D1447075

Switch Mode Power Conversion

ELECTRICAL ENGINEERING AND ELECTRONICS

A Series of Reference Books and Textbooks

Editors

Marlin O. Thurston
Department of Electrical Engineering
The Ohio State University
Columbus, Ohio

William Middendorf
Department of Electrical and Computer Engineering
University of Cincinnati
Cincinnati, Ohio

1. Rational Fault Analysis, *edited by Richard Saeks and S. R. Liberty*
2. Nonparametric Methods in Communications, *edited by P. Papantoni-Kazakos and Dimitri Kazakos*
3. Interactive Pattern Recognition, *Yi-tzuu Chien*
4. Solid-State Electronics, *Lawrence E. Murr*
5. Electronic, Magnetic, and Thermal Properties of Solid Materials, *Klaus Schröder*
6. Magnetic-Bubble Memory Technology, *Hsu Chang*
7. Transformer and Inductor Design Handbook, *Colonel Wm. T. McLyman*
8. Electromagnetics: Classical and Modern Theory and Applications, *Samuel Seely and Alexander D. Poularikas*
9. One-Dimensional Digital Signal Processing, *Chi-Tsong Chen*
10. Interconnected Dynamical Systems, *Raymond A. DeCarlo and Richard Saeks*
11. Modern Digital Control Systems, *Raymond G. Jacquot*
12. Hybrid Circuit Design and Manufacture, *Roydn D. Jones*
13. Magnetic Core Selection for Tranformers and Inductors: A User's Guide to Practice and Specification, *Colonel Wm. T. McLyman*
14. Static and Rotating Electromagnetic Devices, *Richard H. Engelmann*
15. Energy-Efficient Electric Motors: Selection and Application, *John C. Andreas*
16. Electromagnetic Compossibility, *Heinz M. Schlicke*
17. Electronics: Models, Analysis, and Systems, *James G. Gottling*

18. Digital Filter Design Handbook, *Fred J. Taylor*
19. Multivariable Control: An Introduction, *P. K. Sinha*
20. Flexible Circuits: Design and Applications, *Steve Gurley, with contributions by Carl A. Edstrom, Jr., Ray D. Greenway, and William P. Kelly*
21. Circuit Interruption: Theory and Techniques, *Thomas E. Browne, Jr.*
22. Switch Mode Power Conversion: Basic Theory and Design, *K. Kit Sum*

Other Volumes in Preparation

Switch Mode Power Conversion

BASIC THEORY AND DESIGN

K. Kit Sum

Chief Engineer
LH Research, Inc.
Tustin, California

MARCEL DEKKER, INC. New York and Basel

Library of Congress Cataloging in Publication Data

Sum, K. Kit, [date]
Switch mode power conversion.

(Electrical engineering and electronics ; 22)
Bibliography: p.
Includes index.
1. Electronic apparatus and appliances--Power supply--Design and construction. I. Title. II. Series.
TK7881.15.S86 1984 621.31'7 84-12160
ISBN 0-8247-7234-2

Many of the tables and figures in this book have been reprinted from Colonel William T. McLyman's *Transformer and Inductor Design Handbook* (Marcel Dekker, 1978) and *Magnetic Core Selection for Transformers and Inductors* (Marcel Dekker, 1982) with the permission of both the author and the publisher.

Copyright © 1984 by Marcel Dekker, Inc. All Rights Reserved

Neither this book nor any part may be reproduced or transmitted in any form or by any means, electronic or mechanical, including photocopying, microfilming, and recording, or by any information storage and retrieval system, without permission in writing from the publisher.

MARCEL DEKKER, INC.
270 Madison Avenue, New York, New York 10016

Current printing (last digit):
10 9 8 7 6 5 4 3

PRINTED IN THE UNITED STATES OF AMERICA

Foreword

In recent years the demand for power electronics engineers in industry has far outnumbered the supply. This situation is expected to become even worse in the years ahead, considering the demand for engineers in such areas of national attention as energy sufficiency, robotic technology, and defense. In all of these areas, power electronics technology is being used in the efficient control of electric motors for applications such as electric cars and industrial motor drive control, and in developing more reliable, lightweight switching power supplies for various sophisticated computer and communication equipment. Recent advances in microprocessor technology make it very promising to integrate microelectronics and power electronics technology for efficient control of robotic motor drives in factory automation.

Work in the area of switching power supplies has been spurred by the need for higher performance, smaller volume, and lighter weight power sources to be compatible with the shrinking size of all forms of communication and data handling systems, with portable battery-operated equipment in everything from home appliances and handtools to mobile communication equipment. Static dc-to-dc converters and dc-to-ac inverters provide a natural interface with the new direct energy sources such as solar cells, fuel cells, thermoelectric generators, and the like, and form the central ingredient in most uninterruptible power sources. Such solid-state power conditioners operating with internal conversion frequencies in the tens to hundreds of kilohertz range are emerging as the new mainstay of most power supplies for computers and communication systems, as they have been over the last two decades for the space programs of the world.

Along with the growing market in various computer and communication equipment, switching supplies have created a demand for engineers who can design them. Despite this demand and competitive salaries, only a handful of universities in the United States offer courses in switching power supply design. The lack of relevant coursework in colleges has contributed to the growing shortage of engineers who can design the supplies. Most of the power supply engineers in the United States have gained their know-how through many years of on-the-job training and hands-on experience.

A vast amount of technology in the area of switching power supplies modeling, analysis, and control design has been generated in the past decade. This material is scattered in numerous technical journals, conference proceedings, and trade magazines. There is a great need for a book emphasizing analysis and design of switching power supplies that will assist experienced engineers in their design and help to bring the novice up to speed.

I am very pleased to see that Mr. Sum has successfully brought these disciplines together in a well-organized manner. The book is focused on switching mode power supply design, analysis and measurement. The book contains information on the design of magnetics and regulator control loops and is a very good reference source for power supply design engineers. The author also provides a comprehensive listing of references containing most recent publications. Congratulations, Mr. Sum.

Fred C. Lee
Department of Electrical Engineering
Virginia Polytechnic Institute and State University
Blacksburg, Virginia

Preface

This book presents the fundamentals of switch mode power converters with insights into design aspects. Elementary explanations of the general practice of design, analysis, testing, and measurements of switch mode converters are given. The discussions have been confined to basic concepts in an effort to promote better understanding of subject matter. Examples are used to highlight the principles of operation and must not be regarded as optimum design methods for the various applications.

Inverters are not described within these pages, but the principles provided are quite similar to such designs. All modulators described are of the linear ramp (sawtooth) and comparator type; modulators of the sample and hold type are not dealt with here. All regulators are assumed to operate with a constant frequency and variable ON time for output control.

Calculator programs are listed in their entirety for evaluation and analysis of the basic types of converters. Numerous frequency response graphs are provided for the convenience of the reader to coordinate the calculator programs with design efforts toward a stable loop gain realization. State-space averaging analysis is provided in some detail, sufficient to highlight the pertinent characteristics of the different basic converters.

The chapter on magnetic components design is written with the intention of supplementing certain elementary details that are left out in other discussions.

The reader is encouraged to coordinate the discussion in Chapter 4 closely with the programs provided in Chapter 2 and the graphs provided in Appendix B for a complete and detailed understanding of the mechanisms involved in acquiring a stable regulated converter response.

A method of measuring gain and phase is given in some detail for evaluating the closed-loop converter. The information obtained is pertinent for the assessment of the loop gain characteristic for stable system operation.

An extensive list of references, under separate headings, is provided for an up-to-date perspective of the state of the art and for those who wish to pursue the subject further.

The author wishes to thank William M. Polivka of the California Institute of Technology for checking the calculations in Appendix A and Michael Paupst for translating the HP programs to algebraic operating system language for use with the TI calculators.

The author is particularly indebted to Doctors R. D. Middlebrook and Slobodan M. Ćuk of the California Institute of Technology for inspiration and Colonel William T. McLyman of the Jet Propulsion Laboratory for encouragement, without which this book could never have been written.

Thanks are also due to Kamalesh D. Dwivedi, Terry LaLonde, and Milan Skubnik of the Computer Aided Engineering Department, Peter Ryan of the Test Equipment and Calibrations Department, and Richard Charlebois of the Research and Development Division of Mitel Corporation for computer simulation assistance, computing and measurement equipment support, and assistance in generating the illustrations, respectively, and to the Digital Systems Division of Mitel Corporation for the loan of the HP-9845C desk-top color computer.

K. Kit Sum

Contents

1
Switch Mode Power Converters

1.1 FUNDAMENTAL CONCEPTS

Power processing has always been an essential feature of most electrical equipment. The differences in voltage and current requirements for different applications have led to the design of dedicated power converters to meet their specific requirements.

In general, power conversion involves the process of either converting alternating current (ac) to direct current (dc) or converting dc of one voltage level to dc of another voltage level or both. The process of converting dc to ac, however, is sometimes necessary. A processor of this kind is generally known as an inverter [1, 9-16, 18-21].

The concept of switch mode power conversion is not new, but the technology was not quite ready until the last decade or so. The availability of fast-switching high-voltage transistors, low-loss ferrite and metallic glass materials made the complete implementation of switch mode power converters more reliable and practical.

The most significant differences between the linear and the switch mode regulators involve their efficiency, size, weight, thermal requirement, response time, and noise characteristics.

Ideal components in a linear regulator have little effect on the overall performance of the regulator, whereas in a switching regulator ideal components would make a 100% efficient regulated power converter.

The linear regulator is, therefore, sometimes referred to as a dissipative regulator.

In the case of the switching regulator, the disadvantages of the linear regulator are eliminated. However, the switching regulator is also unable to retain the advantages of the linear regulator; namely, it is noisy with switching transients, has complicated circuitry, is difficult to analyze, and has slow transient response.

The switching converter is difficult to analyze because it is nonlinear. All the linear circuit analysis techniques are rendered useless when it comes to switch mode converters.

Fortunately, a method called the state-space averaging technique was developed [73] and further refined [78, 80] to provide a basis for the analysis of these basic configurations. This state-space averaging method succeeded in transforming a nonlinear circuit model to an averaged linear model, which allows the use of linear circuit analysis techniques for analysis and performance prediction. See Appendix A.

In the implementation of switch mode power conversion electronics, one or more of the three basic configurations are usually employed. Other circuit topologies are usually derived from one or more of these configurations.

The three basic configurations are the buck converter, the boost converter, and the buck-boost converter. They are "basic" in the sense that only one switch is used and that there is no isolation from the output circuits. Many of the popular topologies are buck-derived-type converters: the push-pull, the bridge, the half-bridge, and the forward converters.

A new and unique configuration of the buck and boost combination is the Ćuk converter [42, 43, 45, 46, 49, 192, 194, 196]. This configuration is sometimes known as the optimum topology because of the minimum number of switches and energy storage components required. It is optimum also because of the arrangement of components to obtain low input and output current ripple.

1.2 THE PULSE WIDTH MODULATOR

A switching regulator is a power processor in which the power handling devices are operated as switches in either ON or OFF positions. The arrangement of the switches and storage and filtering components is referred to as the *topology* of the converter. A given topology is used to obtain a specific result, such as voltage step-down, voltage step-up, or voltage step-up and current step-up.

The regulation process of a switch mode converter is performed via the pulse width modulator with a control voltage derived from the output of the converter.

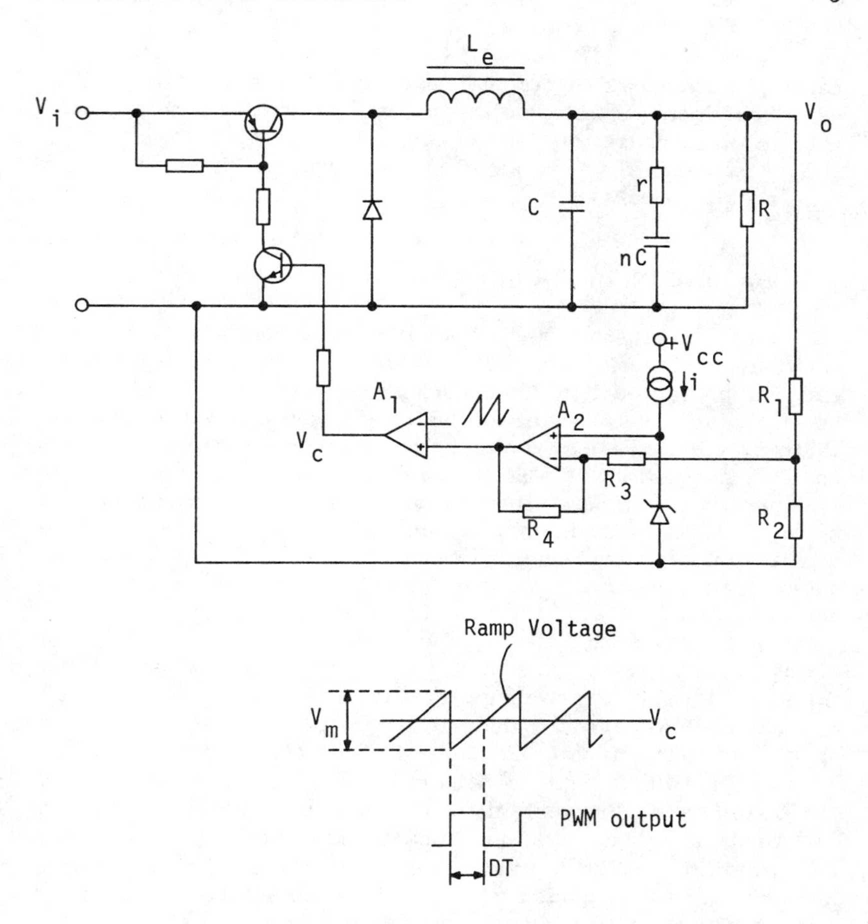

FIG. 1.1 Example of pulse width modulator as used in a buck converter.

The modulators described in this text are assumed to be the linear ramp and comparator type and do not contribute any phase lag (neglecting storage time modulation effects) [71, 76, 90].

This type of modulator operates on the principle of comparing a linear ramp voltage with a control voltage V_c, as shown in Fig. 1.1.

The control voltage is derived from the output voltage V_o. A_1 is a comparator, and A_2 is the error amplifier. If V_o decreases, the output of A_2 (V_c) will be increased. An increase in V_c will cause the comparator to trigger at a higher potential point on the ramp, producing a wider output pulse. The wider output pulse will, in turn, cause the delivery of a higher output voltage V_o. If V_o increases, the reverse effect will occur. The result is a voltage-regulated converter.

1.3 BASIC CONFIGURATIONS

To understand the nature of the three basic configurations, it is convenient to recall Lenz's law. In 1834, Lenz, a Russian physicist, supplemented Faraday's work on electromagnetic induction by pointing out that the direction of the induced emf is the same as that of a current whose magnetic action would neutralize or oppose the flux change. This statement is known as Lenz's law.

When a current flows through an inductor, a magnetic field is set up. Any change in this current will change this field, and an emf is induced. This induced emf will act in such a direction as to maintain the flux at its original density. This effect is known as *self-induction*.

It is evident from Table 1.1 that there is essentially no drastic change of topology for the buck converter for the two switch positions. The small-signal characteristics as provided by Eqs. (A.55) and (A.57 of Appendix A indicate the response of a linear system.

However, this is not the case with the boost and buck-boost converters, which indicate drastic changes in the topology between the ON and the OFF positions of the switches. Figure 1.2 shows the topological changes of the buck-boost converter during ON and OFF conditions. The individual networks of Fig. 1.2b and Fig. 1.2c are linear as shown, but the constant periodic changes between the two states make the converter *nonlinear*. Converters operating in two (ON and OFF) states are referred to as operating in the *continuous conduction* mode. In this operating mode, the buck-boost converter stores energy in the inductor in the form of magnetic field during the ON time and releases the energy to the output circuit during the OFF time. Table 1.1 provides a perspective view of the three basic converters operating in this mode. See also Table 1.20. Figure 1.3 shows the inductor current waveform for the buck-boost converter.

According to Faraday's law, the average voltage across the inductor over a complete period is zero. This means that (*volt-second applied = volt-second released*)

TABLE 1.1 Nature of Energy Transfer in Basic Converters

Converter	Switch position	
	ON	OFF
	(When the current increases, the induced emf across the inductor opposes the applied emf)	(When the current decreases, the induced emf tries to maintain the flux caused by the previously applied emf)
Buck	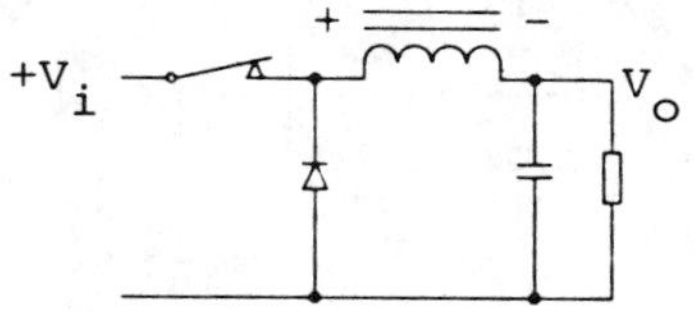 + − $+V_i$ V_o	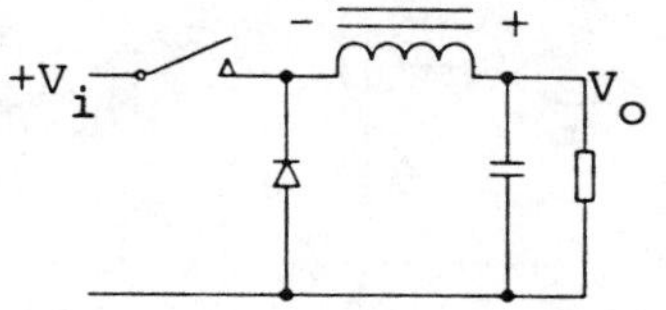 − + $+V_i$ V_o
Boost	+ − $+V_i$ V_o	− + $+V_i$ V_o
Buck-boost	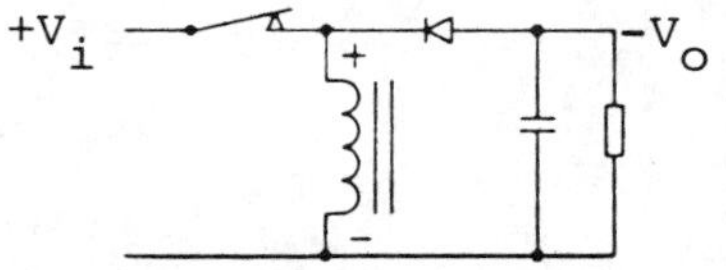 $+V_i$ $-V_o$ + −	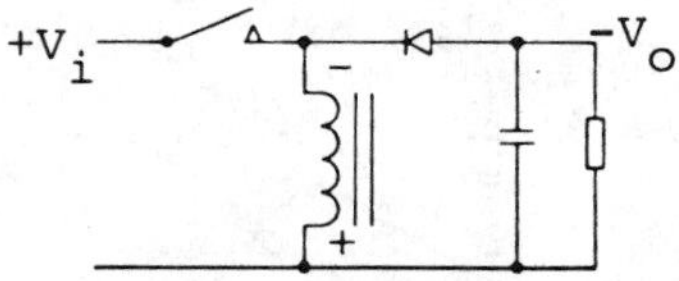 $+V_i$ $-V_o$ − +

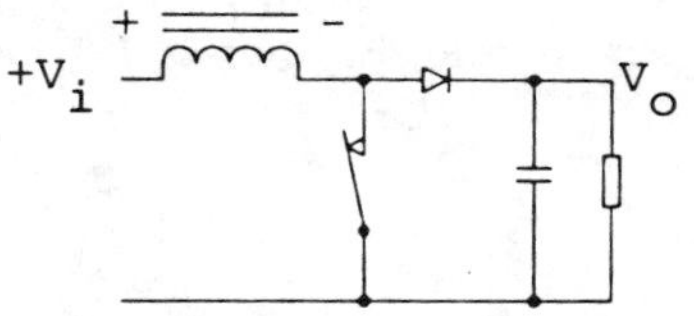

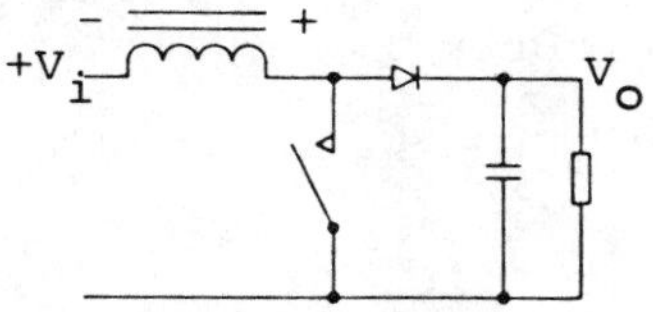

$$V_iDT = V_o(1 - D)T \tag{1.1}$$

from which

$$\frac{V_o}{V_i} = \frac{D}{1 - D} \tag{1.2}$$

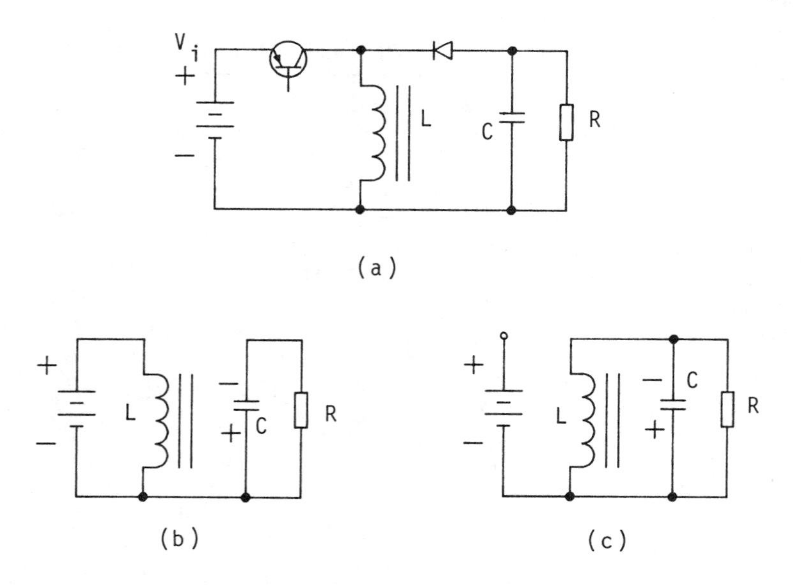

FIG. 1.2 Topological changes with change of switch positions. (a) Buck-boost converter. (b) ON topology. (c) OFF topology.

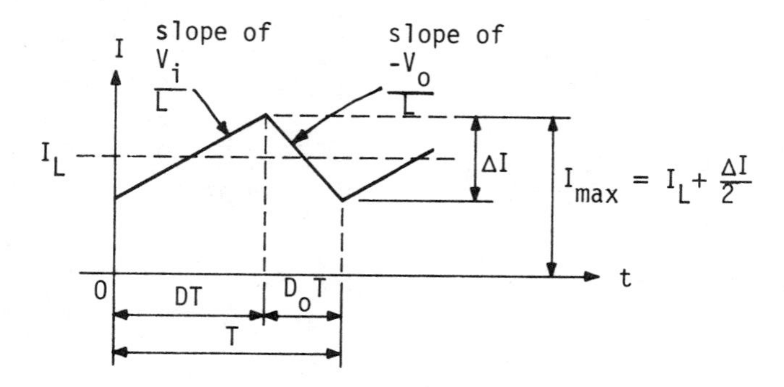

FIG. 1.3 Steady-state buck-boost converter inductor current waveform.

Equation (1.2) is an idealized case, since it indicates an infinite voltage gain for D equal to 1. In practice, a slightly lossy inductor would have limited this gain to a realistically finite value.

In the design of converters, it is usually necessary to accommodate a given range of input voltage. Equation (1.2) can be rewritten to give

$$\frac{V_o}{V_{i_{max}}} = \frac{D_L}{1 - D_L} \tag{1.3}$$

and

$$\frac{V_o}{V_{i_{min}}} = \frac{D_H}{1 - D_H} \tag{1.4}$$

To maintain continuous conduction, it is necessary to assess the boundary condition in which the converter is allowed to operate. Figure 1.4a shows the boundary condition for I_L, and Fig. 1.4b shows the *discontinuous conduction* condition.

Returning to Fig. 1.3, we see that the current ripple is

$$\Delta I = \frac{DTV_i}{L} = \frac{V_o(1 - D)T}{L} \tag{1.5}$$

Note that Eq. (1.5) defines the requirement of B_{ac} which is part of the requirement of the maximum flux density B_{max} for the magnetic design of the inductor. The significance of the boundary condition is that it dictates the minimum inductance L_b required to maintain continuous conduction for a given minimum load; therefore,

$$P_{o_{min}} = \frac{1}{2} L_b(\Delta I)^2 f = \frac{D_L^2 V_i^2 T}{2L_b} = \frac{V_o^2(1 - D_L)^2 T}{2L_b} = \frac{V_o^2}{R_{max}} \tag{1.6}$$

Therefore,

$$L_b = \frac{R_{max} T(1 - D_L)^2}{2} \tag{1.7}$$

TABLE 1.2 DC Gain of Basic Converters with Practical Passive Components

Converter	Function $\frac{V_o}{V_i}$	η
Buck-boost	$\frac{D}{D_o} \frac{D_o^2 R}{D_o^2 R + r_L + DD_o(r//R)}$	$\frac{D_o^2 R}{D_o^2 + r_L + DD_o(r//R)}$
Boost	$\frac{1}{D_o} \frac{D_o^2 R}{D_o^2 R + r_L + DD_o(r//R)}$	$\frac{D_o^2 R}{D_o^2 R + r_L + DD_o(r//R)}$
Buck	$D \frac{R}{R + r_L}$	$\frac{R}{R + r_L}$

From Fig. 1.3 and Eq. (1.5),

$$I_{max} = I_L + \frac{\Delta I}{2}$$

$$I_{max} = \frac{I_{o_{max}}}{(1 - D_H)} + \frac{V_o(1 - D_H)T}{2L_b} \tag{1.8}$$

During the switch ON time, the filter capacitor C discharges through the load resistor R. See Fig. 1.2. The slope of the output voltage within this duration is directly related to the average voltage across the capacitor, V(t):

$$i = C\frac{dV(t)}{dt} = -\frac{V(t)}{R}$$

or

$$\frac{dV(t)}{dt} = -\frac{V(t)}{RC} \tag{1.9}$$

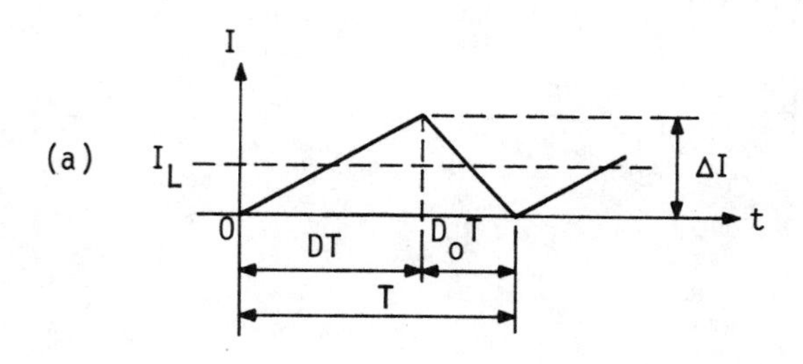

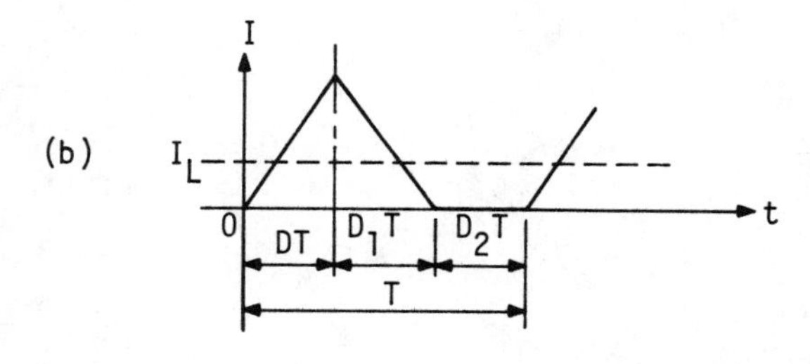

FIG. 1.4 (a) Boundary and (b) discontinuous conduction conditions.

The negative sign indicates a discharging slope.

For an almost constant V(t), the following approximation is valid:

$$-\frac{V(t)}{RC} \simeq \frac{\Delta V}{\Delta t} \simeq \frac{\Delta V_o}{DT} \tag{1.10}$$

$$\frac{\Delta V_o}{V_o} = \frac{D_H T}{R_{min} C} \tag{1.11}$$

Equation (1.11) defines the output voltage ripple.

Similarly, for the buck converter operating in the continuous conduction mode in the steady-state condition with constant volt-second relations,

$$(V_i - V_o)DT = V_o(1 - D)T \tag{1.12}$$

$$V_o = DV_i \tag{1.13}$$

Under the boundary condition $V_o = D_L V_{i_{max}}$,

$$I_{min} = \frac{V_o}{R_{max}} = \frac{D_L V_i}{R_{max}} = \frac{\Delta I}{2} \tag{1.14}$$

$$L\frac{di}{dt} = V_i - V_o = V_i(1 - D_L) \tag{1.15}$$

$$\frac{\Delta I}{\Delta T} = \frac{V_i}{L_b}(1 - D_L) \tag{1.16}$$

The current ripple is

$$\Delta I = \frac{V_i D_L T(1 - D_L)}{L_b} \tag{1.17}$$

$$\frac{\Delta I}{2} = \frac{D_L V_i}{R_{max}} = \frac{V_i D_L T(1 - D_L)}{2L_b} \tag{1.18}$$

$$L_b = \frac{R_{max} T(1 - D_L)}{2} \tag{1.19}$$

$$I_{max} = \frac{V_o}{R_{min}} + \frac{\Delta I}{2} = \frac{V_o}{R_{min}} + \frac{V_{i_{max}} D_L T(1 - D_L)}{2L_b} \tag{1.20}$$

The output voltage ripple is derived as follows: The change in charge ΔQ of the output capacitor C is represented by the shaded area in Fig. 1.5:

$$\Delta Q = \frac{1}{2}\left(\frac{DT}{2} + \frac{D_o T}{2}\right)\frac{\Delta I}{2} = \frac{T \Delta I}{8} \tag{1.21}$$

$$\Delta v_o = \frac{\Delta Q}{C} = \frac{T\ \Delta I}{8C} = \frac{V_o T^2 (1 - D_L)}{8LC} \tag{1.22}$$

For the boost converter, the steady-state continuous conduction relations are as follows: For constant volt-second,

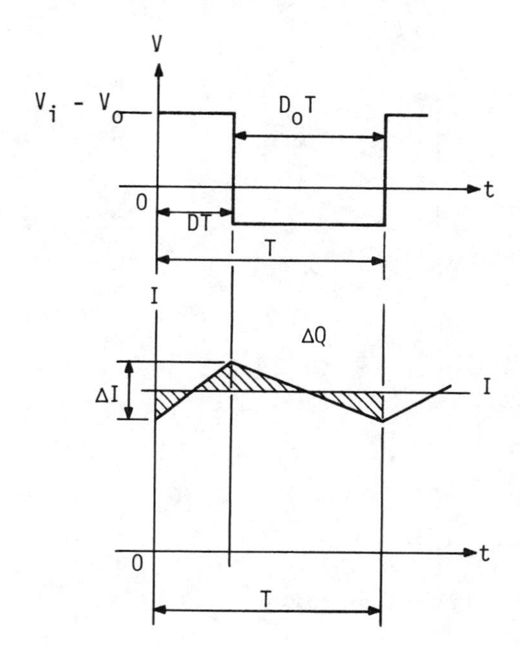

FIG. 1.5 Steady-state buck converter continuous conduction mode waveforms. Over one period, ΔQ is represented by shaded area.

$$V_i DT = (V_o - V_i)(1 - D)T \tag{1.23}$$

$$V_i T = V_o T(1 - D) \tag{1.24}$$

$$\frac{V_o}{V_i} = \frac{1}{1 - D}, \quad \text{a voltage step-up condition} \tag{1.25}$$

$$\frac{I_i}{I_o} = \frac{1}{1 - D}, \quad \text{a current step-down condition} \tag{1.26}$$

Multiplying Eq. (1.25) by Eq. (1.26) gives

$$\frac{V_o}{V_i}\frac{I_i}{I_o} = \frac{1}{(1 - D)^2} \tag{1.27}$$

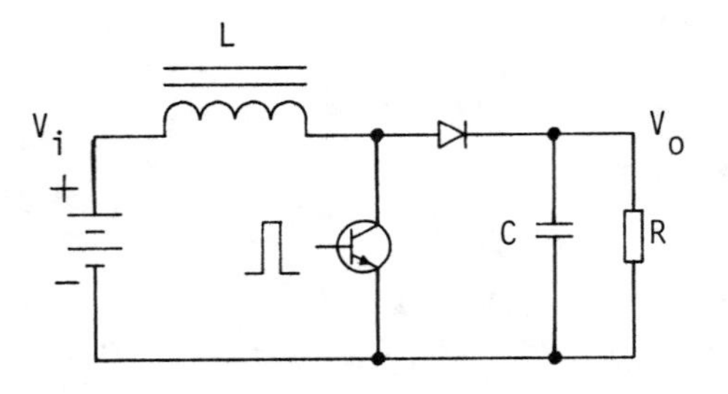

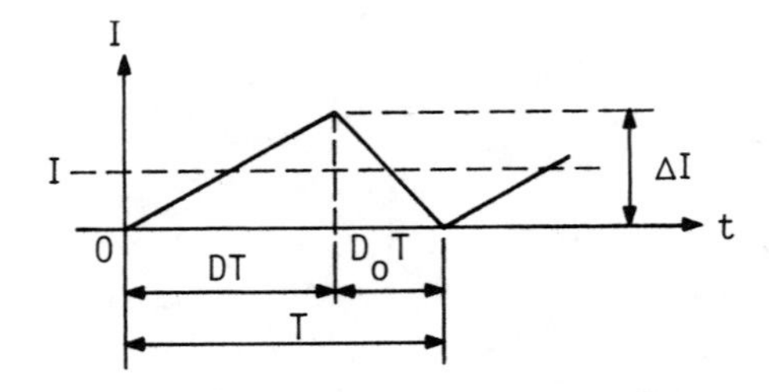

FIG. 1.6 Boost converter boundary condition.

The voltage across the inductor is

$$L\frac{di}{dt} = V_i$$

From Fig. 1.6, the current ripple is

$$\Delta I = \frac{V_i D_L T}{L_b} \tag{1.28}$$

For the boundary condition, the minimum average input current is

$$I_i = \frac{\Delta I}{2} = \frac{V_i D_L T}{2L_b} \tag{1.29}$$

The maximum current is

$$I_{max} = \frac{I_{o_{max}}}{1 - D_H} + \frac{\Delta I}{2} \tag{1.30}$$

$$I_{max} = \frac{V_o}{R_{min}(1 - D_H)} + \frac{V_{i_{min}} TD_H}{2L_b} \tag{1.31}$$

Substituting Eq. (1.29) into Eq. (1.27) for I_i gives

$$\frac{V_o D_L T}{2I_o L_b} = \frac{1}{(1 - D_L)^2} \tag{1.32}$$

or

$$L_b = \frac{V_o TD_L(1 - D_L)^2}{2I_o} \tag{1.33}$$

Substituting R_{max} for V_o/I_o gives

$$L_b = \frac{R_{max} TD_L(1 - D_L)^2}{2} \tag{1.34}$$

The output peak-to-peak ripple voltage Δv_o is the same as that derived for the buck-boost converter and is given by Eq. (1.11), owing to similar output circuit topologies.

In the case of the boost and the buck-boost converters, the losses in the converter elements play a very significant role in reducing the efficiency of the circuit. This is evident from the values of the elements given in Fig. A.4 in which the losses due to the inductor, represented by

$$\frac{r_L}{(1 - D)^2}$$

increase dramatically with the increase of duty ratio D. By means of Eq. (A.58) in Appendix A, it is possible to plot a number of curves to demonstrate the effect of increasing r_L on the converter efficiency. Figure 1.7 shows this effect on a boost converter.

Mathematically, if Eq. (A.58) is used with r equal to zero, so that the effect of r_L on the steady-state dc gain is

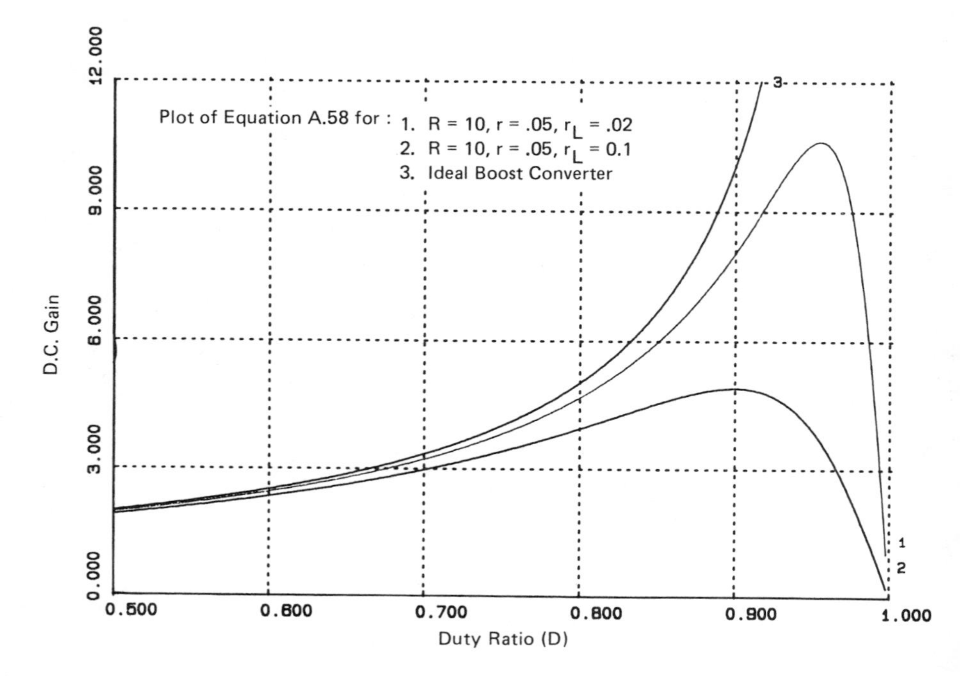

FIG. 1.7 Boost converter dc gain characteristics.

$$\frac{V_o}{V_i} = \frac{RD_o}{r_L + RD_o^2} \tag{1.35}$$

differentiating V_o/V_i with respect to D and equating to zero for the maximum gives

$$\frac{d}{dD}\frac{V_o}{V_i} = \frac{d}{dD}\frac{R(1 - D)}{r_L + R(1 - D)^2} \tag{1.36}$$

$$0 = \frac{d}{dD}\frac{1 - D}{(r_L/R) + (1 - D)^2} \tag{1.37}$$

After some manipulation and simplification, maximum gain D_m is obtained:

$$D_m = 1 - \sqrt{\frac{r_L/(R + r_L)}{1 - r_L/(R + r_L)}} \tag{1.38}$$

$$D_m = 1 - \sqrt{\frac{r_L}{R}} \quad \text{for boost converter} \tag{1.39}$$

Similarly, for the buck-boost converter,

$$D_m = \frac{1 - \sqrt{r_L/(r_L + R)}}{1 - r_L/(r_L + R)} \quad \text{buck-boost converter} \tag{1.40}$$

Substitution of Eq. (1.39) into Eq. (1.35) for D gives the boost converter maximum gain (neglecting other losses):

$$\left.\frac{V_o}{V_i}\right|_{max} = \frac{1}{2}\sqrt{\frac{R}{r_L}} \quad \text{boost converter} \tag{1.41}$$

Similarly,

$$\left.\frac{V_o}{V_i}\right|_{max} = \frac{1 - \sqrt{r_L/(r_L + R)}}{2\sqrt{r_L/(r_L + R)}} \quad \text{buck-boost converter} \tag{1.42}$$

For example, by using Eq. (1.39) and with $R = 10\ \Omega$ $r_L = 0.1\ \Omega$, the calculated maximum gain for the boost converter is 5. This result compares favorably with the curve in Fig. 1.7, which includes the effect of r of the filter capacitor.

The following is an example of the buck-boost converter, showing the previously derived relations as applied to a design case.

The converter is required to accept an input voltage within the range of +12 V dc to +22 V dc. A regulated output voltage of -28 V dc is required. The load current will fall within the range of 200 mA to 2 A. The output voltage ripple is to be within ±0.5%. The switching frequency is to be 33 kHz.

Using Eq. (1.3),

$$\frac{V_o}{V_{i_{max}}} = \frac{D_L}{1 - D_L}$$

Neglecting losses,

$$\frac{28}{22} = \frac{D_L}{1 - D_L}$$

$$D_L = 0.56$$

or

$$R_{max} = \frac{V_o}{I_{o_{min}}} = \frac{28}{0.2} = 140\ \Omega$$

$$T = 1/33\ \text{kHz} = 30.30\ \mu\text{sec}$$

Using Eq. (1.7),

$$L_b = \frac{R_{max} T (1 - D_L)^2}{2}$$

$$= \frac{1}{2}[140 \times 30.3 \times 10^{-6}(1 - 0.56)^2]$$

$$= 410.67\ \mu\text{H} \quad \text{design } \underline{420\ \mu\text{H}}$$

By using Eq. (1.4), D_H is calculated and is equal to 0.7. By using Eq. (1.11), the output voltage ripple is evaluated:

$$\frac{\Delta v_o}{V_o} = \frac{D_H T}{R_{min} C}$$

where $R_{min} = V_o/I_{o_{max}} = 28/2 = 14\ \Omega$. For 0.5% voltage ripple,

$$\frac{0.5}{100} = \frac{0.7 \times 30.3 \times 10^{-6}}{14C}$$

$$C = 303\ \mu\text{F} \quad \text{use available C of } \underline{380\ \mu\text{F}}$$

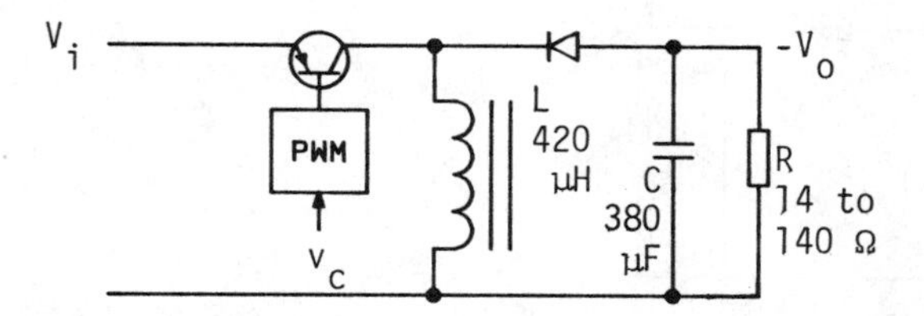

FIG. 1.8 Basic buck-boost converter.

The voltage ripple, however, is often dominated by the esr of the output capacitor, and

$$\Delta V_{esr} = \Delta I \times esr$$

$$= esr \frac{V_o T(1 - D_L)}{L_b}$$

The designer should choose a capacitor of negligible esr to ensure that a low output voltage ripple is obtainable. See the example given at the end of Chapter 2. Figure 1.8 shows the basic buck-boost converter, open loop. Further treatment of this circuit with closed-loop consideration will be given in Chapter 4.

Another example showing the use of formulas derived for the buck converter is given in Section 2.6. The example converter is analyzed with the calculator program provided in Section 2.5, and responses are plotted for the cases of an output filter with and without damping. Further discussion in relation to compensation and stability will be detailed in Chapter 4.

1.4 DERIVED CONVERTERS

As the section title implies, some variations of the basic converters are possible, but the basic principles of operation remain the same. One example of a derived converter is the isolated flyback converter. This is basically a buck-boost derived converter with an isolation winding, so that the input circuit is isolated from the output circuit, and the output voltage can be either positive or negative, depending on the winding and diode connected polarities. See Fig. 1.9.

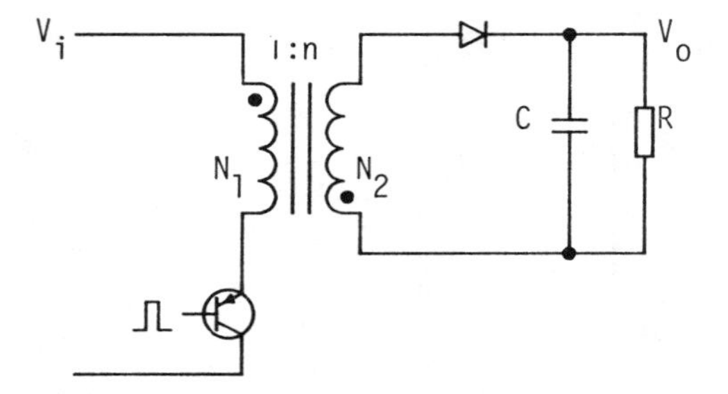

FIG. 1.9 Basic isolated flyback converter.

The circuit relations, without derivation, are as follows:

$$\frac{V_o}{V_i} = \frac{nD}{1 - D} \tag{1.43}$$

$$\Delta I = \frac{V_o(1 - D)T}{nL} \tag{1.44}$$

The primary inductance

$$L = \frac{V_o^2(1 - D)^2 T}{2P_o n^2} = \frac{R_{max}(1 - D)^2 T}{2n^2} \tag{1.45}$$

$$I_{max} = \frac{I_{o\,max}}{1 - D_H} + \frac{\Delta I}{2} = \frac{I_{o\,max}}{1 - D_H} + \frac{V_o(1 - D)T}{2nL}$$

$$I_{max} = \frac{P_{o\,max}\, n}{V_o(1 - D_H)} + \frac{V_o(1 - D_H)T}{2nL} \tag{1.46}$$

From these relations, it would appear that with a suitable choice of turns ratio n a lower operating duty ratio is possible. This is not exactly the case, because if a nonisolated buck-boost converter is compared with the isolated flyback, both having the same element values (with the exception of the isolated winding), then the gain and phase characteristics of the two converters would be very close to each other. This means that if the buck-boost converter is running out of phase margin at a given high-duty ratio, the isolated counterpart of the buck-boost converter (though operating with a lower-duty ratio) is not gaining any advantage, because the turns ratio n in the transfer function serves to modify the overall behavior of the converter to the same extent, as if there were no isolated winding. The reader can easily check out this effect with the programs provided in Section 2.5.

So far, much attention is focused on the continuous conduction mode of operation, because the nature of this mode of operation, using the basic components described up to this point, is that of a second-order control system, whereas the discontinuous conduction mode is only a first-order system. It is, therefore, quite easy to produce a discontinuous conduction converter, close the feedback loop, and realize a stable regulator. The penalty here is a higher input current surge due to the lower energy storage capability of the inductor.

In the case of the continuous conduction converter, the buck-type converters roll off at -12 dB/octave. Any additional phase shift within the feedback loop will be potentially hazardous to the stability of the system. This effect is graphically depicted in Appendix B. Further discussions about stability are given in Chapter 4.

In the discontinuous conduction mode, the inductor current does not behave like a true state variable, because during the dwell time D_2T, it has a definite value of zero. See Fig. 1.4b. As a result, the order of the state-space averaged model is reduced by one. The significance of this order reduction is on the loop stability dynamics.

For a converter designed to operate in the discontinuous conduction mode, care must be taken to check out the behavior of the converter during turn on and under step load conditions. These are the conditions in which the converter is likely to "drop" into the continuous conduction mode momentarily, sometimes long enough to start a parasitic oscillation.

If the converter is found to go into continuous conduction during turn on, or during step load change conditions, then the converter should be compensated for stability for both continuous conduction and discontinuous conduction modes to ensure unconditional stability.

It is also quite common to design converters to operate in both conduction modes for reasons of efficiency or component size reduction. In this case, the converter is allowed to operate in discontinuous conduction mode at light load and in continuous conduction mode at heavier load.

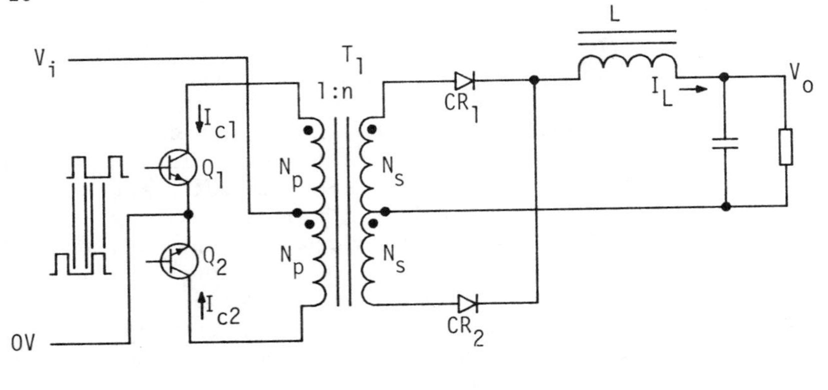

(a)

FIG. 1.10 Push-pull converter. (a) Basic schematic of converter. (b) Idealized waveforms.

Note that the analysis outlined in this volume is a small-signal approximation technique and is not suitable for large-signal response predictions.

Another buck-boost derived converter is the Ćuk converter. This converter has been very well documented elsewhere [42, 43, 45, 46, 49, 192, 194] and is a patented invention.

This converter is a unique topology, just as the buck-boost converter is a unique configuration. Recalling the buck-boost converter, the input current ripple is high, since the switching transient is not isolated by an inductor like the boost converter. The output current ripple is also high, since the switching noise is not isolated from the output like the buck converter. From the foregoing statements, it is apparent that if the buck-boost configuration is reversed to produce a boost-buck configuration, it would have the advantage of both low input current ripple and low output current ripple. This is exactly the case with the Ćuk converter. In short, the boost-buck configuration was produced with simplification of switching arrangement to produce the optimum topology converter.

Note that in the case of the isolated flyback converter the transformer is still doing the same job as the inductor in the buck-boost converter. This means that the so-called transformer is actually an inductor with a secondary winding, and extra window area must be allowed for accommodating the secondary winding. It is, therefore, rather important for the designer to choose the correct core geometry for an isolated flyback converter.

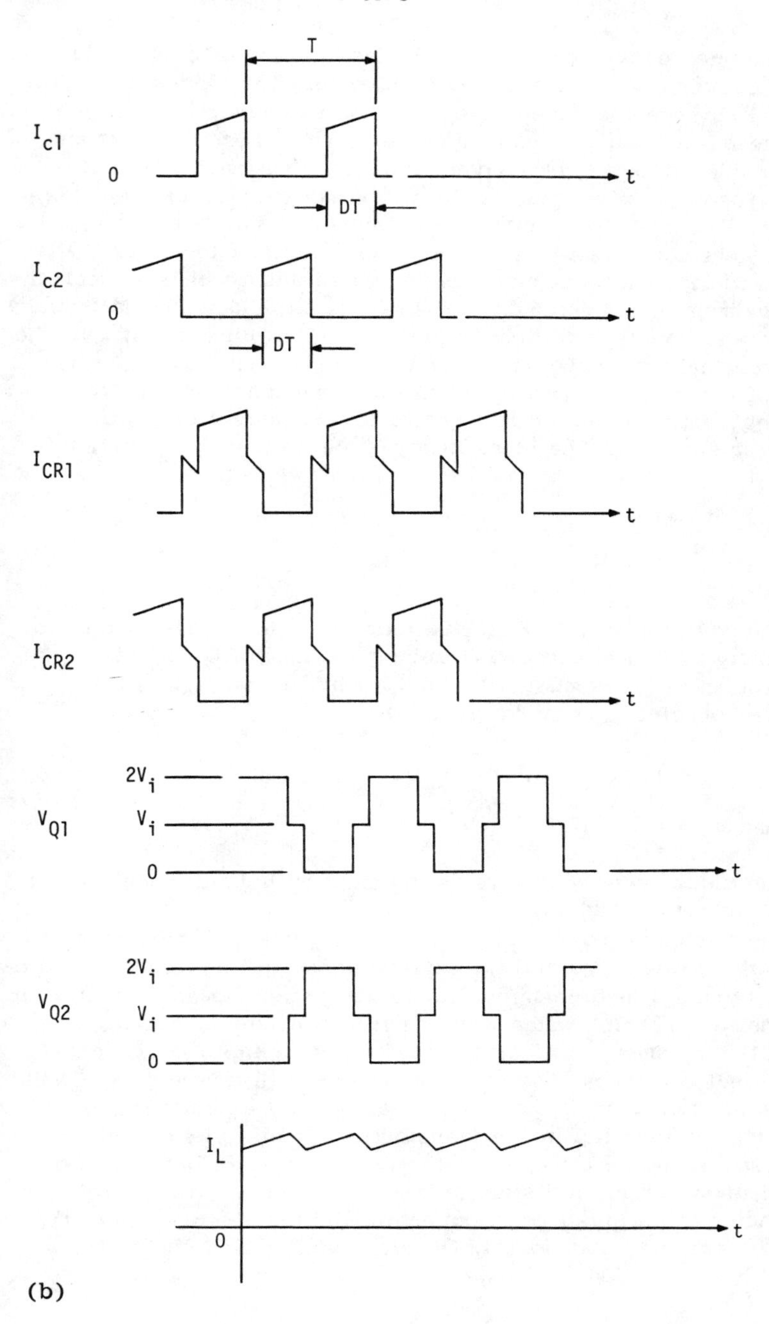

(b)

The buck converter provides the next group of common derived converters: the push-pull converter, the bridge converter, the half-bridge converter, and the forward converter.

The basic push-pull converter is shown in Fig. 1.10. It can be considered as a buck converter with transformer isolation. And by virtue of this isolated coupling, voltage step-up or step-down or polarity reversed outputs are obtainable. The transistors Q_1 and Q_2 are controlled to switch on and off alternately. The ON duty ratio for each transistor is regulated and reaches a theoretical maximum of 0.5 (or 50%). When Q_1 is on, diode CR_1 is forward biased and conducts to deliver part of the output current. In the following half cycle, Q_2 is on, and diode CR_2 delivers the other part of the output current. When both transistors are not conducting, both diodes are forward biased to maintain a continuous inductor current. The ideal steady-state dc transfer function is given by

$$\frac{V_o}{V_i} = 2Dn \tag{1.47}$$

The collector voltage of Q_1 switches from 0 during ON time to V_i during OFF time and to $2V_i$ during ON time of Q_2. This implies that the transistors must withstand a minimum voltage of $2V_{i_{max}}$.

The collector current is given by

$$I_{c_{max}} = n\left(I_o + \frac{\Delta I_L}{2}\right) + I_m \tag{1.48}$$

The diode reverse voltage rating must withstand a minimum of $2nV_i$.

The push-pull converter is suitable for delivering heavy loads, since the switches Q_1 and Q_2 share the load quite effecitvely. However, during a load change, the sudden power demand tends to upset the load-sharing balance of the two transistors, causing one side of the primary winding to "see" a higher input current. If this effect continues, the transformer core will magnetically "walk" to a saturation point. This is the reason why a nonmagnetic gap is sometimes inserted in the transformer of this type of converter. The same situation can arise if the two transistors have somewhat different switching and storage times, or even saturation voltages.

Similar to the push-pull converter, the half-bridge converter has two switches that operate ON alternately. See Fig. 1.11.

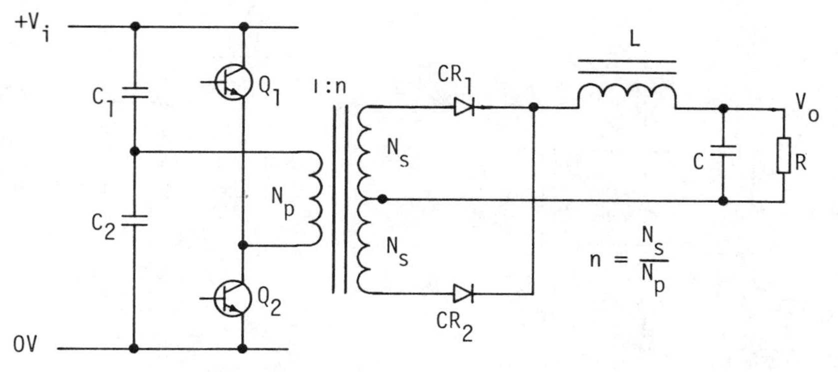

FIG. 1.11 Basic half-bridge converter.

The input voltage is divided between the two capacitors C_1 and C_2, and C_1 equals C_2 in capacitance. The common connecting point of these two capacitors thus has an average voltage of $V_i/2$. This arrangement allows the collector of the switching transistors to see a peak voltage of only V_i, rather than $2V_i$ as in the case of the push-pull converter. However, for a given output power, the half-bridge converter has twice the average primary current seen by the push-pull configuration.

Although one end of the primary is capacitively coupled to the input, there is still danger of saturation for the transformer due to differences in switching times and storage time effects. The blocking capacitor does, however, prevent saturation of the transformer due to mismatches in transistor saturation voltages. When Q_1 is on, CR_1 is forward biased, and power is transferred to the output circuit. In the next half cycle, Q_2 will be on, and CR_2 will be forward biased. When both transistors are off, the diodes are forward biased to maintain a continuous output current. Since one of the two transistor switches is effectively in cascade with the other, an isolated drive transformer is necessary to operate this converter. The ideal steady-state dc transfer function for the half-bridge converter is given by

$$\frac{V_o}{V_i} = nD \tag{1.49}$$

The maximum collector current is given by

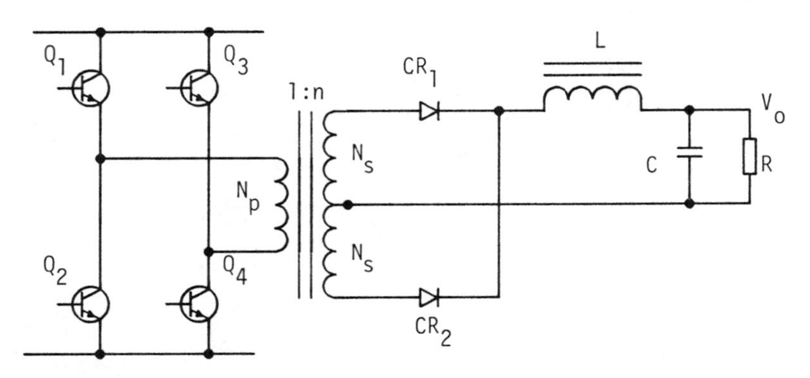

FIG. 1.12 Basic bridge converter.

$$I_{c_{max}} = n\left(I_o + \frac{I_L}{2}\right) + I_m \qquad (1.50)$$

where I_m is the magnetizing current.

Since the capacitors C_1 and C_2 are required to handle high rms currents, they are usually bulky and costly.

The basic bridge converter is shown in Fig. 1.12. In this configuration, the transistors Q_1 and Q_4 turn on at the same time.

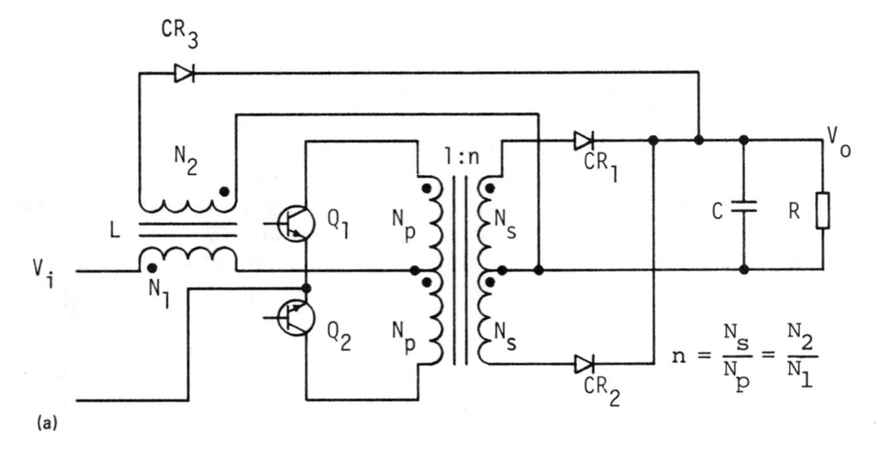

FIG. 1.13 Basic Weinberg converter. (a) Basic schematic of converter. (b) Idealized waveforms.

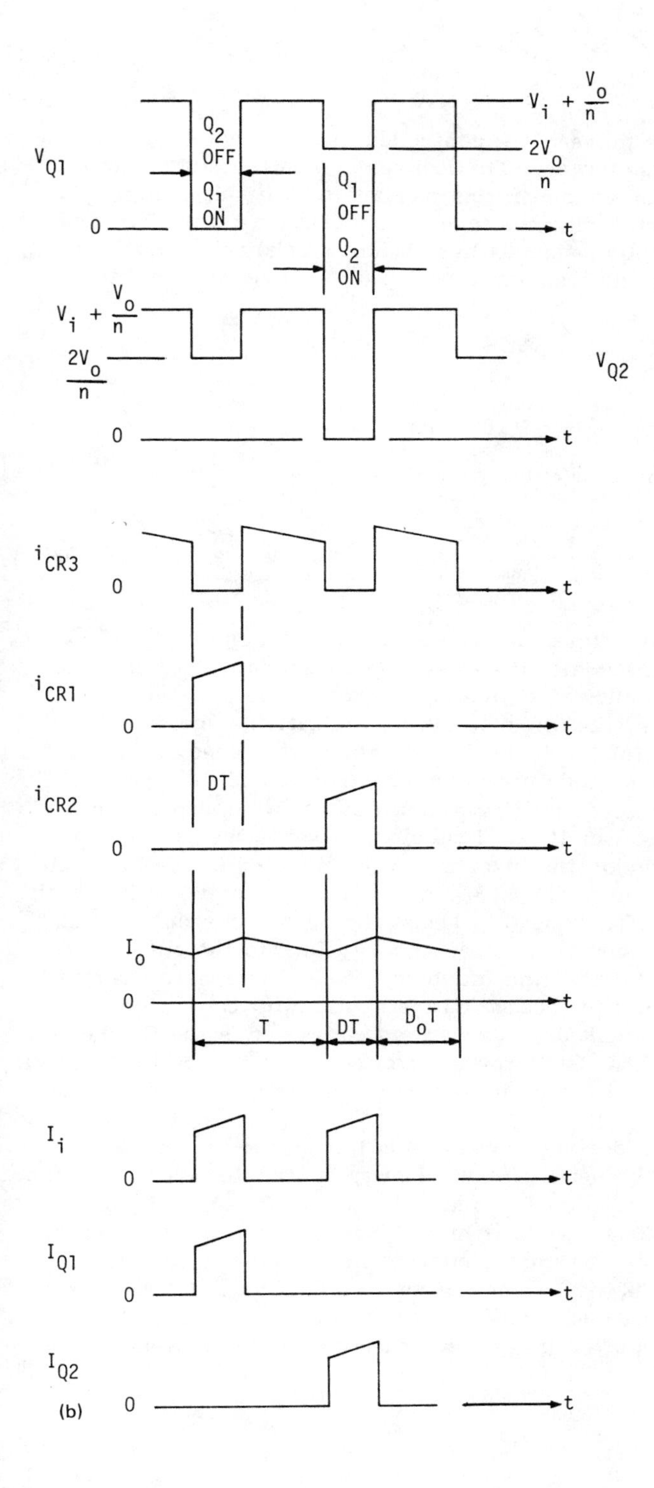

V_{Q1}
$V_i + \frac{V_o}{n}$
$\frac{2V_o}{n}$
Q_2 OFF
Q_1 ON
Q_1 OFF
Q_2 ON
0
t
V_{Q2}
i_{CR3}
i_{CR1}
i_{CR2}
DT
I_o
T
D_oT
I_i
I_{Q1}
I_{Q2}
(b)

Their ON time is pulse width controlled and alternates with Q_2 and Q_3 to process power. This converter has the advantage of the same collector voltage rating as that for the half-bridge converter, V_i, but the disadvantage of transformer saturation of the push-pull converter, since it has similar imbalance conditions. The steady-state dc transfer function of the bridge converter is given by

$$\frac{V_o}{V_i} = 2nD \tag{1.51}$$

The maximum collector current is given by

$$I_{c_{max}} = n\left(I_o + \frac{I_L}{2}\right) + I_m \tag{1.52}$$

In all the converters with two or more switches described above, the alternating nature of the switch operation must be maintained and must not be allowed to overlap on duty ratio. If overlap switching occurs, the effect would be the same as shorting the primary winding of the transformer, which will cause an infinitely high input current and subsequent destruction of the switches.

A relatively recent topology known as the Weinberg converter was first disclosed in 1974. This circuit avoids the problem of overlap switching by the insertion of an inductor in the input circuit as shown in Fig. 1.13. For nonoverlap switching, when both transistors are off, diode CR_3 keeps the output current nonpulsating by transferring the stored energy in L to the output circuit. It is evident that the inductor L could have more than one secondary winding to accommodate a multioutput converter.

Another rather similar current-fed converter is the Clarke converter [190], which feeds the stored energy in L back to the input circuit instead. The current-fed converter is quite well documented elsewhere [67-69].

Another buck derived converter is the forward converter. See Fig. 1.14. As the name implies, energy is transferred during the ON time of Q_1 as diode CR_1 is simultaneously forward biased. As Q_1 turns off, diode CR_1 is reverse biased and CR_2 is forward biased to maintain a continuous current in the output circuit. At the same time, CR_3 is forward biased to allow magnetic resetting of the core. This demagnetizing winding is usually wound with the same number of turns as the primary winding, and as a result

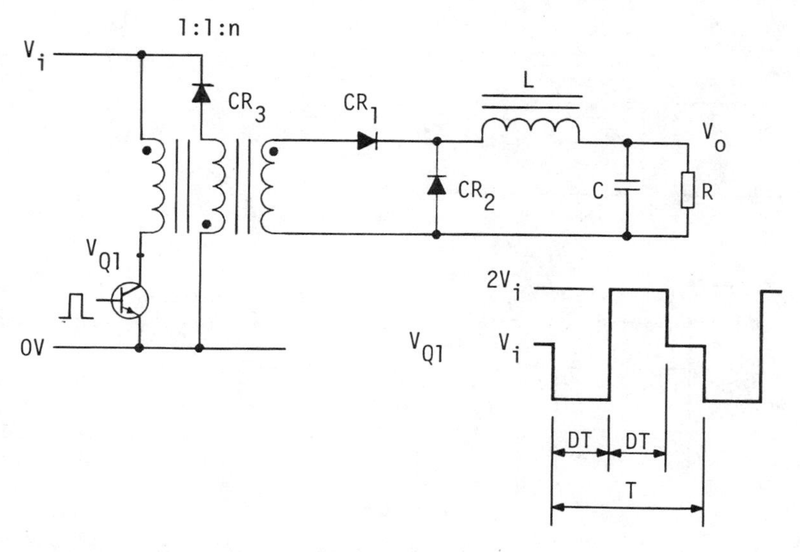

FIG. 1.14 Basic forward converter.

the collector of the transistor must be rated at least twice V_i. To maintain the transformer volt-second balance, the transistor duty cycle must not exceed 0.5 (or 50%); otherwise the transformer will go into saturation. If the ratio of the demagnetization winding to the primary winding is varied, so that the demagnetization winding is less than the 1:1 ratio, the transistor duty cycle may exceed 0.5, but the collector flyback voltage rating becomes higher.

The steady-state dc transfer function of the forward converter is given by

$$\frac{V_o}{V_i} = nD \tag{1.53}$$

1.5 CONVERTERS IN DISCONTINUOUS CONDUCTION MODE

This section will briefly outline the mathematical relations of the three basic types of converters so that a qualitative comparison of the two modes of operation can be sensibly made.

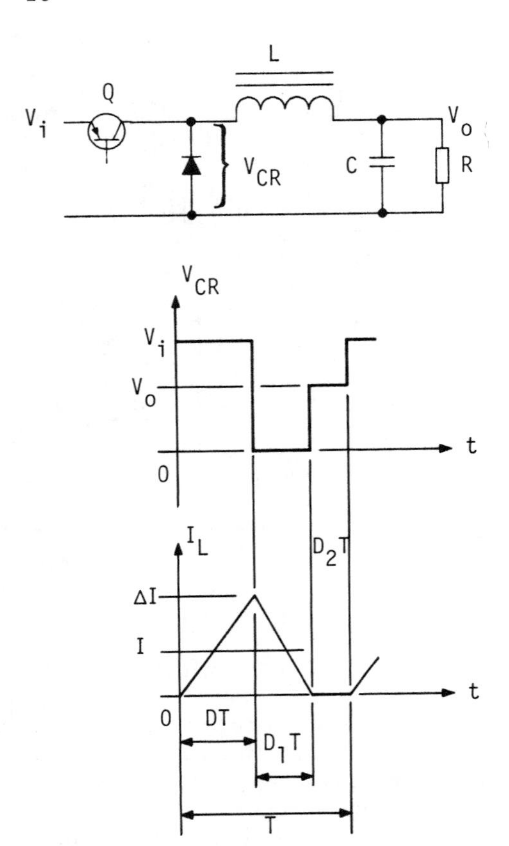

FIG. 1.15 Basic buck converter in discontinuous conduction mode.

The buck converter operating in the discontinuous conduction mode also maintains the same volt-second balance as in the continuous conduction mode; see Fig. 1.15:

$$(V_i - V_o)DT = V_oD_1T \tag{1.54}$$

$$V_iD = V_o(D + D_1) \tag{1.55}$$

$$\frac{V_o}{V_i} = \frac{D}{D + D_1} \tag{1.56}$$

$$\Delta I = \frac{(V_i - V_o)DT}{L} = \frac{V_o D_1 T}{L} \tag{1.57}$$

but

$$I = \frac{\Delta I}{2} \tag{1.58}$$

and

$$I_o = \frac{\Delta I}{2}(D + D_1) \tag{1.59}$$

Therefore,

$$I_o = \frac{V_o T D_1}{2L}(D + D_1) \tag{1.60}$$

$$\frac{V_o}{I_o} = R = \frac{2L}{D_1 T(D + D_1)} \tag{1.61}$$

$$DD_1RT + D_1^2RT = 2L \tag{1.62}$$

$$D_1^2 + DD_1 - \frac{2L}{RT} = 0 \tag{1.63}$$

$$D_1 = \frac{-D + \sqrt{D^2 + (8L/RT)}}{2} \tag{1.64}$$

Substitution of Eq. (1.64) into Eq. (1.56) for D_1 gives

$$\frac{V_o}{V_i} = \frac{2D}{D + \sqrt{D^2 + (8L/RT)}} \tag{1.65}$$

For the boost converter operating in the discontinuous conduction mode (refer to Fig. 1.16),

$$V_iDT = (V_o - V_i)D_1T \tag{1.66}$$

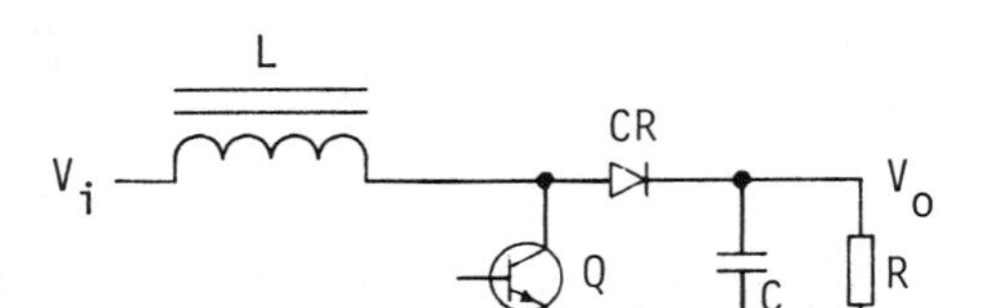

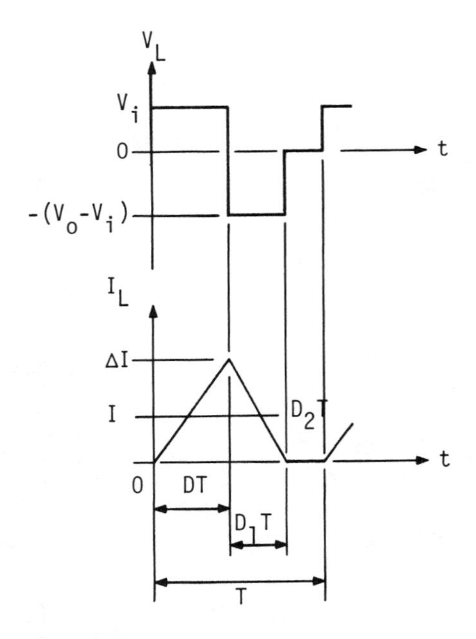

FIG. 1.16 Basic boost converter in discontinuous conduction mode.

$$V_iD = (V_o - V_i)D_1 \tag{1.67}$$

$$\frac{V_o}{V_i} = \frac{D + D_1}{D_1} \tag{1.68}$$

$$\Delta I = \frac{V_iDT}{L} \tag{1.69}$$

$$I = \frac{\Delta I}{2} = \frac{V_iDT}{2L} \tag{1.70}$$

but

$$I = \frac{V_o}{D_1R} \tag{1.71}$$

Equating Eqs. (1.70) and (1.71) gives

$$\frac{V_o}{V_i} = \frac{DD_1TR}{2L} \tag{1.72}$$

or

$$D_1 = \frac{2L}{DTR}\frac{V_o}{V_i} \tag{1.73}$$

Substituting Eq. (1.73) into Eq. (1.68) gives

$$\frac{V_o}{V_i} = \frac{(2L/DTR)(V_o/V_i) + D}{(2L/DTR)(V_o/V_i)} \tag{1.74}$$

$$\frac{V_o^2}{V_i^2} = \frac{V_o}{V_i} + \frac{D^2TR}{2L} \tag{1.75}$$

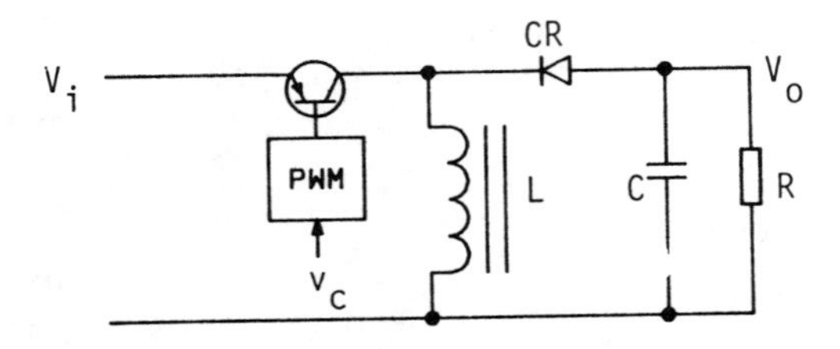

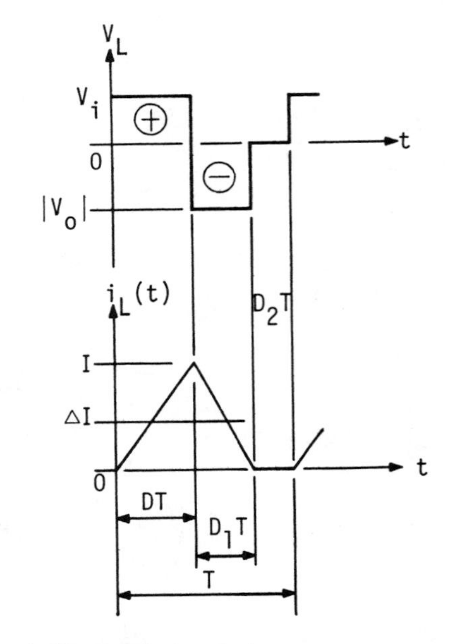

FIG. 1.17 Basic buck-boost converter.

$$\frac{V_o^2}{V_i^2} - \frac{V_o}{V_i} - \frac{D^2TR}{2L} = 0 \tag{1.76}$$

$$\frac{V_o}{V_i} = \frac{1 + \sqrt{1 + (2D^2TR/L)}}{2} \tag{1.77}$$

Therefore, from Eq. (1.73),

$$D_1 = \frac{2L}{DTR} \frac{1 + \sqrt{1 + (2D^2TR/L)}}{2} \tag{1.78}$$

For the buck-boost converter operating in the discontinuous conduction mode, Fig. 1.17 depicts the inductor voltage and current behavior. As before, the constant volt-second relationship holds:

$$V_iDT + V_oD_1T = 0 \tag{1.79}$$

$$\frac{V_o}{V_i} = -\frac{D}{D_1} \tag{1.80}$$

$$P_i = V_iI_i \tag{1.81}$$

$$I_i = DI \tag{1.82}$$

From Fig. 1.17,

$$\Delta I = \frac{V_iDT}{L} \tag{1.83}$$

and

$$I = \frac{\Delta I}{2} \tag{1.84}$$

Therefore,

$$I_i = \frac{D^2 V_i T}{2L} \tag{1.85}$$

Therefore,

$$P_i = \frac{D^2 V_i^2 T}{2L} \tag{1.86}$$

But

$$P_o = \frac{V_o^2}{R} \tag{1.87}$$

For a 100% efficient system,

$$P_i = P_o \quad \text{(or, otherwise, } \eta P_i = P_o) \tag{1.88}$$

$$\frac{D^2 V_i^2 T}{2L} = \frac{V_o^2}{R} \tag{1.89}$$

$$\frac{D^2 TR}{2L} = \frac{V_o^2}{V_i^2} \tag{1.90}$$

$$\frac{V_o}{V_i} = D\sqrt{\tfrac{1}{2}RT/L} \tag{1.91}$$

Comparing Eq. (1.91) with Eq. (1.79) gives

$$D_1 = \sqrt{\frac{2L}{RT}} \tag{1.92}$$

APPENDIX*

Table 1.3 is a summary of some of the most common switching regulator types. The most popular converters depend on the application variables. For a series switching regulator where isolation is not required and $V_{OUT} < V_{IN}$, the buck regulator is most common. The boost regulator performs a similar function for application where $V_{OUT} > V_{IN}$. In applications where isolation between input and output is required, transformer coupling is employed. The half-bridge converter is by far the most widely used regulator in the industry today. The final choice for a converter must be based on output current and voltage, frequency of operation, component cost, and expected performance levels.
The appendix table begins on the following page.

---

*Reprinted with permission of General Electric and Canadian General Electric, Toronto, Ontario, Canada.

TABLE 1.3 Transistor and Diode Requirements for Switching Converters

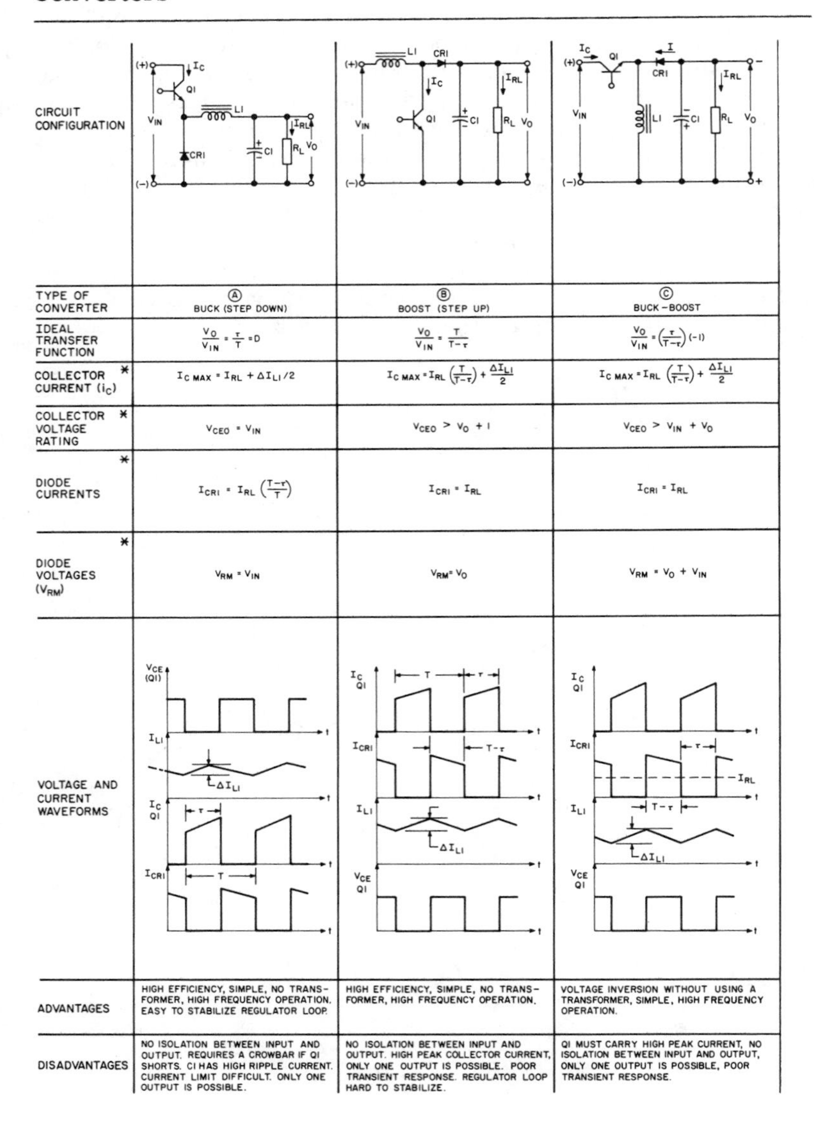

CIRCUIT CONFIGURATION			
TYPE OF CONVERTER	Ⓐ BUCK (STEP DOWN)	Ⓑ BOOST (STEP UP)	Ⓒ BUCK–BOOST
IDEAL TRANSFER FUNCTION	$\frac{V_O}{V_{IN}} = \frac{\tau}{T} = D$	$\frac{V_O}{V_{IN}} = \frac{T}{T-\tau}$	$\frac{V_O}{V_{IN}} = \left(\frac{\tau}{T-\tau}\right)(-1)$
COLLECTOR CURRENT (i_C) *	$I_{C\,MAX} = I_{RL} + \Delta I_{L1}/2$	$I_{C\,MAX} = I_{RL}\left(\frac{T}{T-\tau}\right) + \frac{\Delta I_{L1}}{2}$	$I_{C\,MAX} = I_{RL}\left(\frac{T}{T-\tau}\right) + \frac{\Delta I_{L1}}{2}$
COLLECTOR VOLTAGE RATING *	$V_{CEO} = V_{IN}$	$V_{CEO} > V_O + 1$	$V_{CEO} > V_{IN} + V_O$
DIODE CURRENTS *	$I_{CR1} = I_{RL}\left(\frac{T-\tau}{T}\right)$	$I_{CR1} = I_{RL}$	$I_{CR1} = I_{RL}$
DIODE VOLTAGES (V_{RM}) *	$V_{RM} = V_{IN}$	$V_{RM} = V_O$	$V_{RM} = V_O + V_{IN}$
VOLTAGE AND CURRENT WAVEFORMS			
ADVANTAGES	HIGH EFFICIENCY, SIMPLE, NO TRANSFORMER, HIGH FREQUENCY OPERATION, EASY TO STABILIZE REGULATOR LOOP.	HIGH EFFICIENCY, SIMPLE, NO TRANSFORMER, HIGH FREQUENCY OPERATION.	VOLTAGE INVERSION WITHOUT USING A TRANSFORMER, SIMPLE, HIGH FREQUENCY OPERATION.
DISADVANTAGES	NO ISOLATION BETWEEN INPUT AND OUTPUT. REQUIRES A CROWBAR IF Q1 SHORTS. C1 HAS HIGH RIPPLE CURRENT. CURRENT LIMIT DIFFICULT. ONLY ONE OUTPUT IS POSSIBLE.	NO ISOLATION BETWEEN INPUT AND OUTPUT. HIGH PEAK COLLECTOR CURRENT, ONLY ONE OUTPUT IS POSSIBLE. POOR TRANSIENT RESPONSE. REGULATOR LOOP HARD TO STABILIZE.	Q1 MUST CARRY HIGH PEAK CURRENT, NO ISOLATION BETWEEN INPUT AND OUTPUT, ONLY ONE OUTPUT IS POSSIBLE, POOR TRANSIENT RESPONSE.

TABLE 1.3 (Continued)

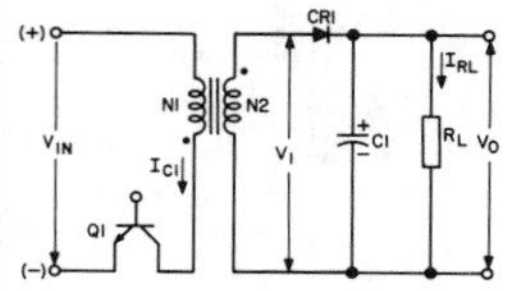	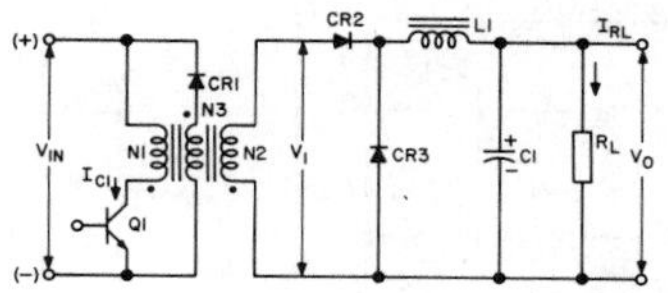
Ⓓ FLYBACK	Ⓔ FORWARD
$\frac{V_O}{V_{IN}} = \frac{N2}{N1}\left(\frac{\tau}{T-\tau}\right)$	$\frac{V_O}{V_{IN}} = \frac{N2}{N1}\left(\frac{\tau}{T}\right)$
$I_{C\,MAX} = I_{RL} \cdot \frac{N2}{N1}\left(\frac{T}{T-\tau}\right) + \frac{\Delta I_{L1}}{2}$	$I_{C\,MAX} = \frac{N2}{N1}\left(I_{RL} + \frac{\Delta I_{L1}}{2}\right) + \hat{I}_{MAG}$
$V_{CEO} > V_{IN} + \left(\frac{N1}{N2}\right) V_{OUT}$	$V_{CEO} > V_{IN}\left(1 + \frac{N1}{N3}\right)$
$I_{CR1} = I_{RL}$	$I_{CR1} = \frac{\hat{I}_{MAG}}{2}\left(\frac{\tau}{T}\right)$ $I_{CR2} = I_{RL}\left(\frac{\tau}{T}\right)$ $I_{CR3} = I_{RL}\left(\frac{T-\tau}{T}\right)$
$V_{RM} = V_{IN}\left(\frac{N2}{N1}\right)$	$V_{CR1} = V_{IN}\left(1 + \frac{N3}{N1}\right)$ $V_{CR2} = V_{IN}\left(\frac{N2}{N3}\right)$ $V_{CR3} = V_{IN}\left(\frac{N2}{N1}\right)$ } V_{RM}
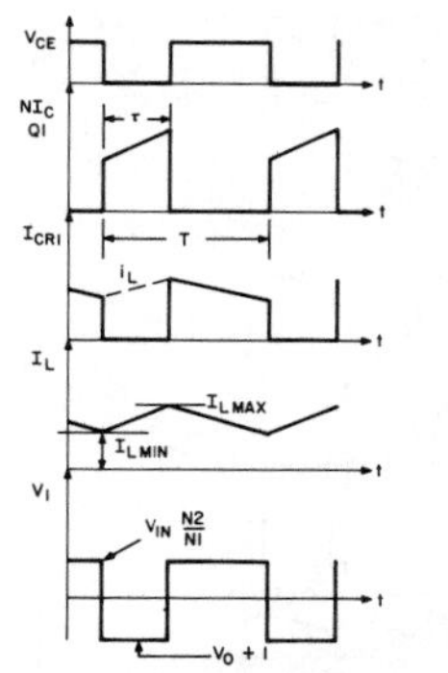	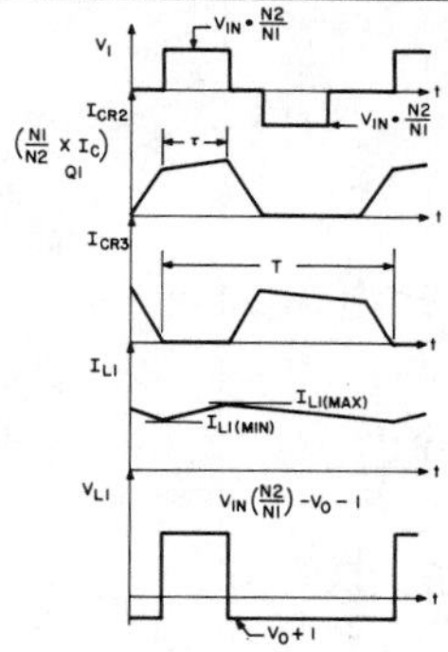
SIMPLE, MULTIPLE OUTPUTS ARE POSSIBLE. COLLECTOR CURRENT REDUCED BY TURNS RATIO OF TRANSFORMER. LOW PARTS COUNT, ISOLATION.	SIMPLE, MULTIPLE OUTPUTS ARE POSSIBLE, COLLECTOR CURRENT REDUCED BY RATIO OF $\frac{N2}{N1}$. LOW OUTPUT RIPPLE.
POOR TRANSFORMER UTILIZATION, TRANSFORMER DESIGN CRITICAL, HIGH OUTPUT RIPPLE.	POOR TRANSFORMER UTILIZATION, POOR TRANSIENT RESPONSE, PARTS COUNT HIGH, TRANSFORMER DESIGN IS CRITICAL.

TABLE 1.3 (Continued)

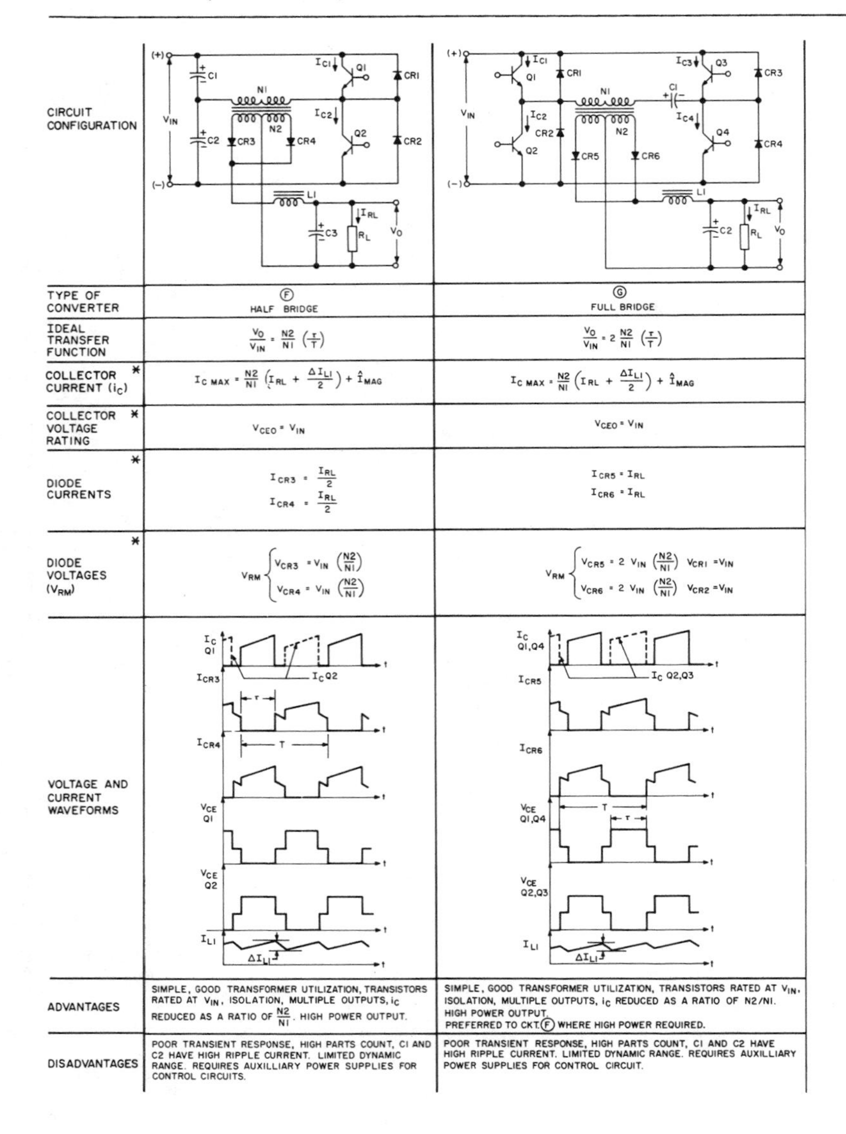

CIRCUIT CONFIGURATION		
TYPE OF CONVERTER	(F) HALF BRIDGE	(G) FULL BRIDGE
IDEAL TRANSFER FUNCTION	$\frac{V_O}{V_{IN}} = \frac{N2}{N1}\left(\frac{\tau}{T}\right)$	$\frac{V_O}{V_{IN}} = 2\,\frac{N2}{N1}\left(\frac{\tau}{T}\right)$
COLLECTOR CURRENT (i_C) *	$I_{C\,MAX} = \frac{N2}{N1}\left(I_{RL} + \frac{\Delta I_{L1}}{2}\right) + \hat{I}_{MAG}$	$I_{C\,MAX} = \frac{N2}{N1}\left(I_{RL} + \frac{\Delta I_{L1}}{2}\right) + \hat{I}_{MAG}$
COLLECTOR VOLTAGE RATING *	$V_{CEO} = V_{IN}$	$V_{CEO} = V_{IN}$
DIODE CURRENTS *	$I_{CR3} = \frac{I_{RL}}{2}$ $I_{CR4} = \frac{I_{RL}}{2}$	$I_{CR5} = I_{RL}$ $I_{CR6} = I_{RL}$
DIODE VOLTAGES (V_{RM}) *	V_{RM}: $V_{CR3} = V_{IN}\left(\frac{N2}{N1}\right)$ $V_{CR4} = V_{IN}\left(\frac{N2}{N1}\right)$	V_{RM}: $V_{CR5} = 2\,V_{IN}\left(\frac{N2}{N1}\right)$ $V_{CR1} = V_{IN}$ $V_{CR6} = 2\,V_{IN}\left(\frac{N2}{N1}\right)$ $V_{CR2} = V_{IN}$
VOLTAGE AND CURRENT WAVEFORMS		
ADVANTAGES	SIMPLE, GOOD TRANSFORMER UTILIZATION, TRANSISTORS RATED AT V_{IN}, ISOLATION, MULTIPLE OUTPUTS, i_C REDUCED AS A RATIO OF $\frac{N2}{N1}$. HIGH POWER OUTPUT.	SIMPLE, GOOD TRANSFORMER UTILIZATION, TRANSISTORS RATED AT V_{IN}, ISOLATION, MULTIPLE OUTPUTS, i_C REDUCED AS A RATIO OF N2/N1. HIGH POWER OUTPUT. PREFERRED TO CKT. (F) WHERE HIGH POWER REQUIRED.
DISADVANTAGES	POOR TRANSIENT RESPONSE, HIGH PARTS COUNT, C1 AND C2 HAVE HIGH RIPPLE CURRENT. LIMITED DYNAMIC RANGE. REQUIRES AUXILLIARY POWER SUPPLIES FOR CONTROL CIRCUITS.	POOR TRANSIENT RESPONSE, HIGH PARTS COUNT, C1 AND C2 HAVE HIGH RIPPLE CURRENT. LIMITED DYNAMIC RANGE. REQUIRES AUXILLIARY POWER SUPPLIES FOR CONTROL CIRCUIT.

TABLE 1.3 (Continued)

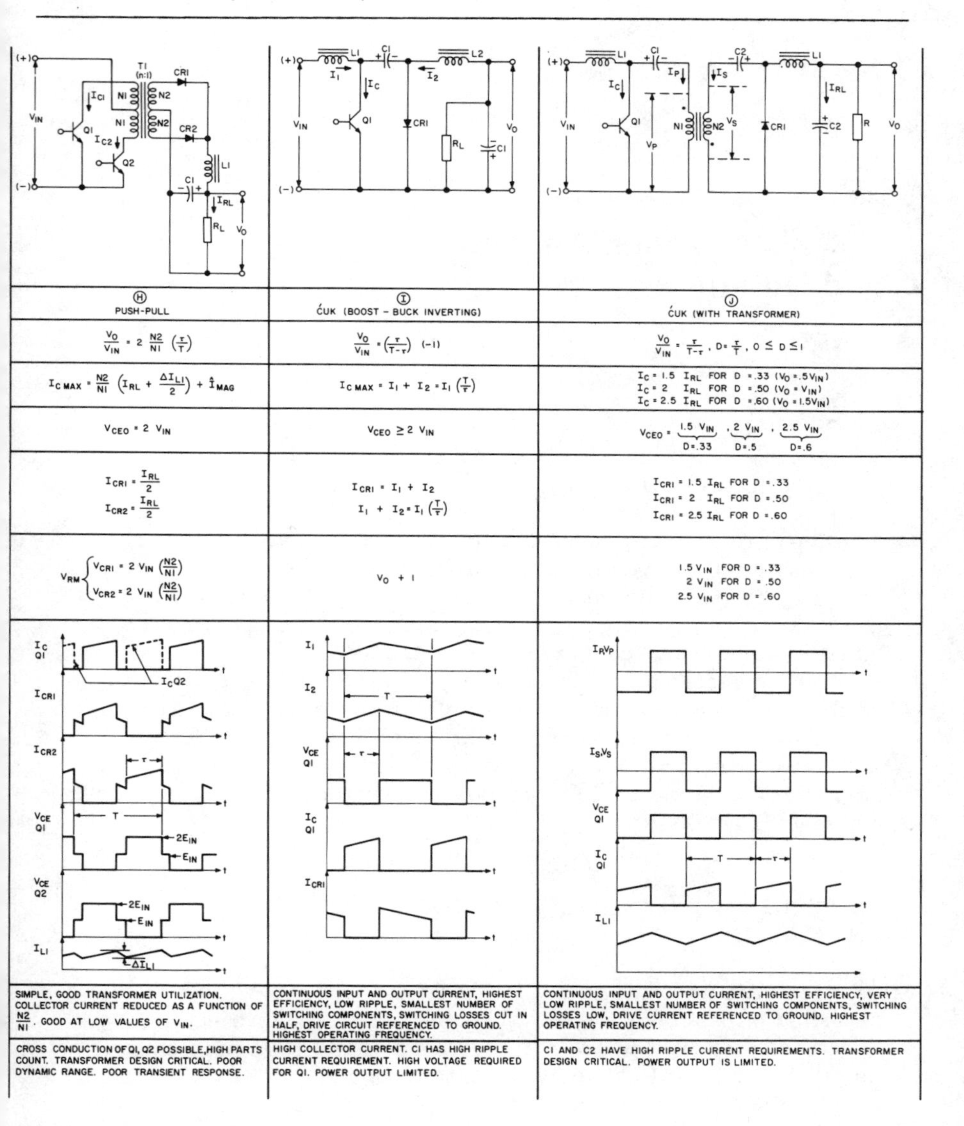

(H) PUSH-PULL	(I) ĆUK (BOOST - BUCK INVERTING)	(J) ĆUK (WITH TRANSFORMER)
$\frac{V_O}{V_{IN}} = 2\frac{N2}{N1}\left(\frac{\tau}{T}\right)$	$\frac{V_O}{V_{IN}} = \left(\frac{\tau}{T-\tau}\right)(-1)$	$\frac{V_O}{V_{IN}} = \frac{\tau}{T-\tau}, D = \frac{\tau}{T}, 0 \le D \le 1$
$I_{C\,MAX} = \frac{N2}{N1}\left(I_{RL} + \frac{\Delta I_{L1}}{2}\right) + \hat{I}_{MAG}$	$I_{C\,MAX} = I_1 + I_2 = I_1\left(\frac{T}{\tau}\right)$	$I_C = 1.5\ I_{RL}$ FOR D = .33 ($V_O = .5V_{IN}$) $I_C = 2\ I_{RL}$ FOR D = .50 ($V_O = V_{IN}$) $I_C = 2.5\ I_{RL}$ FOR D = .60 ($V_O = 1.5V_{IN}$)
$V_{CEO} = 2\ V_{IN}$	$V_{CEO} \ge 2\ V_{IN}$	$V_{CEO} = \underbrace{1.5\ V_{IN}}_{D=.33}, \underbrace{2\ V_{IN}}_{D=.5}, \underbrace{2.5\ V_{IN}}_{D=.6}$
$I_{CR1} = \frac{I_{RL}}{2}$ $I_{CR2} = \frac{I_{RL}}{2}$	$I_{CR1} = I_1 + I_2$ $I_1 + I_2 = I_1\left(\frac{T}{\tau}\right)$	$I_{CR1} = 1.5\ I_{RL}$ FOR D = .33 $I_{CR1} = 2\ I_{RL}$ FOR D = .50 $I_{CR1} = 2.5\ I_{RL}$ FOR D = .60
$V_{RM} \begin{cases} V_{CR1} = 2\ V_{IN}\left(\frac{N2}{N1}\right) \\ V_{CR2} = 2\ V_{IN}\left(\frac{N2}{N1}\right) \end{cases}$	$V_O + 1$	1.5 V_{IN} FOR D = .33 2 V_{IN} FOR D = .50 2.5 V_{IN} FOR D = .60
SIMPLE, GOOD TRANSFORMER UTILIZATION. COLLECTOR CURRENT REDUCED AS A FUNCTION OF $\frac{N2}{N1}$. GOOD AT LOW VALUES OF V_{IN}.	CONTINUOUS INPUT AND OUTPUT CURRENT, HIGHEST EFFICIENCY, LOW RIPPLE, SMALLEST NUMBER OF SWITCHING COMPONENTS, SWITCHING LOSSES CUT IN HALF, DRIVE CIRCUIT REFERENCED TO GROUND. HIGHEST OPERATING FREQUENCY.	CONTINUOUS INPUT AND OUTPUT CURRENT, HIGHEST EFFICIENCY, VERY LOW RIPPLE, SMALLEST NUMBER OF SWITCHING COMPONENTS, SWITCHING LOSSES LOW, DRIVE CURRENT REFERENCED TO GROUND. HIGHEST OPERATING FREQUENCY.
CROSS CONDUCTION OF Q1, Q2 POSSIBLE, HIGH PARTS COUNT. TRANSFORMER DESIGN CRITICAL. POOR DYNAMIC RANGE. POOR TRANSIENT RESPONSE.	HIGH COLLECTOR CURRENT. C1 HAS HIGH RIPPLE CURRENT REQUIREMENT. HIGH VOLTAGE REQUIRED FOR Q1. POWER OUTPUT LIMITED.	C1 AND C2 HAVE HIGH RIPPLE CURRENT REQUIREMENTS. TRANSFORMER DESIGN CRITICAL. POWER OUTPUT IS LIMITED.

*For reliable operation, it is suggested and recommended that all voltage and current ratings be increased to 125% of the required maximum.

2
Calculator-Aided Analysis and Design

2.1 INTRODUCTION

The design and analysis of power converters have been considerably simplified by means of the programmable calculators. In this chapter, complete listings of programs for the analysis of the output filter with damping, the buck converter, the boost converter, and the buck-boost converter are included.

The converter programs analyze converter circuits consisting of the power control switch, the pulse width modulator (PWM), and the related output filter components with damping. The equations used are based on the canonic models derived by Ćuk [78]. The mathematical approach to state-space averaging analysis of these converters is given in Appendix A.

2.2 GENERAL COMMENTS ON PROGRAMS AND CALCULATORS

All programs were originally written for the HP-97* calculator and recorded on magnetic cards. The magnetic cards were then read and translated by the HP-41* card reader and then printed out for use with the HP-41 calculator. An appropriate alpha abbreviated title was added to each program to facilitate easy alpha address.

---

*Registered trademark of the Hewlett-Packard Company.

These programs are easily adaptable to other Hewlett-Packard calculators, i.e., the HP-67,* HP-34C,* and HP-11C,* with 16 data storage registers or more.

The programs have also been converted to the algebraic operating system (AOS)† language for use on the TI-58† calculator. These converted programs are also adaptable to other Texas Instruments calulators, that is, the SR-52† and TI-59† calculators.

Another recent introduction is the HP-15C* calculator from Hewlett-Packard. It has a complex stack for complex number manipulations. This complex stack feature made the HP-15C a very versatile calculator, and the circuits herein could be easily analyzed with this tool.

No attempt was made to minimize the number of steps in any of the programs provided.

2.3 GENERAL USER INSTRUCTIONS

All HP programs assume the use of an optional printer. If no printer is attached, the PRTX command should be changed to R/S in each case, so that the calculated magnitude will be displayed first. Press R/S again to obtain angle.

Component values for the circuit to be analyzed should be calculated in accordance with the procedures outlined in Chapter 1 for continuous conduction mode of operation.

Since the converter programs are analyzing the open-loop response of the power stage with the output filter, the output voltage should be calculated using the designed D_H and D_L values for given input voltages. See the example in Section 2.6.

Load all program steps in "program" mode into the calculator.

Set the calculator to RUN mode; for HP-41, press PROGRAM to come out of program mode and then press the USER key.

Load all data storage registers according to the tabulations given in Table 2.1 for HP calculators or Table 2.2 for TI calculators, as the case may be.

Key in frequency.

Press A (Σ+ for HP-41) to run the program.

Always load D_H in register 6 with the corresponding R_{min} in register 5 for one set of magnitude and phase curves.

---

*Registered trademark of the Hewlett-Packard Company.

†Registered trademark of Texas Instruments Limited.

TABLE 2.1 Allocation of Data Storage Registers for the HP Calculator Programs

Register			Program				
HP-97	HP-67	HP-41	Output filter	Buck	Boost	Buck-boost	Isolated flyback
0	0	00	Reserved for ω in program				
1	1	01	L	L	L	L	L
2	2	02	C	C	C	C	C
3	3	03	r	r	r	r	r
4	4	04	nC	nC	nC	nC	nC
5	5	05	R	R	R	R	R
6	6	06		D	D	D	D
7	7	07		V_o	V_o	V_o	V_o
8	8	08		V_m	V_m	V_m	V_m
9	9	09					
A	A	20					
B	B	21					
C	C	22					
D	D	23					
E	E	24	r_L	r_L			
I	I	25			r_L	r_L	n

TABLE 2.2 Data Register Allocations for TI-58 Programs

	Program			
Register	Output filter	Buck	Boost	Buck-boost
01	L	L	L	L
02	C	C	C	C
03	r	r	r	r
04	nC	nC	nC	nC
05	R	R	R	R
06		D	D	D
07		V_o	V_o	V_o
08		V_m	V_m	V_m
09				
10	r_L			
11				
12				
13		r_L		
14				
15			r_L	r_L

For the next set of curves, load D_L with the corresponding R_{max} accordingly.

The calculated magnitude and phase for the key-in frequency will be printed out as the calculation is completed. If no printer is attached, the initial result displayed is the magnitude. Press R/S to obtain angle.

To run the program for the special case of "no damping," load a very small value for r (say, 1 mΩ) and a small value for nC (say, 1 pF).

Note: It is important to enter a value other than zero for V_m (STO 08) to obtain a rational result.

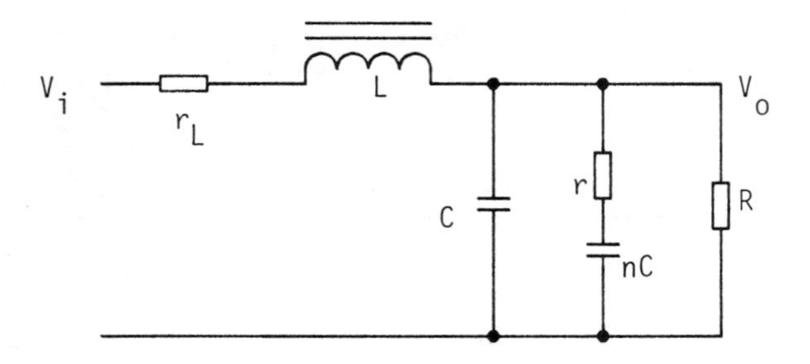

FIG. 2.1 Output filter with damping.

2.4 CIRCUITS AND EQUATIONS

Output Filter with Damping

This circuit is shown in Fig. 2.1. The transfer function of this circuit is represented by $H_e(s) = (V_o/V_i)$.

Buck Converter

This circuit is shown in Fig. 2.2. The small-signal output to control voltage transfer function is given by (based on Ćuk's canonic model [78])

$$\frac{\hat{v}_o}{\hat{v}_c} = \frac{V_o}{DV_m} H_e(s) \tag{2.1}$$

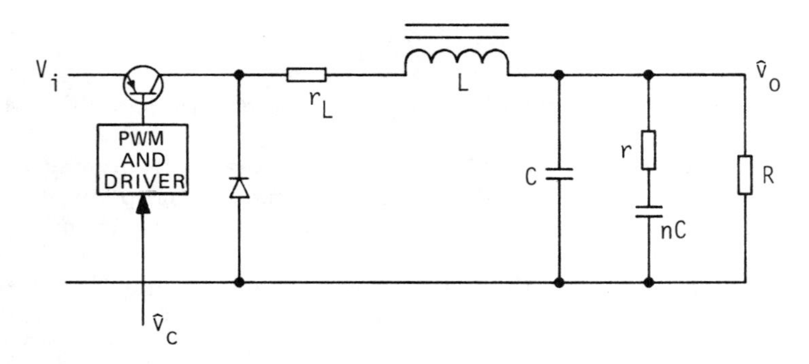

FIG. 2.2 Buck converter.

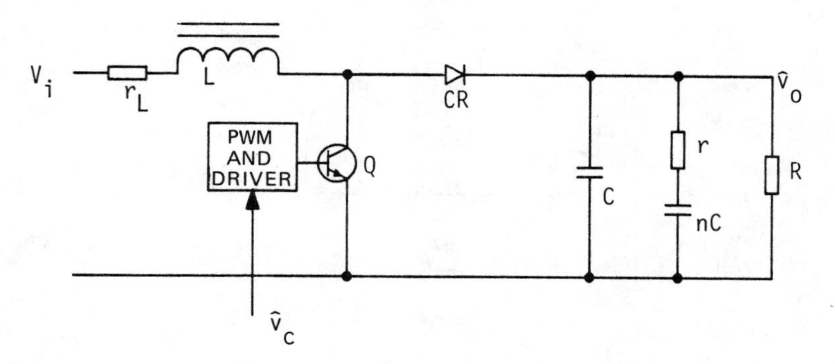

FIG. 2.3 Boost converter.

Boost Converter

This circuit is shown in Fig. 2.3. The small-signal output to control voltage transfer function for this converter is

$$\frac{\hat{v}_o}{\hat{v}_c} = \frac{V_o}{(1 - D)V_m}\left[1 - \frac{j\omega L}{(1 - D)^2 R}\right]H_e(s) \tag{2.2}$$

Buck-Boost Converter

This circuit is shown in Fig. 2.4. The small-signal output to control voltage transfer function for this converter is

$$\frac{\hat{v}_o}{\hat{v}_c} = \frac{V_o}{D(1 - D)V_m}\left[1 - \frac{j\omega DL}{(1 - D)^2 R}\right]H_e(s) \tag{2.3}$$

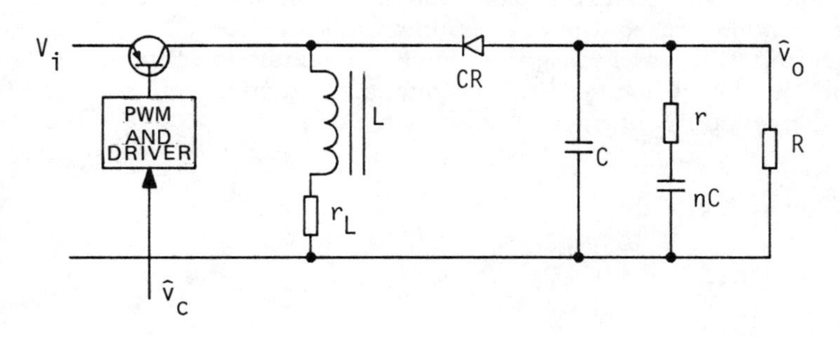

FIG. 2.4 Buck-boost converter.

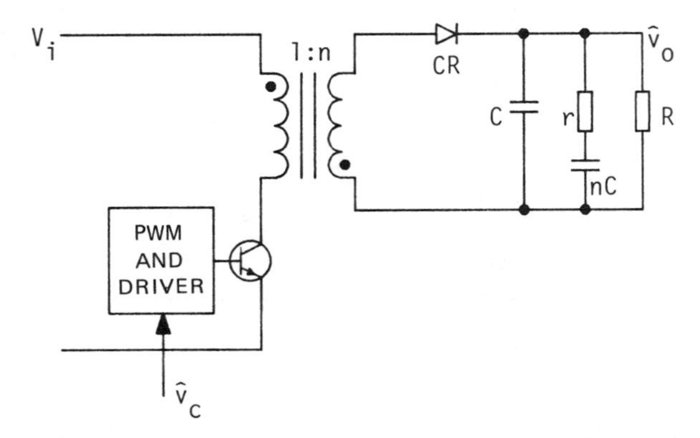

FIG. 2.5 Isolated flyback converter.

Isolated Flyback Converter

This circuit is shown in Fig. 2.5. The small-signal output to control voltage transfer function for this converter is

$$\frac{\hat{v}_o}{\hat{v}_c} = \frac{V_o}{V_m D(1 - D)}\left[1 - \frac{j\omega n^2 DL}{(1 - D)^2 R}\right] H_e(s) \tag{2.4}$$

2.5 PROGRAM LISTINGS

The following programs are listed in their order of appearance:

HP-41 and HP-97: Output filter with damping; buck converter, continuous conduction; boost converter, continuous conduction; buck-boost converter, continuous conduction; isolated flyback converter, continuous conduction mode

TI-58: Output filter with damping; buck converter, continuous conduction; boost converter, continuous conduction; buck-boost converter, continuous conduction mode

Program Listing for HP-41C: Output Filter with Damping

☐ 67 ☐ 97 ☑ 41C

STEP/LINE | KEY ENTRY | KEY CODE (67/97 only) | COMMENTS

```
01◆LBL 67
02◆LBL "O=P
       FIL"
03◆LBL 10
04◆LBL A
05 2
06 *
07 PI
08 *
09 STO 00
10 RCL 04
11 RCL 00
12 *
13 1/X
14 CHS
15 RCL 03
16 R-P
17 1/X
18 X<>Y
19 CHS
20 X<>Y
21 P-R
22 STO 21
23 X<>Y
24 STO 20
25 RCL 05
26 1/X
27 RCL 21
28 +
29 RCL 02
30 RCL 00
31 *
32 RCL 20
33 +
34 X<>Y
35 R-P
36 1/X
37 STO 20
38 X<>Y
39 CHS
40 STO 21
41 X<>Y
42 P-R
43 STO 22
44 RCL 24
45 +
46 STO 22
47 X<>Y
48 STO 23
49 RCL 01
50 RCL 00
51 *
52 RCL 23
53 +
54 RCL 22
55 R-P
56 STO 23
57 X<>Y
58 STO 22
59 RCL 20
60 RCL 23
61 /
62 LOG
63 20
64 *
65 STO 20
66 7PRTX
67 RCL 21
68 RCL 22
69 -
70 STO 23
71 7PRTX
72 RTN
73 STOP
74 .END.
```

Program Listing for HP-41C: Buck Converter, Continuous Conduction

☐ 67 ☐ 97 ■ 41C

STEP/LINE	KEY ENTRY	KEY CODE (67/97 only)	COMMENTS

```
01♦LBL 67              BUCK
02♦LBL "BUC
K"
03♦LBL 10
04♦LBL A
05 2
06 *
07 PI
08 *
09 STO 00
10 RCL 04
11 *
12 1/X
13 CHS
14 RCL 03
15 R-P
16 1/X
17 X<>Y
18 CHS
19 X<>Y
20 P-R
21 STO 21
22 X<>Y
23 STO 22
24 RCL 05
25 1/X
26 RCL 21
27 +
28 RCL 02
29 RCL 00
30 *
31 RCL 22
32 +
33 X<>Y
34 R-P
35 1/X
36 STO 22
37 X<>Y
38 CHS
39 STO 20
40 X<>Y
41 P-R
42 STO 21
43 X<>Y
44 STO 23
45 RCL 01
46 RCL 00
47 *
48 RCL 23
49 +
50 RCL 21
51 RCL 24
52 +
53 R-P
54 STO 23
55 X<>Y
56 STO 21
57 RCL 22
58 RCL 23
59 /
60 LOG
61 20
62 *
63 STO 22
64 RCL 20
65 RCL 21
66 -
67 STO 23
68 RCL 07
69 RCL 08
70 /
71 RCL 06
72 /
73 LOG
74 20
75 *
76 RCL 22
77 +
78 7PRTX
79 RCL 23
80 7PRTX
81 ADV
82 RTN
83 STOP
84 .END.
```

STEP/LINE	KEY ENTRY	KEY CODE (67/97 only)
90		
00		

Program Listing for HP-41C: Boost Converter, Continuous Conduction

☐ 67 ☐ 97 ■ 41C

STEP/LINE	KEY ENTRY	KEY CODE (67/97 only)	COMMENTS
01	◆LBL 67		BOOST
02	◆LBL "BOOST"		
03	◆LBL 10		
04	◆LBL A		
05	2		
06	*		
07	PI		
08	*		
09	STO 00		
10	RCL 01		
11	*		
12	RCL 05		
13	/		
14	RCL 06		
15	CHS		
16	1		
17	+		
18	X↑2		
19	/		
20	RCL 07		
21	*		
22	RCL 08		
23	/		
24	RCL 06		
25	CHS		
26	1		
27	+		
28	/		
29	CHS		
30	RCL 07		
31	RCL 08		
32	/		
33	RCL 06		
34	CHS		
35	1		
36	+		
37	/		
38	R-P		
39	LOG		
40	20		
41	*		
42	STO 09		
43	X<>Y		
44	STO 20		
45	RCL 00		
46	RCL 04		
47	*		
48	1/X		
49	CHS		
50	RCL 03		
51	RCL 06		
52	CHS		
53	1		
54	+		
55	1/X		
56	*		
57	R-P		
58	1/X		
59	X<>Y		
60	CHS		
61	X<>Y		
62	P-R		
63	STO 21		
64	X<>Y		
65	STO 22		
66	RCL 05		
67	1/X		
68	RCL 21		
69	+		
70	RCL 02		
71	RCL 00		
72	*		
73	RCL 22		
74	+		
75	X<>Y		
76	R-P		
77	1/X		
78	STO 22		
79	X<>Y		
80	CHS		
81	STO 24		
82	X<>Y		
83	P-R		
84	STO 21		
85	X<>Y		
86	STO 23		
87	RCL 01		
88	RCL 06		
89	CHS		
90	1		
91	+		
92	X↑2		
93	/		
94	RCL 00		
95	*		
96	RCL 23		
97	+		

HP-41C: Boost Converter (Continued)

☐ 67 ☐ 97 ■ 41C

STEP/LINE	KEY ENTRY	KEY CODE (67/97 only)	COMMENTS	STEP/LINE	KEY ENTRY	KEY CODE (67/97 only)	COMMENTS

```
 98 RCL 25
 99 RCL 06
100 CHS
101 1
102 +
103 X↑2
104 /
105 RCL 21
106 +
107 R-P
108 STO 23
109 X<>Y
110 STO 21
111 RCL 22
112 RCL 23
113 /
114 LOG
115 20
116 *
117 STO 22
118 RCL 24
119 RCL 21
120 -
121 STO 23
122 RCL 22
123 RCL 09
124 +
125 7PRTX
126 RCL 23
127 RCL 20
128 +
129 7PRTX
130 ADV
131 RTN
132 STOP
133 .END.
```

40

50

51

60

70

80

90

00

Program Listing for HP-41C: Buck-Boost Converter, Continuous Conduction

☐ 67 ☐ 97 ☑ 41C

STEP/LINE | KEY ENTRY | KEY CODE (67/97 only) | COMMENTS

```
01♦LBL 67
02♦LBL "BKB
         ST"
03♦LBL 10
04♦LBL A
05 2
06 *
07 PI
08 *
09 STO 00
10 RCL 06
11 *
12 RCL 01
13 *
14 RCL 05
15 /
16 RCL 06
17 CHS
18 1
19 +
20 X↑2
21 /
22 RCL 07
23 *
24 RCL 06
25 /
26 RCL 08
27 /
28 RCL 06
29 CHS
30 1
31 +
32 /
33 CHS
34 RCL 07
35 RCL 06
36 /
37 RCL 08
38 /
39 RCL 06
40 CHS
41 1
42 +
43 /
44 R-P
45 LOG
46 20
47 *
48 STO 09
49 X<>Y
50 STO 20
51 RCL 00
52 RCL 04
53 *
54 1/X
55 CHS
56 RCL 03
57 RCL 06
58 CHS
59 1
60 +
61 1/X
62 *
63 R-P
64 1/X
65 X<>Y
66 CHS
67 X<>Y
68 P-R
69 STO 21
70 X<>Y
71 STO 22
72 RCL 05
73 1/X
74 RCL 21
75 +
76 RCL 02
77 RCL 00
78 *
79 RCL 22
80 +
81 X<>Y
82 R-P
83 1/X
84 STO 22
85 X<>Y
86 CHS
87 STO 24
88 X<>Y
89 P-R
90 STO 21
91 X<>Y
92 STO 23
93 RCL 21
94 RCL 25
95 RCL 06
96 CHS
97 1
```

HP-41C: Buck-Boost Converter (Continued)

☐ 67 ☐ 97 ☑ 41C

STEP/LINE	KEY ENTRY	KEY CODE (67/97 only)	COMMENTS	STEP/LINE	KEY ENTRY	KEY CODE (67/97 only)	COMMENTS

```
 98 +
 99 X↑2
100 1/X
101 *
102 +
103 STO 21
104 RCL 01
105 RCL 06
106 CHS
107 1
108 +
109 X↑2
110 /
111 RCL 00
112 *
113 RCL 23
114 +
115 RCL 21
116 R-P
117 STO 23
118 X<>Y
119 STO 21
120 RCL 22
121 RCL 23
122 /
123 LOG
124 20
125 *
126 STO 22
127 RCL 24
128 RCL 21
129 -
130 STO 23
131 RCL 22
132 RCL 09
133 +
134 7PRTX
135 RCL 23
136 RCL 20
137 +
138 7PRTX
139 ADV
140 RTN
141 STOP
142 .END.
```

50 51 60 70 80 90 00

Program Listing for HP-41C: Isolated Flyback Converter, Continuous Conduction

☐ 67 ☐ 97 ☑ 41C

STEP/LINE | KEY ENTRY | KEY CODE (67/97 only) | COMMENTS

```
01♦LBL 67
02♦LBL "FLY
        BK"
03♦LBL 10
04♦LBL A
05 2
06 *
07 PI
08 *
09 STO 00
10 RCL 06
11 *
12 RCL 01
13 *
14 RCL 25
15 X↑2
16 *
17 RCL 05
18 /
19 RCL 06
20 CHS
21 1
22 +
23 X↑2
24 /
25 RCL 07
26 *
27 RCL 06
28 /
29 RCL 08
30 /
31 RCL 06
32 CHS
33 1
34 +
35 /
36 CHS
37 RCL 07
38 RCL 06
39 /
40 RCL 08
41 /
42 RCL 06
43 CHS
44 1
45 +
46 /
47 R-P
48 LOG
49 20
50 *
51 STO 09
52 X<>Y
53 STO 20
54 RCL 00
55 RCL 04
56 *
57 1/X
58 CHS
59 RCL 03
60 R-P
61 1/X
62 X<>Y
63 CHS
64 X<>Y
65 P-R
66 STO 21
67 X<>Y
68 STO 22
69 RCL 05
70 1/X
71 RCL 21
72 +
73 RCL 02
74 RCL 00
75 *
76 RCL 22
77 +
78 X<>Y
79 R-P
80 1/X
81 STO 22
82 X<>Y
83 CHS
84 STO 24
85 X<>Y
86 P-R
87 STO 21
88 X<>Y
89 STO 23
90 RCL 01
91 RCL 06
92 CHS
93 1
94 +
95 X↑2
96 /
97 RCL 00
```

HP-41C: Isolated Flyback Converter (Continued)

☐ 67 ☐ 97 ☑ 41C

STEP/LINE	KEY ENTRY	KEY CODE (67/97 only)	COMMENTS

```
 98 *
 99 RCL 23
100 +
101 RCL 21
102 R-P
103 STO 23
104 X<>Y
105 STO 21
106 RCL 22
107 RCL 23
108 /
109 LOG
110 20
111 *
112 STO 22
113 RCL 24
114 RCL 21
115 -
116 STO 23
117 RCL 22
118 RCL 09
119 +
120 7PRTX
121 RCL 23
122 RCL 20
123 +
124 7PRTX
125 ADV
126 RTN
127 STOP
128 .END.
```

40

50

STEP/LINE	KEY ENTRY	KEY CODE (67/97 only)	COMMENTS
51			
60			
70			
80			
90			
00			

Program Listing for HP-97: Output Filter with Damping

STEP	KEY ENTRY	KEY CODE	COMMENTS
001	*LBLA	21 11	HP-97
002	2	02	Output Filter
003	×	-35	
004	Pi	16-24	
005	×	-35	
006	STO0	35 00	
007	RCL4	36 04	
008	RCL0	36 00	
009	×	-35	
010	1/X	52	
011	CHS	-22	
012	RCL3	36 03	
013	→P	34	
014	1/X	52	
015	X⇄Y	-41	
016	CHS	-22	
017	X⇄Y	-41	
018	→R	44	
019	STOB	35 12	
020	X⇄Y	-41	
021	STOA	35 11	
022	RCL5	36 05	
023	1/X	52	
024	RCLB	36 12	
025	+	-55	
026	RCL2	36 02	
027	RCL0	36 00	
028	×	-35	
029	RCLA	36 11	
030	+	-55	
031	X⇄Y	-41	
032	→P	34	
033	1/X	52	
034	STOA	35 11	
035	X⇄Y	-41	
036	CHS	-22	
037	STOB	35 12	
038	X⇄Y	-41	
039	→R	44	
040	STOC	35 13	
041	RCLE	36 15	
042	+	-55	
043	STOC	35 13	
044	X⇄Y	-41	
045	STOD	35 14	
046	RCL1	36 01	
047	RCL0	36 00	
048	×	-35	
049	RCLD	36 14	
050	+	-55	
051	RCLC	36 13	
052	→P	34	
053	STOD	35 14	
054	X⇄Y	-41	
055	STOC	35 13	
056	RCLA	36 11	
057	RCLD	36 14	
058	÷	-24	
059	LOG	16 32	
060	2	02	
061	0	00	
062	×	-35	
063	STOA	35 11	
064	PRTX	-14	
065	RCLB	36 12	
066	RCLC	36 13	
067	−	-45	
068	STOD	35 14	
069	PRTX	-14	
070	RTN	24	
071	R/S	51	
080			
090			
100			
110			

REGISTERS

0	1	2	3	4	5	6	7	8	9
	L	C	r	nC	R				

S0	S1	S2	S3	S4	S5	S6	S7	S8	S9

A	B	C	D	E	I
				r_L	

Program Listing for HP-97: Buck Converter, Continuous Conduction

STEP	KEY ENTRY	KEY CODE	COMMENTS
001	*LBLA	21 11	HP-97
002	2	02	Buck Converter
003	×	-35	
004	Pi	16-24	
005	×	-35	
006	STO0	35 00	
007	RCL4	36 04	
008	×	-35	
009	1/X	52	
010	CHS	-22	
011	RCL3	36 03	
012	→P	34	
013	1/X	52	
014	X⇄Y	-41	
015	CHS	-22	
016	X⇄Y	-41	
017	→R	44	
018	STOB	35 12	
019	X⇄Y	-41	
020	STOC	35 13	
021	RCL5	36 05	
022	1/X	52	
023	RCLB	36 12	
024	+	-55	
025	RCL2	36 02	
026	RCL0	36 00	
027	×	-35	
028	RCLC	36 13	
029	+	-55	
030	X⇄Y	-41	
031	→P	34	
032	1/X	52	
033	STOC	35 13	
034	X⇄Y	-41	
035	CHS	-22	
036	STOA	35 11	
037	X⇄Y	-41	
038	→R	44	
039	STOB	35 12	
040	X⇄Y	-41	
041	STOD	35 14	
042	RCL1	36 01	
043	RCL0	36 00	
044	×	-35	
045	RCLD	36 14	
046	+	-55	
047	RCLB	36 12	
048	RCLE	36 15	
049	+	-55	
050	→P	34	
051	STOD	35 14	
052	X⇄Y	-41	
053	STOB	35 12	
054	RCLC	36 13	
055	RCLD	36 14	
056	÷	-24	
057	LOG	16 32	
058	2	02	
059	0	00	
060	×	-35	
061	STOC	35 13	
062	RCLA	36 11	
063	RCLB	36 12	
064	-	-45	
065	STOD	35 14	
066	RCL7	36 07	
067	RCL8	36 08	
068	÷	-24	
069	RCL6	36 06	
070	÷	-24	
071	LOG	16 32	
072	2	02	
073	0	00	
074	×	-35	
075	RCLC	36 13	
076	+	-55	
077	PRTX	-14	
078	RCLD	36 14	
079	PRTX	-14	
080	SPC	16-11	
081	RTN	24	
082	R/S	51	
090			
100			
110			

REGISTERS

0	1	2	3	4	5	6	7	8	9
	L	C	r	nC	R	D	V_o	V_m	

S0	S1	S2	S3	S4	S5	S6	S7	S8	S9

A	B	C	D	E	I
				r_L	

Program Listing for HP-97: Boost Converter, Continuous Conduction

STEP	KEY ENTRY	KEY CODE	COMMENTS	STEP	KEY ENTRY	KEY CODE	COMMENTS
001	*LBLA	21 11	HP-97	057	X⇄Y	-41	
002	2	02	Boost Converter	058	CHS	-22	
003	x	-35		059	X⇄Y	-41	
004	Pi	16-24		060	→R	44	
005	x	-35		061	STOB	35 12	
006	STO0	35 00		062	X⇄Y	-41	
007	RCL1	36 01		063	STOC	35 13	
008	x	-35		064	RCL5	36 05	
009	RCL5	36 05		065	1/X	52	
010	÷	-24		066	RCLB	36 12	
011	RCL6	36 06		067	+	-55	
012	CHS	-22		068	RCL2	36 02	
013	1	01		069	RCL0	36 00	
014	+	-55		070	x	-35	
015	X²	53		071	RCLC	36 13	
016	÷	-24		072	+	-55	
017	RCL7	36 07		073	X⇄Y	-41	
018	x	-35		074	→P	34	
019	RCL8	36 08		075	1/X	52	
020	÷	-24		076	STOC	35 13	
021	RCL6	36 06		077	X⇄Y	-41	
022	CHS	-22		078	CHS	-22	
023	1	01		079	STOE	35 15	
024	+	-55		080	X⇄Y	-41	
025	÷	-24		081	→R	44	
026	CHS	-22		082	STOB	35 12	
027	RCL7	36 07		083	X⇄Y	-41	
028	RCL8	36 08		084	STOD	35 14	
029	÷	-24		085	RCL1	36 01	
030	RCL6	[illegible] 06		086	RCL6	36 06	
031	CHS	-22		087	CHS	-22	
032	1	01		088	1	01	
033	+	-55		089	+	-55	
034	÷	-24		090	X²	53	
035	→P	34		091	÷	-24	
036	LOG	16 32		092	PCL0	36 00	
037	2	02		093	x	-35	
038	0	00		094	RCLD	36 14	
039	x	-35		095	+	-55	
040	STO9	35 09		096	RCLI	36 46	
041	X⇄Y	-41		097	RCL6	36 06	
042	STOA	35 11		098	CHS	-22	
043	RCL0	36 00		099	1	01	
044	RCL4	36 04		100	+	-55	
045	x	-35		101	X²	53	
046	1/X	52		102	÷	-24	
047	CHS	-22		103	RCLB	36 12	
048	RCL3	36 03		104	+	-55	
049	RCL6	36 06		105	→P	34	
050	CHS	-22		106	STOD	35 14	
051	1	01		107	X⇄Y	-41	
052	+	-55		108	STOB	35 12	
053	1/X	52		109	RCLC	36 13	
054	x	-35		110	RCLD	36 14	
055	→P	34		111	÷	-24	
056	1/X	52		112	LOG	16 32	

REGISTERS

0	1	2	3	4	5	6	7	8	9
	L	C	r	nC	R	D	V_o	V_m	

S0	S1	S2	S3	S4	S5	S6	S7	S8	S9

A	B	C	D	E	I
					r_L

HP-97: Boost Converter (Continued)

STEP	KEY ENTRY	KEY CODE	COMMENTS
113	2	02	
114	0	00	
115	x	-35	
116	STOC	35 13	
117	RCLE	36 15	
118	RCLB	36 12	
119	-	-45	
120	STOD	35 14	
121	RCLC	36 13	
122	RCL9	36 09	
123	+	-55	
124	PRTX	-14	
125	RCLD	36 14	
126	RCLA	36 11	
127	+	-55	
128	PRTX	-14	
129	SPC	16-11	
130	RTN	24	
131	R/S	51	
140			
150			
160			
170			
180			
190			
200			
210			
220			

LABELS					FLAGS
A	B	C	D	E	0
a	b	c	d	e	1
0	1	2	3	4	2
5	6	7	8	9	3

SET STATUS		
FLAGS	TRIG	DISP
ON OFF		
0 ☐ ☐	DEG ☐	FIX ☐
1 ☐ ☐	GRAD ☐	SCI ☐
2 ☐ ☐	RAD ☐	ENG ☐
3 ☐ ☐		n ___

Program Listing for HP-97: Buck-Boost Converter, Continuous Conduction

STEP	KEY ENTRY	KEY CODE	COMMENTS	STEP	KEY ENTRY	KEY CODE	COMMENTS
001	*LBLA	21 11	HP-97	057	1	01	
002	2	02	Buck-Boost	058	+	-55	
003	×	-35	Converter	059	1/X	52	
004	Pi	16-24		060	×	-35	
005	×	-35		061	→P	34	
006	STO0	35 00		062	1/X	52	
007	RCL6	36 06		063	X⇄Y	-41	
008	×	-35		064	CHS	-22	
009	RCL1	36 01		065	X⇄Y	-41	
010	×	-35		066	→R	44	
011	RCL5	36 05		067	STOB	35 12	
012	÷	-24		068	X⇄Y	-41	
013	RCL6	36 06		069	STOC	35 13	
014	CHS	-22		070	RCL5	36 05	
015	1	01		071	1/X	52	
016	+	-55		072	RCLB	36 12	
017	X²	53		073	+	-55	
018	÷	-24		074	RCL2	36 02	
019	RCL7	36 07		075	RCL0	36 00	
020	×	-35		076	×	-35	
021	RCL6	36 06		077	RCLC	36 13	
022	÷	-24		078	+	-55	
023	RCL8	36 08		079	X⇄Y	-41	
024	÷	-24		080	→P	34	
025	RCL6	36 06		081	1/X	52	
026	CHS	-22		082	STOC	35 13	
027	1	01		083	X⇄Y	-41	
028	+	-55		084	CHS	-22	
029	÷	-24		085	STOE	35 15	
030	CHS	-22		086	X⇄Y	-41	
031	RCL7	36 07		087	→R	44	
032	RCL6	36 06		088	STOB	35 12	
033	÷	-24		089	X⇄Y	-41	
034	RCL8	36 08		090	STOD	35 14	
035	÷	-24		091	RCLB	36 12	
036	RCL6	36 06		092	RCLI	36 46	
037	CHS	-22		093	RCL6	36 06	
038	1	01		094	CHS	-22	
039	+	-55		095	1	01	
040	÷	-24		096	+	-55	
041	→P	34		097	X²	53	
042	LOG	16 32		098	1/X	52	
043	2	02		099	×	-35	
044	0	00		100	+	-55	
045	×	-35		101	STOB	35 12	
046	STO9	35 09		102	RCL1	36 01	
047	X⇄Y	-41		103	RCL6	36 06	
048	STOA	35 11		104	CHS	-22	
049	RCL0	36 00		105	1	01	
050	RCL4	36 04		106	+	-55	
051	×	-35		107	X²	53	
052	1/X	52		108	÷	-24	
053	CHS	-22		109	RCL0	36 00	
054	RCL3	36 03		110	×	-35	
055	RCL6	36 06		111	RCLD	36 14	
056	CHS	-22		112	+	-55	

REGISTERS

0	1 L	2 C	3 r	4 nC	5 R	6 D	7 V_o	8 V_m	9
S0	S1	S2	S3	S4	S5	S6	S7	S8	S9

A	B	C	D	E	I r_L

HP-97: Buck-Boost Converter (Continued)

STEP	KEY ENTRY	KEY CODE	COMMENTS
113	RCLB	36 12	
114	→P	34	
115	STOD	35 14	
116	X⇄Y	-41	
117	STOB	35 12	
118	RCLC	36 13	
119	RCLD	36 14	
120	÷	-24	
121	LOG	16 32	
122	2	02	
123	0	00	
124	×	-35	
125	STOC	35 13	
126	RCLE	36 15	
127	RCLB	36 12	
128	-	-45	
129	STOD	35 14	
130	RCLC	36 13	
131	RCL9	36 09	
132	+	-55	
133	PRTX	-14	
134	RCLD	36 14	
135	RCLA	36 11	
136	+	-55	
137	PRTX	-14	
138	SPC	16-11	
139	RTN	24	
140	R/S	51	

Program Listing for HP-97: Isolated Flyback Converter, Continuous Conduction

STEP	KEY ENTRY	KEY CODE	COMMENTS	STEP	KEY ENTRY	KEY CODE	COMMENTS
001	*LBLA	21 11	HP-97	057	RCL3	36 03	
002	2	02	Isolated	058	→P	34	
003	x	-35	Flyback	059	1/X	52	
004	Pi	16-24	Converter	060	X⇄Y	-41	
005	x	-35		061	CHS	-22	
006	STO0	35 00		062	X⇄Y	-41	
007	RCL6	36 06		063	→R	44	
008	x	-35		064	STOB	35 12	
009	RCL1	36 01		065	X⇄Y	-41	
010	x	-35		066	STOC	35 13	
011	RCLI	36 46		067	RCL5	36 05	
012	X²	53		068	1/X	52	
013	x	-35		069	RCLB	36 12	
014	RCL5	36 05		070	+	-55	
015	÷	-24		071	RCL2	36 02	
016	RCL6	36 06		072	RCL0	36 09	
017	CHS	-22		073	x	-35	
018	1	01		074	RCLC	36 13	
019	+	-55		075	+	-55	
020	X²	53		076	X⇄Y	-41	
021	÷	-24		077	→P	34	
022	RCL7	36 07		078	1/X	52	
023	x	-35		079	STOC	35 13	
024	RCL6	36 06		080	X⇄Y	-41	
025	÷	-24		081	CHS	-22	
026	RCL8	36 06		082	STOE	35 15	
027	÷	-24		083	X⇄Y	-41	
028	RCL6	36 06		084	→R	44	
029	CHS	-22		085	STOB	35 12	
030	1	01		086	X⇄Y	-41	
031	+	-55		087	STOD	35 14	
032	÷	-24		088	RCL1	36 01	
033	CHS	-22		089	RCL6	36 06	
034	RCL7	36 07		090	CHS	-22	
035	RCL6	36 06		091	1	01	
036	÷	-24		092	+	-55	
037	RCL8	36 08		093	X²	53	
038	÷	-24		094	÷	-24	
039	RCL6	36 06		095	RCL0	36 00	
040	CHS	-22		096	x	-35	
041	1	01		097	RCLD	36 14	
042	+	-55		098	+	-55	
043	÷	-24		099	RCLB	36 12	
044	→P	34		100	→P	34	
045	LOG	16 32		101	STOD	35 14	
046	2	02		102	X⇄Y	-41	
047	0	00		103	STOB	35 12	
048	x	-35		104	RCLC	36 13	
049	STO9	35 09		105	RCLD	36 14	
050	X⇄Y	-41		106	÷	-24	
051	STOA	35 11		107	LOG	16 32	
052	RCL0	36 00		108	2	02	
053	RCL4	36 04		109	0	00	
054	x	-35		110	x	-35	
055	1/X	52		111	STOC	35 13	
056	CHS	-22		112	RCLE	36 15	

REGISTERS

0	1 L	2 C	3 r	4 nC	5 R	6 D	7 V_o	8 V_m	9
S0	S1	S2	S3	S4	S5	S6	S7	S8	S9

A	B	C	D	E	I n

HP-97: Isolated Flyback Converter (Continued)

STEP	KEY ENTRY	KEY CODE	COMMENTS
113	RCLB	36 12	
114	-	-45	
115	STOD	35 14	
116	RCLC	36 13	
117	RCL9	36 09	
118	+	-55	
119	PRTX	-14	
120	RCLD	36 14	
121	RCLA	36 11	
122	+	-55	
123	PRTX	-14	
124	SPC	16-11	
125	RTN	24	
126	R/S	51	
140			
150			
160			
170			
180			
190			
200			
210			
220			

LABELS					FLAGS
A	B	C	D	E	0
a	b	c	d	e	1
0	1	2	3	4	2
5	6	7	8	9	3

SET STATUS

FLAGS	ON	OFF	TRIG		DISP	
0	☐	☐	DEG	☐	FIX	☐
1	☐	☐	GRAD	☐	SCI	☐
2	☐	☐	RAD	☐	ENG	☐
3	☐	☐			n ____	

Program Listing for TI-58: Output Filter with Damping

TITLE Output Filter With Damping PAGE ____ OF ____ **TI Programmable Coding Form**

PROGRAMMER ____________ DATE ____________

LOC	CODE	KEY	COMMENTS

```
000  76  LBL        055  07   07        110  09   09
001  11   A         056  32  X:T        111  91  R/S
002  65   ×         057  35  1/X        112  92  RTN
003  02   2         058  42  STO
004  65   ×         059  06   06
005  89   π         060  32  X:T
006  95   =         061  37  P/R
007  42  STO        062  42  STO
008  00   00        063  09   09
009  43  RCL        064  32  X:T
010  04   04        065  42  STO
011  65   ×         066  08   08
012  43  RCL        067  85   +
013  00   00        068  43  RCL
014  95   =         069  10   10
015  35  1/X        070  95   =
016  94  +/-        071  32  X:T
017  32  X:T        072  43  RCL
018  43  RCL        073  01   01
019  03   03        074  65   ×
020  32  X:T        075  43  RCL
021  22  INV        076  00   00
022  37  P/R        077  95   =
023  94  +/-        078  85   +
024  32  X:T        079  43  RCL
025  35  1/X        080  09   09
026  32  X:T        081  95   =
027  37  P/R        082  22  INV
028  42  STO        083  37  P/R
029  06   06        084  42  STO
030  32  X:T        085  08   08
031  42  STO        086  32  X:T
032  07   07        087  42  STO
033  43  RCL        088  09   09
034  05   05        089  43  RCL
035  35  1/X        090  06   06
036  85   +         091  55   ÷
037  43  RCL        092  43  RCL
038  07   07        093  09   09
039  95   =         094  95   =
040  32  X:T        095  28  LOG
041  43  RCL        096  65   ×
042  02   02        097  02   2
043  65   ×         098  00   0
044  43  RCL        099  95   =
045  00   00        100  42  STO
046  95   =         101  06   06
047  85   +         102  91  R/S
048  43  RCL        103  43  RCL
049  06   06        104  07   07
050  95   =         105  75   -
051  22  INV        106  43  RCL
052  37  P/R        107  08   08
053  94  +/-        108  95   =
054  42  STO        109  42  STO
```

MERGED CODES

62 Pgm Ind	72 STO Ind	83 GTO Ind
63 Exc Ind	73 RCL Ind	84 Op Ind
64 Prd Ind	74 SUM Ind	92 INV SBR

TEXAS INSTRUMENTS INCORPORATED

© 1977 Texas Instruments Incorporated

Program Listing for TI-58: Buck Converter, Continuous Conduction

TITLE Buck Converter, Continuous Conduction PAGE ____ OF ____

PROGRAMMER ____________ DATE ________

TI Programmable Coding Form

LOC	CODE	KEY	COMMENTS

```
000  76  LBL
001  11  A
002  65  ×
003  02  2
004  95  =
005  65  ×
006  89  π
007  95  =
008  42  STO
009  00  00
010  65  ×
011  43  RCL
012  04  04
013  95  =
014  35  1/X
015  94  +/-
016  32  X:T
017  43  RCL
018  03  03
019  32  X:T
020  22  INV
021  37  P/R
022  94  +/-
023  32  X:T
024  35  1/X
025  32  X:T
026  37  P/R
027  42  STO
028  11  11
029  32  X:T
030  42  STO
031  10  10
032  43  RCL
033  05  05
034  35  1/X
035  85  +
036  43  RCL
037  10  10
038  95  =
039  32  X:T
040  43  RCL
041  02  02
042  65  ×
043  43  RCL
044  00  00
045  95  =
046  85  +
047  43  RCL
048  11  11
049  95  =
050  22  INV
051  37  P/R
052  94  +/-
053  42  STO
054  09  09
055  32  X:T
056  35  1/X
057  42  STO
058  11  11
059  32  X:T
060  37  P/R
061  42  STO
062  12  12
063  32  X:T
064  42  STO
065  10  10
066  43  RCL
067  01  01
068  65  ×
069  43  RCL
070  00  00
071  95  =
072  85  +
073  43  RCL
074  12  12
075  95  =
076  32  X:T
077  43  RCL
078  10  10
079  85  +
080  43  RCL
081  13  13
082  95  =
083  32  X:T
084  22  INV
085  37  P/R
086  42  STO
087  10  10
088  32  X:T
089  42  STO
090  12  12
091  43  RCL
092  11  11
093  55  ÷
094  43  RCL
095  12  12
096  95  =
097  28  LOG
098  65  ×
099  02  2
100  00  0
101  95  =
102  42  STO
103  11  11
104  43  RCL
105  09  09
106  75  -
107  43  RCL
108  10  10
109  95  =
110  42  STO
111  12  12
112  43  RCL
113  07  07
114  55  ÷
115  43  RCL
116  08  08
117  95  =
118  55  ÷
119  43  RCL
120  06  06
121  95  =
122  28  LOG
123  65  ×
124  02  2
125  00  0
126  95  =
127  85  +
128  43  RCL
129  11  11
130  95  =
131  91  R/S
132  43  RCL
133  12  12
134  91  R/S
```

MERGED CODES

62 Pgm Ind	72 STO Ind	83 GTO Ind
63 Exc Ind	73 RCL Ind	84 Op Ind
64 Prd Ind	74 SUM Ind	92 INV SBR

TEXAS INSTRUMENTS INCORPORATED

© 1977 Texas Instruments Incorporated

Program Listing for TI-58: Boost Converter, Continuous Conduction

TITLE Boost Converter,Continuous Conduction PAGE ____ OF ____ TI Programmable Coding Form

PROGRAMMER ________________ DATE ________

LOC	CODE	KEY	COMMENTS
000	76	LBL	
001	11	A	
002	65	×	
003	02	2	
004	95	=	
005	65	×	
006	89	π	
007	95	=	
008	42	STO	
009	00	00	
010	65	×	
011	43	RCL	
012	01	01	
013	95	=	
014	55	÷	
015	43	RCL	
016	05	05	
017	55	÷	
018	53	(	
019	43	RCL	
020	06	06	
021	94	+/-	
022	85	+	
023	01	1	
024	54	)	
025	33	X²	
026	95	=	
027	65	×	
028	43	RCL	
029	07	07	
030	95	=	
031	55	÷	
032	43	RCL	
033	08	08	
034	95	=	
035	55	÷	
036	53	(	
037	43	RCL	
038	06	06	
039	94	+/-	
040	85	+	
041	01	1	
042	54	)	
043	95	=	
044	94	+/-	
045	32	X⇌T	
046	43	RCL	
047	07	07	
048	55	÷	
049	43	RCL	
050	08	08	
051	95	=	
052	55	÷	
053	53	(	
054	43	RCL	
055	06	06	
056	94	+/-	
057	85	+	
058	01	1	
059	54	)	
060	95	=	
061	32	X⇌T	
062	22	INV	
063	37	P/R	
064	42	STO	
065	10	10	
066	32	X⇌T	
067	28	LOG	
068	65	×	
069	02	2	
070	00	0	
071	95	=	
072	42	STO	
073	09	09	
074	43	RCL	
075	00	00	
076	65	×	
077	43	RCL	
078	04	04	
079	95	=	
080	35	1/X	
081	94	+/-	
082	32	X⇌T	
083	43	RCL	
084	03	03	
085	65	×	
086	53	(	
087	43	RCL	
088	06	06	
089	94	+/-	
090	85	+	
091	01	1	
092	54	)	
093	35	1/X	
094	95	=	
095	32	X⇌T	
096	22	INV	
097	37	P/R	
098	94	+/-	
099	32	X⇌T	
100	35	1/X	
101	32	X⇌T	
102	37	P/R	
103	42	STO	
104	12	12	
105	32	X⇌T	
106	42	STO	
107	11	11	
108	43	RCL	
109	05	05	
110	35	1/X	
111	85	+	
112	43	RCL	
113	11	11	
114	95	=	
115	32	X⇌T	
116	43	RCL	
117	02	02	
118	65	×	
119	43	RCL	
120	00	00	
121	95	=	
122	85	+	
123	43	RCL	
124	12	12	
125	95	=	
126	22	INV	
127	37	P/R	
128	94	+/-	
129	42	STO	
130	14	14	
131	32	X⇌T	
132	35	1/X	
133	42	STO	
134	12	12	
135	32	X⇌T	
136	37	P/R	
137	42	STO	
138	13	13	
139	32	X⇌T	
140	42	STO	
141	11	11	
142	43	RCL	
143	01	01	
144	55	÷	
145	53	(	
146	43	RCL	
147	06	06	
148	94	+/-	
149	85	+	
150	01	1	
151	54	)	
152	33	X²	
153	95	=	
154	65	×	
155	43	RCL	
156	00	00	
157	95	=	

MERGED CODES

62 Pgm Ind	72 STO Ind	83 GTO Ind
63 Exc Ind	73 RCL Ind	84 Op Ind
64 Prd Ind	74 SUM Ind	92 INV SBR

TEXAS INSTRUMENTS INCORPORATED

© 1977 Texas Instruments Incorporated

TI-58: Boost Converter (Continued)

TITLE Boost Converter, Continuous Conduction PAGE ___ OF ___ TI Programmable Coding Form

PROGRAMMER ___ DATE ___

LOC	CODE	KEY	COMMENTS
158	85	+	
159	43	RCL	
160	13	13	
161	95	=	
162	32	X⇄T	
163	43	RCL	
164	15	15	
165	55	÷	
166	53	(	
167	43	RCL	
168	06	06	
169	94	+/-	
170	85	+	
171	01	1	
172	54	)	
173	33	X²	
174	95	=	
175	85	+	
176	43	RCL	
177	11	11	
178	95	=	
179	32	X⇄T	
180	22	INV	
181	37	P/R	
182	42	STO	
183	11	11	
184	32	X⇄T	
185	42	STO	
186	13	13	
187	43	RCL	
188	12	12	
189	55	÷	
190	43	RCL	
191	13	13	
192	95	=	
193	28	LOG	
194	65	×	
195	02	2	
196	00	0	
197	95	=	
198	42	STO	
199	12	12	
200	43	RCL	
201	14	14	
202	75	-	
203	43	RCL	
204	11	11	
205	95	=	
206	42	STO	
207	13	13	
208	43	RCL	
209	12	12	
210	85	+	
211	43	RCL	
212	09	09	
213	95	=	
214	91	R/S	
215	43	RCL	
216	13	13	
217	85	+	
218	43	RCL	
219	10	10	
220	95	=	
221	91	R/S	
222	92	RTN	
223	00	0	

MERGED CODES

62 Pgm Ind	72 STO Ind	83 GTO Ind
63 Exc Ind	73 RCL Ind	84 Op Ind
64 Prd Ind	74 SUM Ind	92 INV SBR

TEXAS INSTRUMENTS INCORPORATED

© 1977 Texas Instruments Incorporated

Program Listing for TI-58: Buck-Boost Converter, Continuous Conduction

TITLE Buck-Boost Converter Continuous Conduction PAGE ____ OF ____ **TI Programmable Coding Form**

PROGRAMMER ____ DATE ____

LOC	CODE	KEY
000	76	LBL
001	11	A
002	65	×
003	02	2
004	65	×
005	89	π
006	95	=
007	42	STO
008	00	00
009	65	×
010	43	RCL
011	06	06
012	95	=
013	65	×
014	43	RCL
015	01	01
016	55	÷
017	43	RCL
018	05	05
019	95	=
020	55	÷
021	53	(
022	43	RCL
023	06	06
024	94	+/-
025	85	+
026	01	1
027	54	)
028	33	X²
029	95	=
030	65	×
031	43	RCL
032	07	07
033	95	=
034	55	÷
035	43	RCL
036	06	06
037	95	=
038	55	÷
039	43	RCL
040	08	08
041	95	=
042	55	÷
043	53	(
044	43	RCL
045	06	06
046	94	+/-
047	85	+
048	01	1
049	54	)
050	95	=
051	94	+/-
052	32	X⇄T
053	43	RCL
054	07	07
055	55	÷
056	43	RCL
057	06	06
058	95	=
059	55	÷
060	43	RCL
061	08	08
062	95	=
063	55	÷
064	53	(
065	43	RCL
066	06	06
067	94	+/-
068	85	+
069	01	1
070	54	)
071	95	=
072	32	X⇄T
073	22	INV
074	37	P/R
075	42	STO
076	10	10
077	32	X⇄T
078	28	LOG
079	65	×
080	02	2
081	00	0
082	95	=
083	42	STO
084	09	09
085	43	RCL
086	00	00
087	65	×
088	43	RCL
089	04	04
090	95	=
091	35	1/X
092	94	+/-
093	32	X⇄T
094	43	RCL
095	03	03
096	65	×
097	53	(
098	43	RCL
099	06	06
100	94	+/-
101	85	+
102	01	1
103	54	)
104	35	1/X
105	95	=
106	32	X⇄T
107	22	INV
108	37	P/R
109	94	+/-
110	32	X⇄T
111	35	1/X
112	32	X⇄T
113	37	P/R
114	42	STO
115	12	12
116	32	X⇄T
117	42	STO
118	11	11
119	43	RCL
120	05	05
121	35	1/X
122	85	+
123	43	RCL
124	11	11
125	95	=
126	32	X⇄T
127	43	RCL
128	02	02
129	65	×
130	43	RCL
131	00	00
132	95	=
133	85	+
134	43	RCL
135	12	12
136	95	=
137	22	INV
138	37	P/R
139	94	+/-
140	42	STO
141	14	14
142	32	X⇄T
143	35	1/X
144	42	STO
145	12	12
146	32	X⇄T
147	37	P/R
148	42	STO
149	13	13
150	32	X⇄T
151	42	STO
152	11	11
153	43	RCL
154	15	15
155	65	×
156	53	(
157	43	RCL
158	06	06
159	94	+/-

MERGED CODES

62 Pgm Ind	72 STO Ind	83 GTO Ind
63 Exc Ind	73 RCL Ind	84 Op Ind
64 Prd Ind	74 SUM Ind	92 INV SBR

TEXAS INSTRUMENTS INCORPORATED

© 1977 Texas Instruments Incorporated

TI-58: Buck-Boost Converter (Continued)

TITLE Buck-Boost Converter Continuous Conduction PAGE ____ OF ____ TI Programmable Coding Form

PROGRAMMER ____________ DATE ____________

LOC	CODE	KEY	LOC	CODE	KEY
160	85	+	215	12	12
161	01	1	216	43	RCL
162	54	)	217	14	14
163	33	X²	218	75	-
164	35	1/X	219	43	RCL
165	95	=	220	11	11
166	85	+	221	95	=
167	43	RCL	222	42	STO
168	11	11	223	13	13
169	95	=	224	43	RCL
170	42	STO	225	12	12
171	11	11	226	85	+
172	43	RCL	227	43	RCL
173	01	01	228	09	09
174	55	÷	229	95	=
175	53	(	230	91	R/S
176	43	RCL	231	43	RCL
177	06	06	232	13	13
178	94	+/-	233	85	+
179	85	+	234	43	RCL
180	01	1	235	10	10
181	54	)	236	95	=
182	33	X²	237	91	R/S
183	95	=	238	92	RTN
184	65	×	239	00	0
185	43	RCL			
186	00	00			
187	95	=			
188	85	+			
189	43	RCL			
190	13	13			
191	95	=			
192	32	X⇌T			
193	43	RCL			
194	11	11			
195	32	X⇌T			
196	22	INV			
197	37	P/R			
198	42	STO			
199	11	11			
200	32	X⇌T			
201	42	STO			
202	13	13			
203	43	RCL			
204	12	12			
205	55	÷			
206	43	RCL			
207	13	13			
208	95	=			
209	28	LOG			
210	65	×			
211	02	2			
212	00	0			
213	95	=			
214	42	STO			

MERGED CODES

62 Pgm Ind 72 STO Ind 83 GTO Ind
63 Exc Ind 73 RCL Ind 84 Op Ind
64 Prd Ind 74 SUM Ind 92 INV SBR

TEXAS INSTRUMENTS INCORPORATED

© 1977 Texas Instruments Incorporated

2.6 BUCK CONVERTER (CONTINUOUS CONDUCTION) EXAMPLE

$V_{i_{min}} = 11\ V \qquad R_{min} = 2\ \Omega, \quad I_o = 4\ A$

$V_{i_{max}} = 20\ V \qquad R_{max} = 8\ \Omega, \quad I_o = 1\ A$

$V_o = 8\ V \qquad f_s = 40\ kHz$

$$\frac{V_o}{V_{i_{min}}} = D_H = \frac{8}{11}, \text{ ideal condition}$$

$$= \frac{8}{11 - 1}, \text{ practical, allowing for 1 V drop across switch}$$

$$= 0.8$$

It is also imperative that the designer select a commutating diode with a low V_f with fast-switching characteristics for high efficiency performance:

$$D_L = \frac{V_o}{V_{i_{max}}} = \frac{8}{20 - 1} = 0.421$$

$$L_{min} = \frac{T_s(1 - D_L)R_{max}}{2}$$

$$= \frac{2.5 \times 10^{-5}(1 - 0.421) \times 8}{2}$$

$$= 57.9\ \mu H, \quad \text{say } 60\ \mu H$$

For 5mV output voltage ripple, neglecting the effect of esr,

$$\Delta v_o = \frac{V_o(1 - D_L)}{8LCf_s^2}$$

$$.005 = \frac{8(1 - 0.421)}{8 \times 60 \times C \times 40^2} \quad \text{or} \quad C = 1206\ \mu F$$

Use 1500 μF

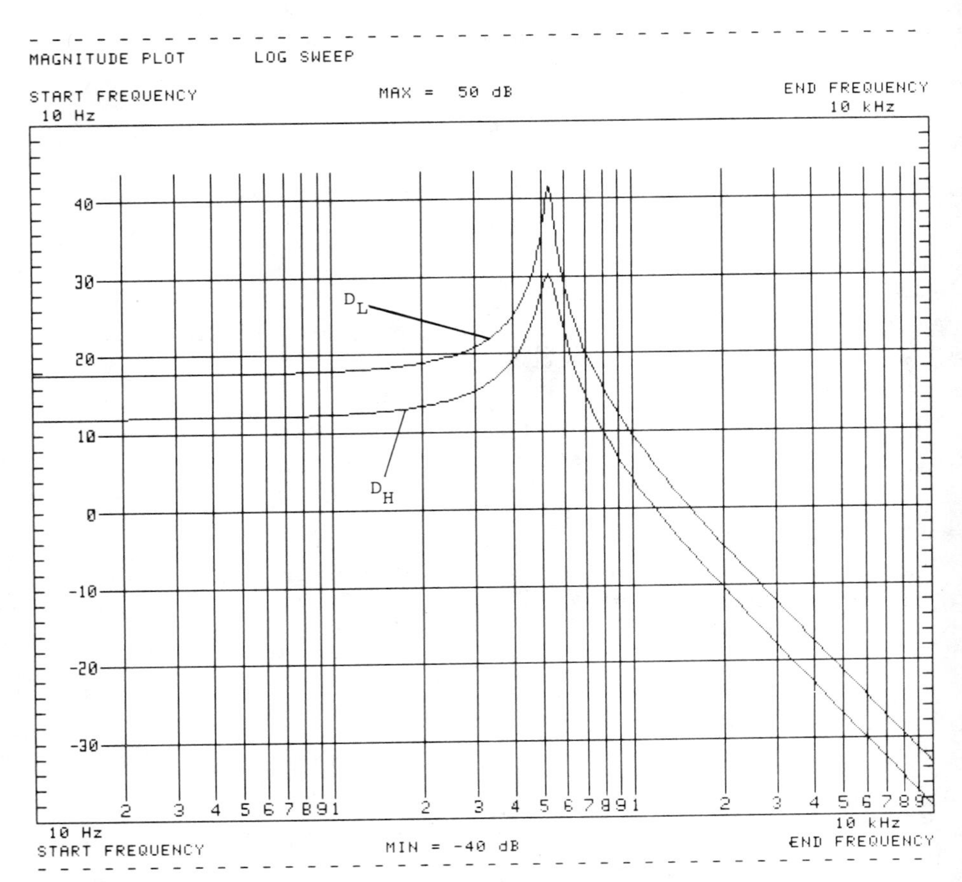

FIG. 2.6 Buck converter (no output filter damping).

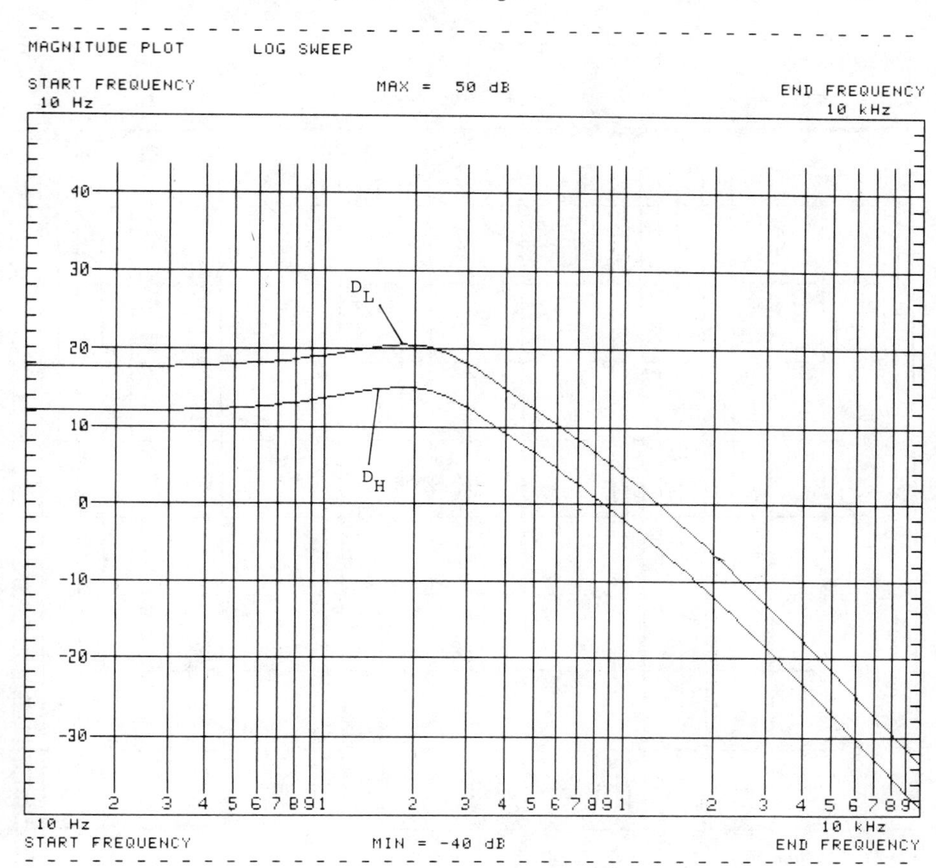

FIG. 2.7 Buck converter with output filter damping.

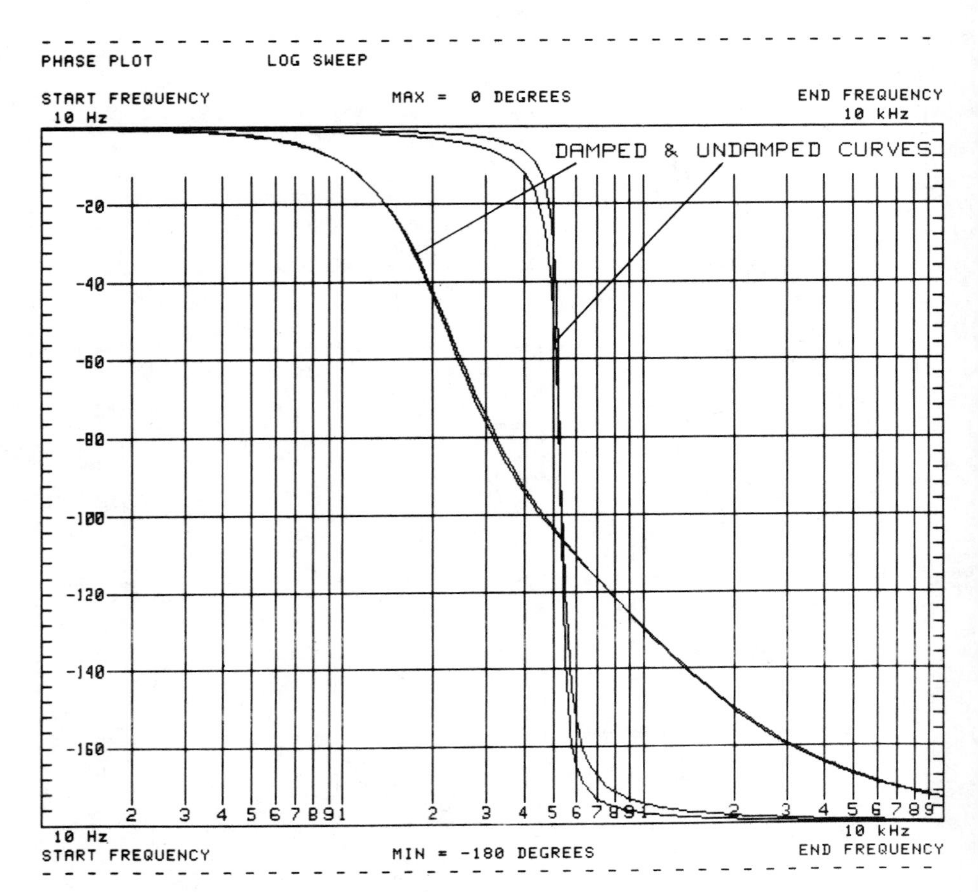

FIG. 2.8 Damped and undamped curves.

```
BUCK EXAMPLE

WITH DAMPING

 60-06 STO 0
           1
 1,500-06 ST
         O 02
  0.1 STO 03
 9,000-06 ST
         O 04
    2 STO 05
  0.8 STO 06
    8 STO 07
  2.5 STO 08

PRESS USER

ENTER FREQUE
NCY
   20  XEQ A
 12.12700+00
         ***
-273.2462-03
         ***

  100  XEQ A
 13.86651+00
         ***
-7.814267+00
         ***

  200  XEQ A
 15.54417+00
         ***
-43.61321+00
         ***

  300  XEQ A
 12.81393+00
         ***
-78.30363+00
         ***

  500  XEQ A
 6.774215+00
         ***
-105.2004+00
         ***
```

```
 1,000  XEQ
           A
-1.940610+00
         ***
-129.7713+00
         ***

 2,000  XEQ
           A
-12.01113+00
         ***
-150.4720+00
         ***

 5,000  XEQ
           A
-27.10840+00
         ***
-167.4073+00
         ***

 10,000  XEQ
           A
-39.01633+00
         ***
-173.6389+00
         ***

 20,000  XEQ
           A
-51.02347+00
         ***
-176.8112+00
         ***

 50,000  XEQ
           A
-66.93148+00
         ***
-178.7235+00
         ***
```

```
LIGHT LOAD

WITH DAMPING
    8 STO 05
 0.421 STO 0
           6


ENTER FREQUE
NCY
   20  XEQ A
 17.70324+00
         ***
-109.6381-03
         ***

  100  XEQ A
 19.46199+00
         ***
-6.821867+00
         ***

  200  XEQ A
 21.37333+00
         ***
-41.80549+00
         ***

  300  XEQ A
 18.79319+00
         ***
-77.73958+00
         ***

  500  XEQ A
 12.67915+00
         ***
-105.8018+00
         ***

 1,000  XEQ
           A
 3.824852+00
         ***
-130.8302+00
         ***

 2,000  XEQ
           A
-6.359778+00
         ***
-151.3638+00
         ***

 5,000  XEQ
           A
-21.51772+00
         ***
-167.8440+00
         ***

 10,000  XEQ
           A
-33.43643+00
         ***
-173.8644+00
         ***

 20,000  XEQ
           A
-45.44637+00
         ***
-176.9248+00
         ***

 50,000  XEQ
           A
-61.35517+00
         ***
-178.7691+00
         ***
```

The allowable esr $= \Delta V_o / \Delta I_o$

$$esr_{max} = \frac{\Delta v_o}{0.15 \times 4} \quad \text{for 15\% current ripple}$$

$$= \frac{0.005}{0.6}$$

$$= 8.33\ m\Omega$$

Note that the voltage ripple contributed by actual esr is

$$\Delta v_{esr} = esr\ \Delta I = \frac{esr\ V_o\ T(1 - D_L)}{L_b}$$

Store 60 μH in data storage register 1.

Store 1500 μF in data storage register 2.

For analysis without damping of the output filter, store small values in registers 3 and 4.

Store 0.001 Ω in data storage register 3.

Store 0.001 μF in data storage register 4.

Store 8 Ω in data storage register 5.

Store the corresponding D_L of 0.421 in data storage register 6.

Store V_o of 8 V in data storage register 7.

Store V_m of 2.5 V (for the SG 1524 control circuit) in data storage register 8.

Neglect r_L for this calculation.

A frequency response plot of the preceding entries was made for both magnitude (decibel) and phase (degree). The curves for D_H and R_{min} were plotted by changing the contents of the appropriate data storage registers. Figures 2.6 to 2.8 show the results of these calculations.

To apply output filter damping, consult Appendix B, Figs. B.22 and B.23. Choose n = 6, r = 0.1 Ω. Load data storage registers 3 and 4 accordingly. Plot D_H and D_L curves as before. See Figs. 2.7 and 2.8. (*Note*: For n = 6, load nC = 9000 μF in register 4.)

3

Design of Magnetic Components

3.1 THE TRANSFORMER

The ideal transformer has no winding resistance, has infinite primary inductance, and has perfect coupling such that all flux due to the ac excitation of the primary will link with the secondary winding. Under no load condition (Fig. 3.1), an infinitely small current will flow. This current lags the applied voltage by 90° and is responsible for setting up the flux linking the primary winding with the secondary winding of the transformer. *This flux is constant for a constant applied voltage.* Since the flux is the same for the primary and the secondary, the induced voltages will be proportional to the primary turns N_1 and secondary turns N_2, respectively. Therefore,

$$\frac{V_2}{V_1} = \frac{N_2}{N_1} \tag{3.1}$$

For a loaded transformer (Fig. 3.2), a secondary current flows:

$$I_1N_1 = I_2N_2 \tag{3.2}$$

or

$$\frac{I_1}{I_2} = \frac{N_2}{N_1} \tag{3.3}$$

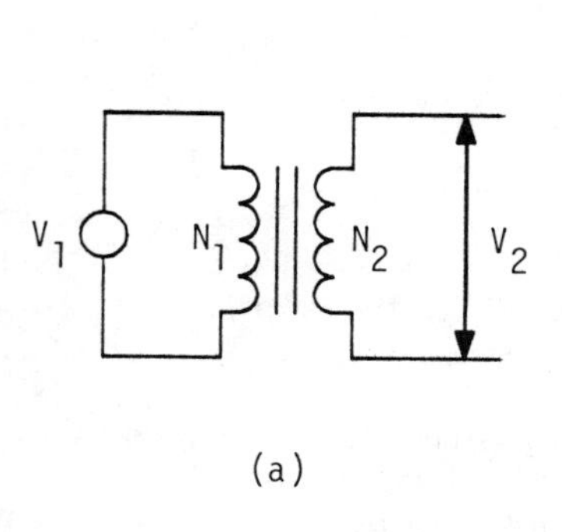

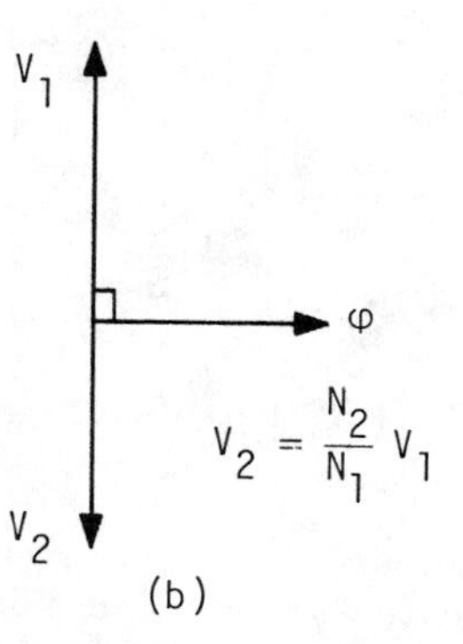

FIG. 3.1 Ideal transformer with no load.

But $N_2/N_1 = V_2/V_1$; therefore, multiplying Eqs. (3.1) and (3.3) gives

$$\frac{N_2^2}{N_1^2} = \frac{I_1}{I_2} \frac{V_2}{V_1} = \frac{V_2/I_2}{V_1/I_1} = \frac{Z_2}{Z_1} \tag{3.4}$$

This is the concept of impedance transformation. See Fig. 3.3.

In general, if $v_1(t)$ is a time-varying voltage source, which, at a given instant, assumes the polarity shown in Fig. 3.4, then

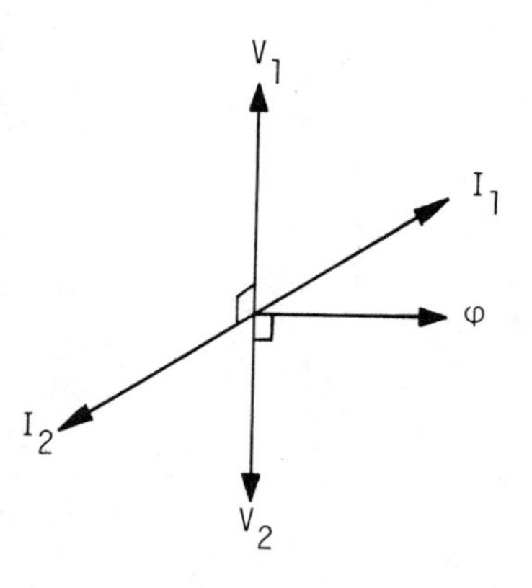

FIG. 3.2 Vector diagram of loaded ideal transformer.

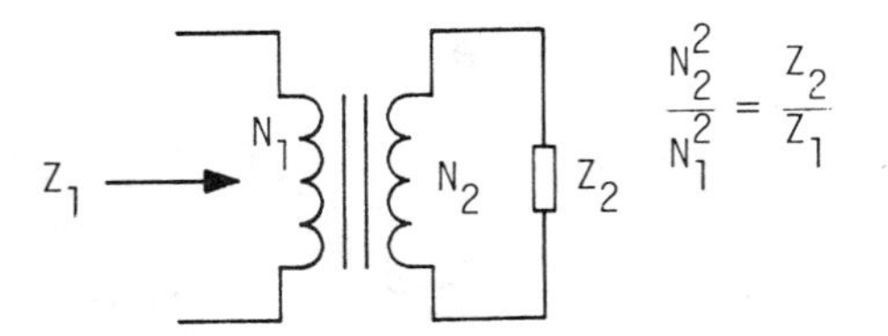

FIG. 3.3 Impedance-transforming property to transformer.

the currents $i_1(t)$, $i_2(t)$, and $i_3(t)$ will assume the directions indicated and are increasing with time. The positive end of the winding is marked with a dot. *This is the dot notation.*

For a practical transformer, the primary inductance is finite, and the winding has finite resistance. A current will flow in the primary circuit even when there is no load on the secondary circuit; this current is called the *magnetizing current*. The magnitude of the magnetizing current is a good indication of the adequacy of the primary inductance for a given number of turns.

Figure 3.5 shows an equivalent circuit of a practical transformer with losses, where r_p and r_s represent winding resistance losses, L_1 and L_2 represent leakage inductances, L_p represents finite primary inductance, r_h represents eddy current and hysteresis losses, and c_p and c_s represent self-capacitances of the windings.

For a sinusoidal input voltage, the flux φ varies alternately:

$$\varphi = \varphi_{max} \sin \omega t \tag{3.5}$$

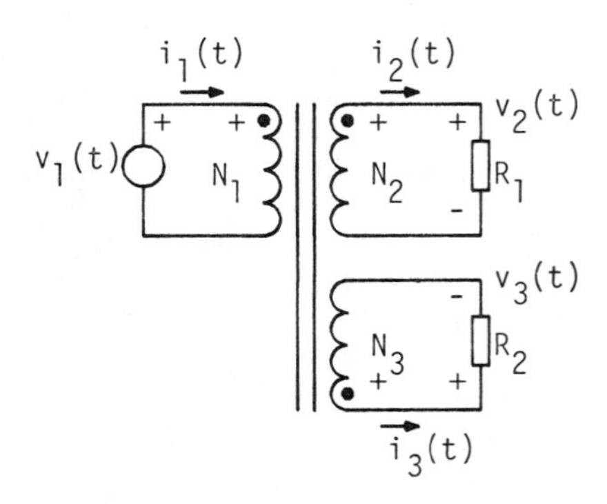

FIG. 3.4 Transformer with dot notation.

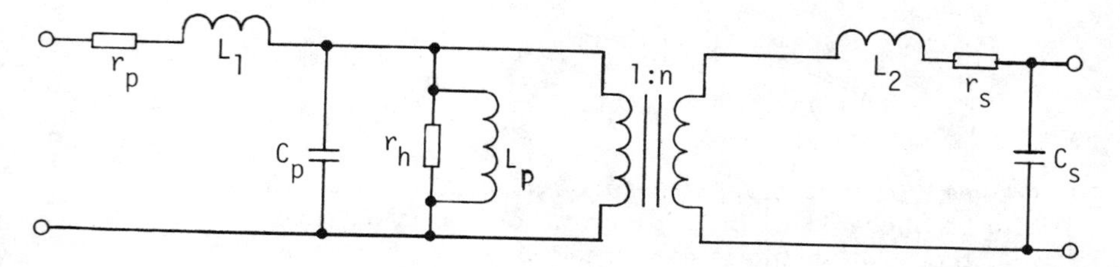

FIG. 3.5 Lumped equivalent circuit of practical transformer.

The instantaneous voltage induced in the primary is, according to Faraday's law,

$$e_1 = -\frac{d\varphi}{dt} N_1, \text{ volts} \tag{3.6}$$

$$e_1 = -N_1 \varphi_{max} \omega \cos \omega t \tag{3.7}$$

$$e_1 = -2\pi f N_1 \varphi_{max} \cos \omega t \tag{3.8}$$

Therefore,

$$E_{1_{max}} = -2\pi f N_1 \varphi_{max} \tag{3.9}$$

or, the rms value of $E_{1_{max}} = E_1$:

$$E_1 = \frac{2\pi}{\sqrt{2}} f N_1 \varphi_{max} \tag{3.10}$$

$$= -4.44 f N_1 \varphi_{max} \tag{3.11}$$

For the general case, the applied voltage

$$V_1 = K f N_1 \varphi_{max} \tag{3.12}$$

where K (= 4.44 for sinusoids, = 4 for rectangular wave) is a constant.

The apparent power [120] handled by the transformer is the sum of the primary volt-amps plus the secondary volt-amps.

For $N_1 = N_2 = N$, $I_1 = I_2 = I$,

$$P_t = V_1 I_1 + V_2 I_2 = KfNBA_c I \tag{3.13}$$

For a given number of amp-turns NI with a given current density J to be allocated within a given window area W_a with a fill factor K_u, the relationship is expressed as

$$NI = K_u W_a J \tag{3.14}$$

Therefore,

$$P_t = KfBA_c K_u W_a J \tag{3.15}$$

and area product

$$A_p = W_a A_c = \frac{P_t}{KfBK_u J} \tag{3.16}$$

The current density J, as derived in Appendix D, is

$$J = K_j A_p^{-0.14} \tag{3.17}$$

or

$$A_p = \frac{P_t}{KfBK_u K_j A_p^{-0.14}} \tag{3.18}$$

Multiplying both sides of Eq. (3.18) by $A_p^{-0.14}$ gives

$$\frac{P_t}{KfBK_u K_j} = A_p^{0.86} \tag{3.19}$$

The rationalized area product (consistent with the units and dimensions given in the list of symbols) is

$$A_p = \left(\frac{P_t \times 10^4}{KBfK_u K_j}\right)^{1.16} \tag{3.20}$$

Equation (3.20) permits the selection of a core on the basis of the area product A_p being proportional to the power handling capability of the transformer. (See Fig. 3.6.) In other words, the amount of copper (wire) and the amount of iron (ferrite or other appropriate core material) determine the total power capability of the transformer.

For a practical transformer, the efficiency is less than 100%, and

$$P_i = \frac{P_o}{\eta} \tag{3.21}$$

where η is the transformer efficiency and

$$P_t = P_i + P_o \tag{3.22}$$

$$P_t = \frac{P_o}{\eta} + P_o \tag{3.23}$$

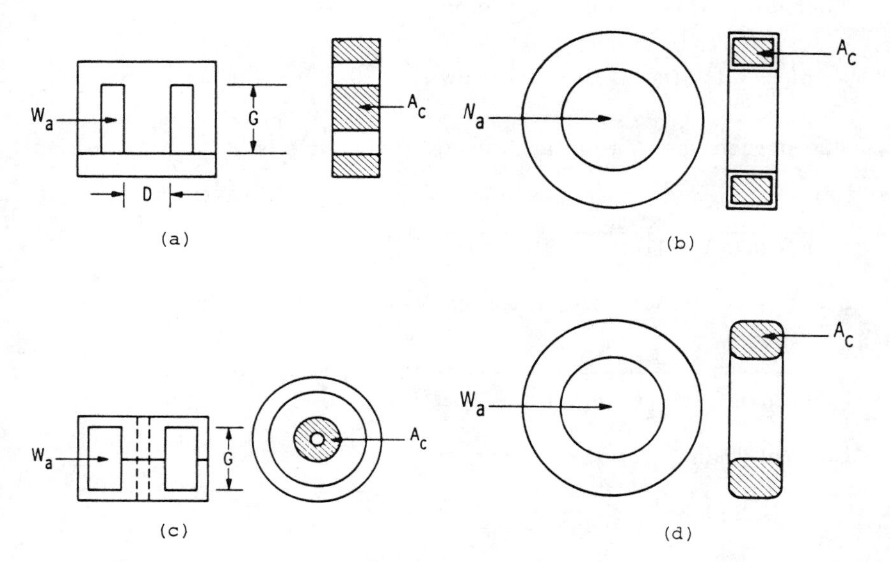

FIG. 3.6 Core areas of various core types. (a) EI lamination. (b) Tape-wound toroidal core. (c) Pot core. (d) Powder core.

$$P_t = P_o\left(\frac{1}{\eta} + 1\right) \tag{3.24}$$

The actual apparent power P_t is dependent on the circuit configuration as shown in Fig. 3.7.

With reference to Eq. (3.20), the flux density B is determined by the core material and operating frequency. The window utilization factor K_u depends on the wire lay of the coil(s), the tightness of the winding for a given wire tension, the shape of the wire, and the amount and type of insulation material used. Normally, K_u varies between 0.4 and 0.6 and is also affected by wire size. For example, if the wire layers are stacked consistently as shown in Fig. 3.8 (neglecting the thickness of the enamel insulation), then the wire area of one turn is πa^2. Therefore, the wire area of n turns is $n\pi a^2$. One side of rectangle, h, is $(m - 1)b + 2a$, where m is the number of layers and a is the radius of the wire. The other side of the rectangle, w, is

$$\frac{2na}{m} + a = a\left(\frac{2n}{m} + 1\right) = w$$

Therefore, the area of the rectangle is

$$a\left(\frac{2n}{m} + 1\right) [(m - 1)b + 2a] = hw$$

The percentage of total area used by the wire is

$$\frac{n\pi a^2}{a[(2n/m) + 1][(m - 1)b + 2a]} \times 100\% \tag{3.25}$$

But $b = a\sqrt{3}$; therefore, the area occupied by wire is

$$\frac{n\pi}{[(2n/m) + 1][(m - 1)\sqrt{3} + 2]} \times 100\% \text{ of available area}$$

But McLyman [120] shows that

$$K_u = S_1S_2S_3S_4 \tag{3.26}$$

and Eq. (3.25) is equal to S_2S_3, or

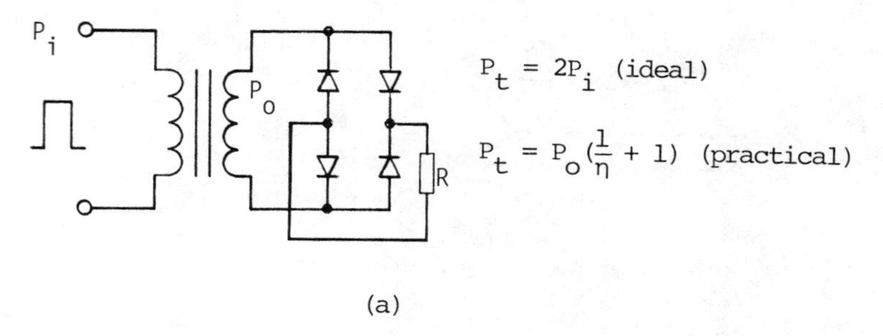

(a)

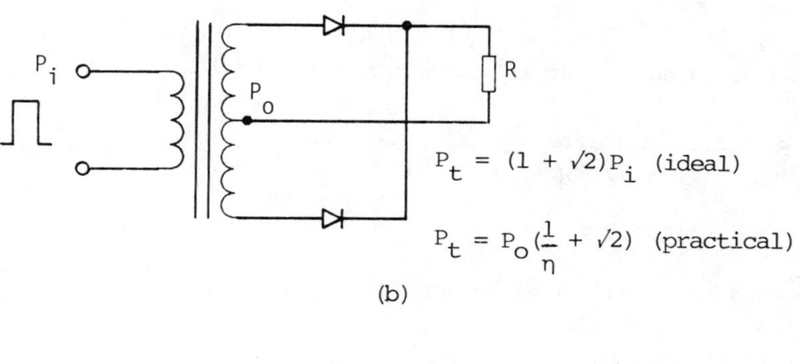

(b)

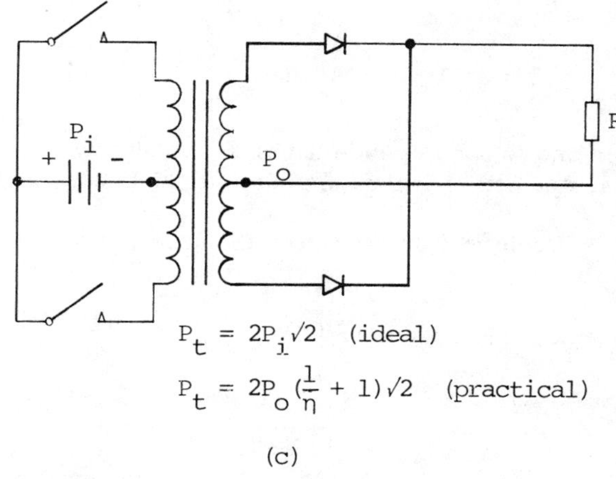

(c)

FIG. 3.7 Transformer apparent power.

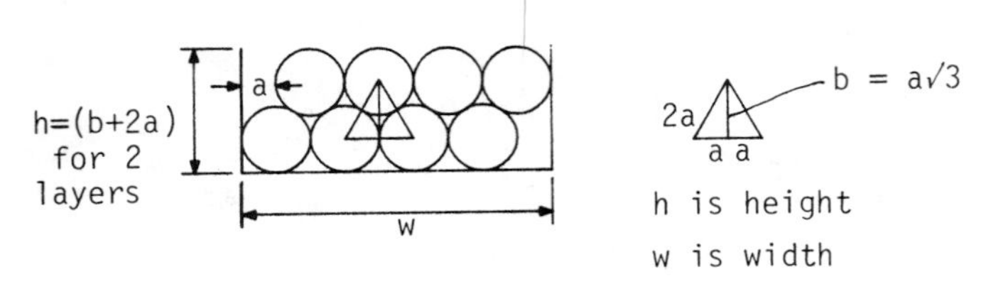

FIG. 3.8 Typical wire arrangement of winding.

$$S_2S_3 = \frac{n\pi}{[(2n/m) + 1][(M - 1)\sqrt{3} + 2]} \quad (3.27)$$

For better space utilization with m = 5, n = 40,

$$S_2S_3 = \frac{40 \times 3.1416}{\{[(2 \times 40)/5] + 1\}[(5 - 1)\sqrt{3} + 2]}$$

$$= 0.828$$

Allocating 35% window area for insulation, we obtain

$$S_1 = \frac{1}{1.35} = 0.74$$

$$K_u = S_1S_2S_3S_4 = 0.74 \times 0.828 \times 0.7 = \underline{0.43}$$

This example shows the dependence of K_u on the wire size, number of turns, number of layers, insulation thickness, and wire lay.

EXAMPLE 3.1 Design of 60-Hz power transformer for off-line step-down application:

$V_i = 115$ V ac, 60-Hz sinusoid

$V_o = 35$ V ac

$I_o = 7.5$ A

Let $\eta = 95\%$ as a design goal.

Assume the configuration of Fig. 3.7a with bridge rectifier output

$$P_t = P_o\left(1 + \frac{1}{\eta}\right)$$

$$= 262.5\left(\frac{1}{0.95} + 1\right)$$

$$= 539 \text{ W}$$

From Eq. (3.20),

$$A_p = \left(\frac{P_t \times 10^4}{KBfK_uK_j}\right)^{1.16} \text{ cm}^4$$

$$= \left(\frac{539 \times 10^4}{4.44 \times 1.4 \times 60 \times 0.4 \times 366}\right)^{1.16}$$

where B is selected for 1.4 tesla for laminations, K_u is 0.4, and K_j is 366 from Table 3.1, column 3. Therefore,

$$A_p = 205.82 \text{ cm}^4$$

Select the core with A_p closest to the calculated A_p value from Table 3.2; core 2-138EI has $A_p = 223.39 \text{ cm}^4$. Core $W_t = 3.901$ kg. Tables 3.3 to 3.7 present other types of core characteristics. A 95% efficient transformer would have a total loss of

$$P_\Sigma = \frac{P_o}{\eta} - P_o$$

$$= 1.052632 \times 262.5$$

$$= 13.816 \text{ W}$$

By using the formula in Fig. 3.9, the core loss is calculated:

$$\frac{P_{fe}}{W_t} = \text{watts/kilogram} = 0.557 \times 10^{-3} f^{1.68} B_m^{\ 1.86}$$

$$= 0.557 \times 10^{-3} \times 60^{1.68} \times 1.4^{1.86}$$

$$= 1.01 \text{ W/kg}$$

TABLE 3.1 Core Configuration Constants

Core	Losses	K_j (25°C)	K_j (50°C)	(x)	K_s	K_w	K_v
Pot core	$P_{cu} = P_{fe}$	433	632	-0.17	33.8	48.0	14.5
Powder core	$P_{cu} \gg P_{fe}$	403	590	-0.12	32.5	58.8	13.1
Lamination	$P_{cu} = P_{fe}$	366	534	-0.12	41.3	68.2	19.7
C-core	$P_{cu} = P_{fe}$	323	468	-0.14	39.2	66.6	17.9
Single-coil	$P_{cu} \gg P_{fe}$	395	569	-0.14	44.5	76.6	25.6
Tape-wound core	$P_{cu} = P_{fe}$	250	365	-0.13	50.9	82.3	25.0

$J = K_j A_p^{(x)}$ $\qquad$ $A_t = K_s A_p^{0.50}$

$W_t = K_w A_p^{0.75}$ $\qquad$ $Vol = K_v A_p^{0.75}$

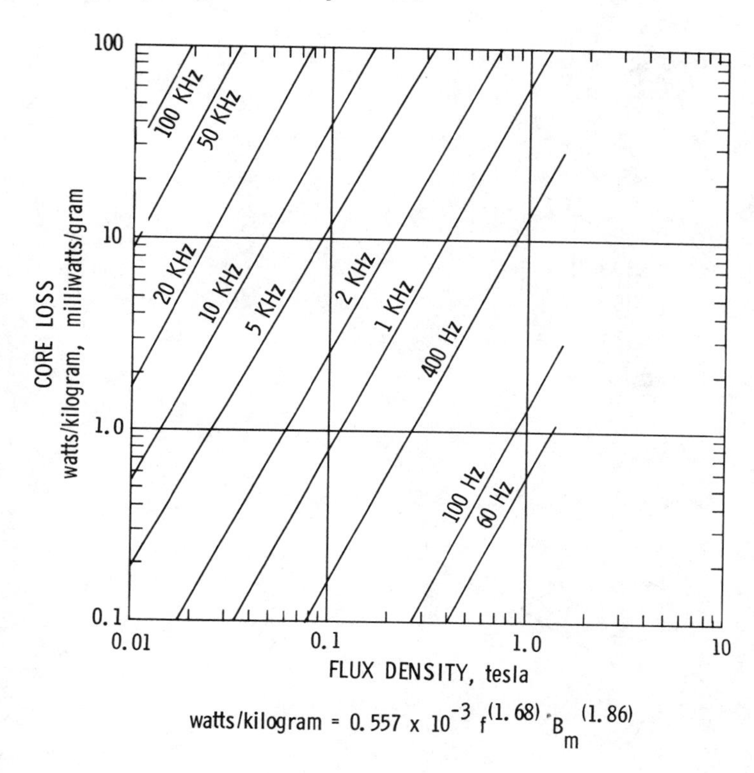

FIG. 3.9 Armco silicon steel 14 mil.

For the core 2-138EI with W_t of 3.901 kg, the core loss is

$$P_{fe} = 3.901 \times 1.01 \text{ W}$$

$$= 3.946 \text{ W}$$

Note that for higher-frequency transformer designs, the core material is selected by means of Figs. 3.11 to 3.19 in a similar manner.

The number of primary turns is derived from Eq. (3.12) as follows:

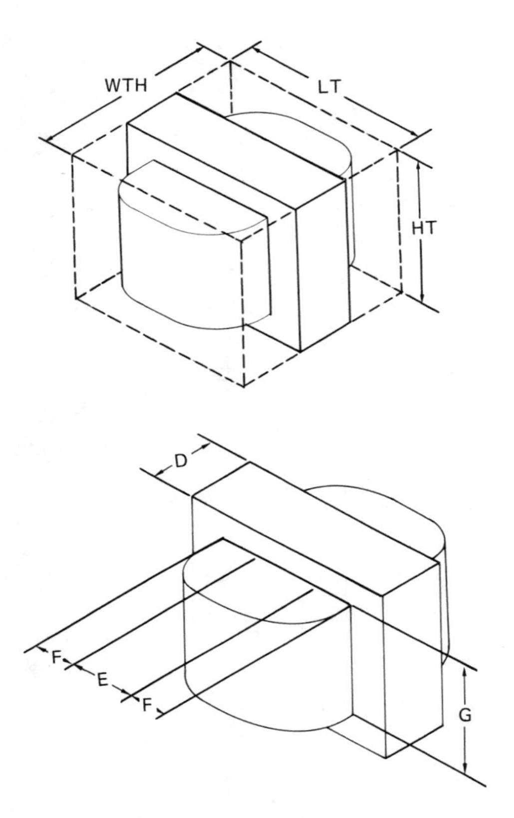

FIG. 3.10 Lamination EI and EE dimensional outline--mechanical outline for Table 3.2.

$$V_i = KfN_p\varphi_{max}$$

$$V_i = KfN_pB_{max}A_c \tag{3.28}$$

Therefore, the rationalized formula for the number of turns is

$$N_p = \frac{V_i \times 10^4}{KfB_{max}A_c} \text{ turns} \tag{3.29}$$

$$= \frac{115 \times 10^4}{4.44 \times 60 \times 1.4 \times 24.4} = 125.37\text{; use 130 turns}$$

TABLE 3.2 Magnetic Metals, EI and EE Laminations

DEFINITIONS

INFORMATION GIVEN IS LISTED BY COLUMN AS,

COLUMN			
1.	D	STACK BUILD UP	(CM)
2.	E	TONGUE WIDTH	(CM)
3.	F	WINDOW WIDTH	(CM)
4.	G	WINDOW LENGTH	(CM)
5.	MPL	MAGNETIC PATH LENGTH	(CM)
6.	HT	FINISHED TRANSFORMER HEIGHT =(E+G)	(CM)
7.	WTH	FINISHED TRANSFORMER WIDTH =(E+2F)	(CM)
8.	LT	FINISHED TRANSFORMER LENGTH =2(E+F)	(CM)
9.	W_{tfe}	IRON WEIGHT =(MPL)(A_c)(7.63)/1000	(KGM)
10.	W_{tcu}	COPPER WEIGHT =(MLT)(W_a)(K_u)(8.89)/1000	(KGM)
11.	MLT	MEAN LENGTH TURN =2(D+2J)+2(E+2J)+(3.14)(F)	(CM)
12.	A_c	IRON AREA =(D)(E)....(GROSS)	(CM SQ)
13.	W_a	WINDOW AREA =(F)(G)....(GROSS)	(CM SQ)
14.	A_p	AREA PRODUCT =(A_c)(W_a)....(GROSS)	(CM 4TH)
15.	K_g	CORE GEOMETRY =(A_p) (A_c) (K_u)/MLT....(GROSS)	(CM 5TH)
16.	A_t	SURFACE AREA	(CM SQ)

For mechanical outline, see Fig. 3.10.
Note: Number in front of part number is times square stack.

TABLE 3.2 (Continued)

MAGNETIC METALS EI AND EE LAMINATIONS

CAT. NO.	D (CM)	E (CM)	F (CM)	G (CM)	MPL (CM)	HT (CM)	WTH (CM)	LT (CM)	GROSS WTFE (KGM)	WTCU (KGM)	MLT (CM)	GROSS AC (CM SQ)	GROSS WA (CM SQ)	GROSS AP (CM 4TH)	GROSS KG (CM 5TH)	AT (CM SQ)
	1	2	3	4	5	6	7	8	9	10	11	12	13	14	15	16
1-94EI	.236	.236	.239	.396	1.7	.6	.7	1.0	.001	.001	2.1	.06	.09	.01	.000056	3.0
1-30-31EE	.236	.236	.239	.714	2.4	1.0	.7	1.0	.001	.001	2.1	.06	.17	.01	.000101	4.1
2-94EI	.472	.236	.239	.396	1.7	.6	.9	1.0	.001	.001	2.6	.11	.09	.01	.000183	3.8
2-30-31EE	.472	.236	.239	.714	2.4	1.0	.9	1.0	.002	.002	2.6	.11	.17	.02	.000330	5.0
3-94EI	.709	.236	.239	.396	1.7	.6	1.2	1.0	.002	.001	3.0	.17	.09	.02	.000348	4.5
1-28-29EE	.317	.317	.317	.795	2.9	1.1	1.0	1.3	.002	.002	2.7	.10	.25	.03	.000384	6.6
1-32-33EE	.356	.356	.381	.698	2.9	1.1	1.1	1.5	.003	.003	3.0	.13	.27	.03	.000563	7.7
3-30-31EE	.709	.236	.239	.714	2.4	1.0	1.2	1.0	.003	.002	3.0	.17	.17	.03	.000627	5.9
2-28-29EE	.635	.317	.317	.795	2.9	1.1	1.3	1.3	.004	.003	3.3	.20	.25	.05	.001240	8.1
1-186EI	.478	.478	.478	.635	3.2	1.1	1.4	1.9	.006	.004	3.8	.23	.30	.07	.001652	11.0
2-32-33EE	.711	.356	.381	.698	2.9	1.1	1.5	1.5	.006	.004	3.7	.25	.27	.07	.001822	9.5
1-185EI	.478	.478	.478	.874	3.7	1.3	1.4	1.9	.006	.006	3.8	.23	.42	.10	.002274	12.6
3-28-29EE	.952	.317	.317	.795	2.9	1.1	1.6	1.3	.007	.004	3.9	.30	.25	.08	.002341	9.6
1-187EI	.478	.478	.478	1.113	4.1	1.6	1.4	1.9	.007	.007	3.8	.23	.53	.12	.002895	14.2
3-32-33EE	1.067	.356	.381	.698	2.9	1.1	1.8	1.5	.008	.004	4.4	.38	.27	.10	.003444	11.3
1-188EI	.478	.478	.478	1.587	5.1	2.1	1.4	1.9	.009	.010	3.8	.23	.76	.17	.004131	17.4
1-186-187EE	.478	.478	.478	1.748	5.4	2.2	1.4	1.9	.009	.011	3.8	.23	.83	.19	.004547	18.5
2-186EI	.955	.478	.478	.635	3.2	1.1	1.9	1.9	.011	.005	4.8	.46	.30	.14	.005286	13.9
1-186-188EE	.478	.478	.478	2.222	6.4	2.7	1.4	1.9	.011	.014	3.8	.23	1.06	.24	.005783	21.7
1-187-188EE	.478	.478	.478	2.697	7.3	3.2	1.4	1.9	.013	.017	3.8	.23	1.29	.29	.007019	24.8
2-185EI	.955	.478	.478	.874	3.7	1.3	1.9	1.9	.013	.007	4.8	.46	.42	.19	.007274	15.7
1-25EIS	.635	.635	.635	.952	4.4	1.6	1.9	2.5	.014	.011	4.9	.40	.60	.24	.007960	20.6
2-187EI	.955	.478	.478	1.113	4.1	1.6	1.9	1.9	.014	.009	4.8	.46	.53	.24	.009262	17.6
3-186EI	1.433	.478	.478	.635	3.2	1.1	2.4	1.9	.017	.006	5.9	.68	.30	.21	.009571	16.8
1-24-25EE	.635	.635	.635	1.270	5.1	1.9	1.9	2.5	.016	.014	4.9	.40	.81	.33	.010614	23.4
3-185EI	1.433	.478	.478	.874	3.7	1.3	2.4	1.9	.019	.009	5.9	.68	.42	.29	.013170	18.9
2-188EI	.955	.478	.478	1.587	5.1	2.1	1.9	1.9	.018	.013	4.8	.46	.76	.35	.013216	21.2
2-186-187EE	.955	.478	.478	1.748	5.4	2.2	1.9	1.9	.019	.014	4.8	.46	.83	.38	.014548	22.4
3-187EI	1.433	.478	.478	1.113	4.1	1.6	2.4	1.9	.022	.011	5.9	.68	.53	.36	.016769	20.9
2-186-188EE	.955	.478	.478	2.222	6.4	2.7	1.9	1.9	.022	.018	4.8	.46	1.06	.48	.018502	26.1
2-187-188EE	.955	.478	.478	2.697	7.3	3.2	1.9	1.9	.025	.022	4.8	.46	1.29	.59	.022457	29.7
3-188EI	1.433	.478	.478	1.587	5.1	2.1	2.4	1.9	.027	.016	5.9	.68	.76	.52	.023928	25.0
2-25EIS	1.270	.635	.635	.952	4.4	1.6	2.5	2.5	.027	.014	6.4	.81	.60	.49	.024529	25.8
3-186-187EE	1.433	.478	.478	1.748	5.4	2.2	2.4	1.9	.028	.018	5.9	.68	.83	.57	.026340	26.4
2-24-25EE	1.270	.635	.635	1.270	5.1	1.9	2.5	2.5	.031	.018	6.4	.81	.81	.65	.032705	29.0
3-186-188EE	1.433	.478	.478	2.222	6.4	2.7	2.4	1.9	.033	.022	5.9	.68	1.06	.73	.033499	30.5
3-187-188EE	1.433	.478	.478	2.697	7.3	3.2	2.4	1.9	.038	.027	5.9	.68	1.29	.88	.040659	34.5
1-26-38EE	.952	.952	.635	1.321	5.8	2.3	2.2	3.2	.040	.019	6.2	.91	.84	.76	.044456	33.8
1-312EI	.795	.795	.952	1.984	7.5	2.8	2.7	3.5	.036	.044	6.6	.63	1.89	1.19	.045893	47.2
3-25EIS	1.905	.635	.635	.952	4.4	1.6	3.2	2.5	.041	.017	7.7	1.21	.60	.73	.046069	31.0

TABLE 5.2 (Continued)

MAGNETIC METALS EI AND EE LAMINATIONS

CAT. NO.	D (CM)	E (CM)	F (CM)	G (CM)	MPL (CM)	HT (CM)	WTH (CM)	LT (CM)	GROSS WTFE (KGM)	WTCU (KGM)	MLT (CM)	GROSS AC. (CM SQ)	GROSS WA (CM SQ)	GROSS AP (CM 4TH)	GROSS KG (CM 5TH)	AT (CM SQ)
	1	2	3	4	5	6	7	8	9	10	11	12	13	14	15	16
1-26-27EE	.952	.952	.635	1.748	6.7	2.7	2.2	3.2	.046	.025	6.2	.91	1.11	1.01	.058819	38.4
3-24-25EE	1.905	.635	.635	1.270	5.1	1.9	3.2	2.5	.047	.022	7.7	1.21	.81	.98	.061425	34.7
1-27-38EE	.952	.952	.635	2.113	7.4	3.1	2.2	3.2	.051	.030	6.2	.91	1.34	1.22	.071129	42.4
1-375EI	.952	.952	.795	1.905	7.3	2.9	2.5	3.5	.051	.036	6.7	.91	1.51	1.37	.074266	46.2
2-26-38EE	1.905	.952	.635	1.321	5.8	2.3	3.2	3.2	.081	.025	8.3	1.81	.84	1.52	.132764	44.2
2-312EI	1.590	.795	.952	1.984	7.5	2.8	3.5	3.5	.072	.056	8.4	1.26	1.89	2.39	.144254	57.2
1-50EI	1.270	1.270	.635	1.905	7.6	3.2	2.5	3.8	.094	.033	7.7	1.61	1.21	1.95	.163800	53.2
2-26-27EE	1.905	.952	.635	1.748	6.7	2.7	3.2	3.2	.092	.033	8.3	1.81	1.11	2.01	.175657	49.6
1-21EI	1.270	1.270	.795	2.065	8.3	3.3	2.9	4.1	.102	.048	8.2	1.61	1.64	2.65	.208652	62.1
2-27-38EE	1.905	.952	.635	2.113	7.4	3.1	3.2	3.2	.102	.040	8.3	1.81	1.34	2.43	.212422	54.3
2-375EI	1.905	.952	.795	1.905	7.3	2.9	3.5	3.5	.101	.048	8.8	1.81	1.51	2.75	.226078	58.3
3-26-38EE	2.857	.952	.635	1.321	5.8	2.3	4.1	3.2	.121	.030	10.2	2.72	.84	2.28	.243063	54.6
3-312EI	2.385	.795	.952	1.984	7.5	2.8	4.3	3.5	.108	.067	10.0	1.90	1.89	3.58	.272769	67.2
3-26-27EE	2.857	.952	.635	1.748	6.7	2.7	4.1	3.2	.139	.040	10.2	2.72	1.11	3.02	.321592	60.8
3-27-38EE	2.857	.952	.635	2.113	7.4	3.1	4.1	3.2	.154	.049	10.2	2.72	1.34	3.65	.388902	66.2
3-375EI	2.857	.952	.795	1.905	7.3	2.9	4.4	3.5	.152	.058	10.7	2.72	1.51	4.12	.418345	70.4
2-50EI	2.540	1.270	.635	1.905	7.6	3.2	3.8	3.8	.188	.044	10.2	3.23	1.21	3.90	.492436	71.0
1-625EI	1.587	1.587	.795	2.383	9.5	4.0	3.2	4.8	.183	.064	9.5	2.52	1.89	4.77	.508804	83.2
2-21EI	2.540	1.270	.795	2.065	8.3	3.3	4.1	4.1	.203	.063	10.7	3.23	1.64	5.30	.636996	81.1
1-68EI	1.748	1.748	.874	2.619	10.5	4.4	3.5	5.2	.244	.084	10.3	3.05	2.29	6.99	.825090	100.7
3-50EI	3.810	1.270	.635	1.905	7.6	3.2	5.1	3.8	.281	.056	13.0	4.84	1.21	5.85	.873603	88.7
3-21EI	3.810	1.270	.795	2.065	8.3	3.3	5.4	4.1	.305	.079	13.5	4.84	1.64	7.94	1.141374	100.0
1-202EI	1.905	1.905	1.270	2.286	10.9	3.9	4.4	7.0	.302	.126	12.2	3.63	2.90	10.54	1.251557	131.7
1-75EI	1.905	1.905	.952	2.857	11.4	4.8	3.8	5.7	.316	.109	11.2	3.63	2.72	9.88	1.277636	119.8
2-625EI	3.175	1.587	.795	2.383	9.5	4.0	4.8	4.8	.367	.085	12.6	5.04	1.89	9.55	1.523696	110.9
2-68EI	3.495	1.748	.874	2.619	10.5	4.4	5.2	5.2	.488	.113	13.8	6.11	2.29	13.98	2.466918	134.2
3-625EI	4.762	1.587	.795	2.383	9.5	4.0	6.4	4.8	.550	.108	16.0	7.56	1.89	14.32	2.704956	138.6
1-87EI	2.222	2.222	1.113	3.335	13.3	5.6	4.4	6.7	.503	.171	13.0	4.94	3.71	18.33	2.786457	163.0
2-75EI	3.810	1.905	.952	2.857	11.4	4.8	5.7	5.7	.633	.147	15.2	7.26	2.72	19.75	3.764375	159.7
2-202EI	3.810	1.905	1.270	2.286	10.9	3.9	6.3	7.0	.605	.168	16.2	7.26	2.90	21.07	3.768576	173.1
3-68EI	4.762	1.748	.874	2.619	10.5	4.4	6.5	5.2	.665	.135	16.6	8.32	2.29	19.04	3.824033	158.6
1-100EI	2.540	2.540	1.270	3.810	15.2	6.3	5.1	7.6	.750	.254	14.8	6.45	4.84	31.22	5.458068	212.9
3-75EI	5.715	1.905	.952	2.857	11.4	4.8	7.6	5.7	.949	.184	19.0	10.89	2.72	29.63	6.775479	199.6
3-202EI	5.715	1.905	1.270	2.286	10.9	3.9	8.3	7.0	.907	.207	20.0	10.89	2.90	31.61	6.867467	214.6
2-87EI	4.445	2.222	1.113	3.335	13.3	5.6	6.7	6.7	1.006	.233	17.6	9.88	3.71	36.65	8.209421	217.3
1-112EI	2.857	2.857	1.430	4.288	17.2	7.1	5.7	8.6	1.068	.360	16.5	8.17	6.13	50.06	9.890285	269.5
3-87EI	6.667	2.222	1.113	3.335	13.3	5.6	8.9	6.7	1.508	.297	22.5	14.82	3.71	54.98	14.487548	271.7
2-100EI	5.080	2.540	1.270	3.810	15.2	6.3	7.6	7.6	1.500	.345	20.0	12.90	4.84	62.43	16.077472	283.9
1-125EI	3.175	3.175	1.587	4.762	19.0	7.9	6.3	9.5	1.465	.492	18.3	10.08	7.56	76.21	16.795435	332.7
1-138EI	3.492	3.492	1.748	5.240	21.0	8.7	7.0	10.5	1.951	.654	20.1	12.20	9.16	111.69	27.152304	402.5

TABLE 3.2 (Continued)

MAGNETIC METALS EI AND EE LAMINATIONS

CAT. NO.	D (CM)	E (CM)	F (CM)	G (CM)	MPL (CM)	HT (CM)	WTH (CM)	LT (CM)	GROSS WTFE (KGM)	WTCU (KGM)	MLT (CM)	GROSS AC (CM SQ)	GROSS WA (CM SQ)	GROSS AP (CM 4TH)	GROSS KG (CM 5TH)	AT (CM SQ)
	1	2	3	4	5	6	7	8	9	10	11	12	13	14	15	16
3-100EI	7.620	2.540	1.270	3.810	15.2	6.3	10.2	7.6	2.251	.439	25.5	19.35	4.84	93.65	28.400309	354.8
2-112EI	5.715	2.857	1.430	4.288	17.2	7.1	8.6	8.6	2.137	.489	22.5	16.33	6.13	100.13	29.132582	359.3
1-150EI	3.810	3.810	1.905	5.715	22.9	9.5	7.6	11.4	2.532	.853	22.0	14.52	10.89	158.04	41.638281	479.0
2-125EI	6.350	3.175	1.587	4.762	19.0	7.9	9.5	9.5	2.930	.679	25.3	20.16	7.56	152.43	48.669813	443.5
3-112EI	8.572	2.857	1.430	4.288	17.2	7.1	11.4	8.6	3.205	.623	28.6	24.50	6.13	150.19	51.505089	449.1
1-145EI	3.683	3.683	2.349	7.620	27.3	11.4	8.4	12.1	2.826	1.447	22.7	13.56	17.90	242.85	57.985265	600.5
2-138EI	6.985	3.492	1.748	5.240	21.0	8.7	10.5	10.5	3.901	.901	27.7	24.40	9.16	223.39	78.793698	536.7
3-125EI	9.525	3.175	1.587	4.762	19.0	7.9	12.7	9.5	4.396	.850	31.6	30.24	7.56	228.64	87.50[illegible]661	554.4
1-36EI	4.127	4.127	3.175	6.667	27.9	10.8	10.5	14.6	3.632	2.055	27.3	17.04	21.17	360.65	90.0[illegible]224	742.7
1-175EI	4.445	4.445	2.222	6.680	26.7	11.1	8.9	13.3	4.024	1.350	25.6	19.76	14.85	293.34	90.64[illegible]741	652.0
2-150EI	7.620	3.810	1.905	5.715	22.9	9.5	11.4	11.4	5.064	1.164	30.1	29.03	10.89	316.08	122.088304	638.7
3-138EI	10.477	3.492	1.748	5.240	21.0	8.7	14.0	10.5	5.852	1.128	34.6	36.59	9.16	335.08	141.547155	670.9
1-19EI	4.445	4.445	4.445	7.620	33.0	12.1	13.3	17.8	4.978	3.922	32.6	19.76	33.87	669.22	162.443682	1066.9
2-145EI	7.366	3.683	2.349	7.620	27.3	11.4	12.1	12.1	5.652	1.954	30.7	27.13	17.90	485.70	171.683420	773.5
3-150EI	11.430	3.810	1.905	5.715	22.9	9.5	15.2	11.4	7.596	1.459	37.7	43.55	10.89	474.11	219.153503	798.4
1-212EI	5.397	5.397	2.700	8.098	32.4	13.5	10.8	16.2	7.200	2.401	30.9	29.13	21.86	636.95	240.316839	961.6
2-175EI	8.890	4.445	2.222	6.680	26.7	11.1	13.3	13.3	8.049	1.841	34.9	39.52	14.85	586.68	265.924103	869.4
2-36EI	8.255	4.127	3.175	6.667	27.9	10.8	14.6	14.6	7.264	2.707	36.0	34.07	21.17	721.29	273.372444	952.4
3-145EI	11.049	3.683	2.349	7.620	27.3	11.4	15.7	12.1	8.478	2.423	38.1	40.69	17.90	728.54	311.537388	946.6
1-225EI	5.715	5.715	2.857	8.572	34.3	14.3	11.4	17.1	8.545	2.844	32.7	32.66	24.50	800.07	320.126244	1077.8
3-175EI	13.335	4.445	2.222	6.680	26.7	11.1	17.8	13.3	12.073	2.310	43.8	59.27	14.85	880.03	476.782902	1086.7
3-36EI	12.382	4.127	3.175	6.667	27.9	10.8	18.7	14.6	10.895	3.328	44.2	51.11	21.17	1081.94	500.250389	1162.1
2-19EI	8.890	4.445	4.445	7.620	33.0	12.1	17.8	17.8	9.956	5.041	41.9	39.52	33.87	1338.44	505.454826	1332.3
2-212EI	10.795	5.397	2.700	8.098	32.4	13.5	16.2	16.2	14.400	3.272	42.1	58.27	21.86	1273.90	705.431343	1282.0
1-20EI	6.350	6.350	4.762	9.525	41.3	15.9	15.9	22.2	12.699	6.708	41.6	40.32	45.36	1829.14	709.478561	1673.4
3-19EI	13.335	4.445	4.445	7.620	33.0	12.1	22.2	17.8	14.934	6.112	50.7	59.27	33.87	2007.67	938.036354	1597.6
2-225EI	11.430	5.715	2.857	8.572	34.3	14.3	17.1	17.1	17.090	3.875	44.5	65.32	24.50	1600.13	939.811943	1437.1
3-212EI	16.192	5.397	2.700	8.098	32.4	13.5	21.6	16.2	21.599	4.111	52.9	87.40	21.86	1910.85	1263.219345	1602.5
1-3EI	7.620	7.620	3.810	11.430	45.7	19.0	15.2	22.9	20.255	6.763	43.7	58.06	43.55	2528.61	1344.824692	1916.1
3-225EI	17.145	5.715	2.857	8.572	34.3	14.3	22.9	17.1	25.636	4.871	55.9	97.98	24.50	2400.20	1682.339569	1796.4
1-301EI	7.620	7.620	5.715	11.430	49.5	19.0	19.0	26.7	21.943	11.534	49.7	58.06	65.32	3792.91	1774.078934	2409.7
2-20EI	12.700	6.350	4.762	9.525	41.3	15.9	22.2	22.2	25.397	8.756	54.3	80.64	45.36	3658.28	2173.958160	2157.3
3-20EI	19.050	6.350	4.762	9.525	41.3	15.9	28.6	22.2	38.096	10.805	67.0	120.97	45.36	5487.43	3963.993256	2641.1
2-3EI	15.240	7.620	3.810	11.430	45.7	19.0	22.9	22.9	40.511	9.123	58.9	116.13	43.55	5057.21	3987.680969	2554.8
2-301EI	15.240	7.620	5.715	11.430	49.5	19.0	26.7	26.7	43.887	15.074	64.9	116.13	65.32	7585.82	5429.829346	3106.4
1-4EI	10.160	10.160	5.080	15.240	61.0	25.4	20.3	30.5	48.013	15.918	57.8	103.23	77.42	7991.64	5706.912170	3406.4
3-3EI	22.860	7.620	3.810	11.430	45.7	19.0	30.5	22.9	60.766	11.483	74.2	174.19	43.55	7585.82	7128.220459	3193.5
3-301EI	22.860	7.620	5.715	11.430	49.5	19.0	34.3	26.7	65.830	18.614	80.1	174.19	65.32	11378.73	9893.697510	3803.2
2-4EI	20.320	10.160	5.080	15.240	61.0	25.4	30.5	30.5	96.026	21.512	78.1	206.45	77.42	15983.29	16891.450928	4541.9
3-4EI	30.480	10.160	5.080	15.240	61.0	25.4	40.6	30.5	144.038	27.106	98.5	309.68	77.42	23974.93	30162.247070	5677.4

TABLE 3.3 Powder Core Characteristics

	1	2	3	4	5		6	7	8	9	10	11	12	13	14		15	16
	Core	A_t cm^2	A_p cm^4	MLT cm	N	AWG	Ω @ 50°C	P_Σ	$I = \sqrt{\frac{W}{\Omega}}$	ΔT 25°C $J = I/cm^2$	Ω @ 75°C	P_Σ	$I = \sqrt{\frac{W}{\Omega}}$	ΔT 50°C $J = I/cm^2$	Weight fe	Cu	Volume cm^3	A_c cm^2
1	55051	6.569	0.0432	2.16	86	25	0.215	0.216	1.00	617	0.236	0.503	1.46	899	3.1	2.71	1.39	0.113
2	55121	11.24	0.139	2.74	160	25	0.513	0.369	0.848	522	0.563	0.861	1.23	762	6.8	6.3	3.11	0.196
3	55848	15.69	0.264	2.97	257	25	0.897	0.519	0.761	469	0.985	1.211	1.11	683	10	11.3	5.07	0.232
4	55059	20.02	0.460	3.45	316	25	1.27	0.657	0.719	443	1.39	1.533	1.05	647	16	16.3	7.28	0.327
5	55894	28.32	0.997	4.61	351	25	1.87	0.924	0.703	433	2.06	2.16	1.02	631	36	23.2	12.4	0.639
6	55586	44.24	1.83	4.32	902	25	4.69	1.46	0.558	344	5.15	3.40	0.812	500	35	59.9	23.3	0.458
7	55071	40.68	1.95	4.80	656	25	3.70	1.34	0.602	371	4.07	3.13	0.877	540	47	47.4	21.0	0.666
8	55076	46.91	2.44	4.88	815	25	4.71	1.55	0.574	353	5.17	3.61	0.814	518	52	61.0	25.7	0.670
9	55083	61.05	4.53	6.07	959	25	6.84	2.00	0.541	333	7.50	4.68	0.790	487	92	86.0	39.1	1.06
10	55090	81.58	8.06	6.66	1372	25	10.8	2.68	0.498	307	11.8	6.26	0.728	449	131	140	59.5	1.32
11	55439	79.37	8.33	7.62	959	25	8.49	2.60	0.553	341	9.32	6.08	0.807	497	182	109	58.1	1.95
12	55716	91.32	9.32	6.50	1684	25	13.0	3.00	0.480	296	14.3	7.00	0.699	431	133	170	69.0	1.24
13	55110	112.4	13.65	7.00	2125	25	17.8	3.72	0.457	282	19.6	8.68	0.665	410	176	226	93.4	1.44

copper loss ≫ iron loss

TABLE 3.4 Pot Core Characteristics

	1	2	3	4	5		6	7	8	9	10	11	12	13	14		15	16
	Core	A_t cm^2	A_p cm^4	MLT cm	N	AWG	Ω @ 50°C	P_Σ	$I = \sqrt{\frac{W}{\Omega}}$	ΔT 25°C $J = I/cm^2$	Ω @ 75°C	P_Σ	$I = \sqrt{\frac{W}{\Omega}}$	ΔT 50°C $J = I/cm^2$	Weight f_e	Cu	Volume cm^3	A_c cm^2
1	9 x 5	2.93	0.0065	1.85	25	30	0.175	0.098	0.529	1044	0.192	0.230	0.774	1527	0.8	0.32	0.367	0.10
2	11 x 7	4.35	0.0152	2.2	37	30	0.309	0.130	0.458	904	0.339	0.304	0.670	1322	1.7	0.38	0.662	0.16
3	14 x 8	6.96	0.0393	2.8	74	30	0.787	0.208	0.363	716	0.864	0.487	0.531	1048	3.2	0.98	1.35	0.25
4	18 x 11	11.3	0.114	3.56	143	30	1.934	0.339	0.296	584	2.12	0.791	0.432	853	6.0	2.37	2.78	0.43
5	22 x 13	17.0	0.246	4.4	207	30	3.46	0.510	0.271	535	3.80	1.190	0.396	782	13	4.30	5.17	0.63
6	26 x 16	23.9	0.498	5.2	96	25	0.592	0.717	0.778	479	0.650	1.67	1.13	696	21	7.5	8.65	0.94
7	30 x 19	32.8	1.016	6.0	144	25	1.024	0.984	0.693	427	1.12	2.30	1.01	622	36	12.9	13.9	1.36
8	36 x 22	44.8	2.01	7.3	189	25	1.636	1.34	0.639	394	1.79	3.14	0.937	577	57	20.8	22.0	2.01
9	47 x 28	76.0	5.62	9.3	345	25	3.81	2.28	0.547	337	4.18	5.32	0.798	492	125	48.0	48.6	3.12
10	59 x 36	122.0	13.4	12.0	608	25	8.65	3.66	0.459	283	9.50	8.54	0.670	413	270	109	98.3	4.85

copper loss = iron loss

TABLE 3.5 C-Core Characteristics

	1	2	3	4	5		6	7	8	9	10	11	12	13	14		15	16
	Core	A_t cm^2	A_p cm^4	MLT cm	N	AWG	Ω @ 50° C	P_Σ	$I = \sqrt{\frac{W}{\Omega}}$	Δ T 25° C $J = \frac{amps}{cm^2}$	Ω @ 75° C	P_Σ	$I = \sqrt{\frac{W}{\Omega}}$	Δ T 50° C $J = \frac{amps}{cm^2}$	Weight f_e	Cu	Volume cm^3	A_c cm^2
1	AL-2	20.9	0.265	3.55	662	30	8.93	0.627	0.187	370	9.81	1.46	0.273	538	12.2	11.1	7.14	0.265
2	AL-3	23.9	0.410	4.18	662	30	10.5	0.717	0.185	365	11.5	1.67	0.269	531	18.1	13.1	8.92	0.410
3	AL-5	33.6	0.767	4.59	946	30	16.5	1.01	0.174	345	18.1	2.35	0.255	503	31.3	20.5	14.06	0.539
4	AL-6	37.5	1.011	5.23	946	30	18.8	1.13	0.172	341	20.6	2.63	0.253	490	41.7	23.4	16.88	0.716
5	AL-124	45.3	1.44	5.50	1317	30	27.5	1.36	0.157	310	30.2	3.17	0.229	452	46.6	34.2	22.50	0.716
6	AL-8	63.4	2.31	5.74	221	20	0.482	1.90	1.404	271	0.529	4.44	2.05	395	67.9	60.0	35.66	0.806
7	AL-9	69.0	3.09	6.38	221	20	0.535	2.07	1.39	268	0.587	4.83	2.03	391	89.2	66.6	41.62	1.077
8	AL-10	74.5	3.85	7.01	221	20	0.588	2.24	1.38	266	0.646	5.22	2.01	387	110.0	73.2	47.55	1.342
9	AL-12	87.0	4.57	7.09	278	20	0.748	2.61	1.32	255	0.821	6.09	1.93	371	111.0	93.2	61.38	1.26
10	AL-135	93.7	5.14	7.36	325	20	0.908	2.81	1.24	240	0.997	6.56	1.81	345	114.0	113.0	69.63	1.26
11	AL-78	98.1	6.07	7.01	312	20	0.831	2.94	1.33	256	0.912	6.87	1.94	374	155.0	103.0	62.83	1.34
12	AL-18	118	7.92	7.61	510	20	1.47	3.55	1.10	211	1.61	8.26	1.60	308	138.0	183.0	94.79	1.25
13	AL-15	120	9.07	8.05	386	20	1.18	3.58	1.23	237	1.30	8.40	1.79	346	205.0	147.0	94.43	1.80
14	AL-16	127	10.8	8.89	386	20	1.30	3.80	1.20	233	1.43	8.89	1.76	340	235.0	162.0	104.95	2.15
15	AL-17	142	14.4	10.3	386	20	1.51	4.25	1.185	228	1.66	9.94	1.73	333	314.0	188.0	124.94	2.87
16	AL-19	159	18.0	10.8	511	20	2.10	4.77	1.065	205	2.31	11.1	1.55	299	328.0	261.0	155.44	2.87
17	AL-20	182	22.6	11.5	511	20	2.23	5.46	1.106	213	2.45	12.7	1.61	310	437.0	278.0	187.08	3.58
18	AL-22	202	28.0	11.5	637	20	2.78	6.05	1.043	201	3.05	14.1	1.52	293	489.0	346.0	212.04	3.58
19	AL-23	220	34.9	12.7	637	20	3.07	6.60	1.036	200	3.37	15.4	1.51	291	612.0	382.0	244.67	4.48
20	AL-24	245	40.0	12.0	948	20	4.32	7.35	0.922	178	4.74	17.1	1.35	259	552.0	538.0	280.91	3.58

copper loss = iron loss

TABLE 3.6 Single-Coil C-Core Characteristics

	1	2	3	4	5	6	7	8	9	10	11	12	13	14		15	16
	Core	$A_t cm^2$	$A_p cm^4$	MLT_{cm}	$N/_{AWG}$	Ω @ 50°C	P_Σ	$I = \sqrt{\frac{W}{\Omega}}$	ΔT 25°C J = I/cm^2	Ω @ 75°C	P_Σ	$I = \sqrt{\frac{W}{\Omega}}$	ΔT 50°C J = I/cm^2	Weight f_e	Weight Cu	Volume cm^3	$A_c cm^2$
1	AL-2	24.6	0.265	4.47	83/20	0.138	0.737	2.31	445	0.151	1.72	3.37	651	12.2	16.9	10.7	0.264
2	AL-3	27.6	0.410	5.10	83/20	0.158	0.828	2.28	441	0.173	1.93	3.34	644	18.1	19.3	12.5	0.406
3	AL-5	38.1	0.767	5.42	119/20	0.238	1.14	2.18	422	0.262	2.67	3.19	615	31.3	29.2	19.7	0.539
4	AL-6	41.9	1.011	6.06	119/20	0.266	1.26	2.17	420	0.292	2.93	3.16	611	41.7	32.6	21.9	0.716
5	AL-124	51.8	1.44	6.56	175/20	0.426	1.55	1.90	368	0.468	3.63	2.78	537	46.6	52.1	30.8	0.716
6	AL-8	72.8	2.31	7.06	255/20	0.669	2.18	1.80	348	0.734	5.10	2.63	508	67.9	81.7	53.5	0.806
7	AL-9	78.4	3.09	7.69	255/20	0.728	2.35	1.79	346	0.799	5.49	2.62	505	89.2	89.0	59.5	1.08
8	AL-10	83.9	3.85	8.33	255/20	0.788	2.52	1.78	345	0.866	5.87	2.60	502	110.0	96.4	65.4	1.34
9	AL-12	101.0	4.57	9.00	327/20	1.09	3.03	1.66	321	1.20	7.07	2.42	468	111.0	134.4	92.1	1.26
10	AL-135	110.0	5.14	9.50	370/20	1.31	3.30	1.58	306	1.43	7.70	2.32	447	114.0	159.0	107.0	1.26
11	AL-78	110.0	6.08	8.15	406/20	1.23	3.30	1.63	316	1.35	7.70	2.38	460	155.0	150.0	81.3	1.34
12	AL-18	142.0	7.87	7.51	564/20	2.14	4.26	1.41	272	2.35	9.94	2.05	396	138.0	260.0	147.0	1.25
13	AL-15	136.0	9.07	10.1	444/20	1.66	4.08	1.56	302	1.83	9.52	2.28	440	205.0	203.0	136.0	1.80
14	AL-16	143.0	10.8	10.7	444/20	1.77	4.29	1.55	300	1.94	10.0	2.27	438	235.0	216.0	147.0	2.15
15	AL-17	158.0	14.4	12.0	444/20	1.97	4.74	1.55	299	2.20	11.1	2.24	433	314.0	241.0	168.0	2.87
16	AL-19	182.0	•18.1	13.0	563/20	2.71	5.46	1.41	274	2.97	12.7	2.06	399	328.0	332.0	212.0	2.87
17	AL-20	205.0	22.6	13.6	563/20	2.84	6.15	1.47	284	3.12	14.4	2.14	414	437.0	348.0	259.0	3.58
18	AL-22	228.0	28.0	13.6	704/20	3.56	6.84	1.38	267	3.91	16.0	2.02	390	489.0	435.0	294.0	3.58
19	AL-23	246.0	35.0	15.9	704/20	3.89	7.38	1.37	265	4.27	17.2	2.00	387	612.0	479.0	326.0	4.48
20	AL-24	282.0	40.0	14.6	1026	5.57	8.46	1.23	238	6.11	19.7	1.79	346	552.0	680.0	401.0	3.58

TABLE 3.7 Tape-Wound Core Characteristics

	1	2	3	4	5		6	7	8	9	10	11	12	13	14		15	16
	Core	A_t cm^2	A_p cm^4	MLT cm	N	AWG	Ω @ 50°C	P_Σ	$I = \sqrt{\frac{W}{\Omega}}$	ΔT 25°C J = I/cm^2	Ω @ 75°C	P_Σ	$I = \sqrt{\frac{W}{\Omega}}$	ΔT 50°C J = I/cm^2	Weight f_e	Weight Cu	Volume cm^3	A_c cm^2
1	52402	7.26	0.0100	2.05	302	30	2.35	0.218	0.215	425	2.58	0.508	0.313	619	0.63	3.12	1.42	0.022
2	52153	8.29	0.0196	2.22	302	30	2.54	0.249	0.221	436	2.80	0.580	0.322	636	1.31	3.29	1.71	0.053
3	52107	11.1	0.0201	2.21	606	30	5.09	0.333	0.180	357	5.59	0.777	0.263	520	0.80	6.84	2.63	0.022
4	52403	13.5	0.0267	2.30	621	30	5.43	0.405	0.193	381	5.96	0.945	0.281	556	0.88	9.52	3.48	0.022
5	52057	17.4	0.0659	2.53	1017	30	9.78	0.522	0.163	322	10.7	1.22	0.238	471	2.05	13.1	4.98	0.043
6	52000	15.2	0.0787	2.70	606	30	6.22	0.456	0.191	378	6.82	1.06	0.278	550	3.73	7.97	3.99	0.086
7	52063	20.7	0.132	2.85	1017	30	11.0	0.621	0.167	331	12.1	1.45	0.244	483	4.47	14.4	6.20	0.086
8	52002	21.8	0.144	2.88	1114	30	12.2	0.654	0.163	323	13.4	1.53	0.239	472	4.62	16.0	6.72	0.086
9	52007	27.6	0.380	3.87	982	30	14.4	0.828	0.169	334	15.8	1.93	0.246	487	14.5	17.7	9.84	0.257
10	52167	31.5	0.516	4.23	1000	30	16.1	0.945	0.171	338	17.6	2.21	0.250	494	20.9	19.0	11.9	0.343
11	52094	30.4	0.592	4.47	1017	30	17.3	0.912	0.162	321	19.0	2.13	0.237	468	21.8	21.0	12.2	0.386
12	52004	46.1	0.725	4.02	315	20	0.469	1.38	1.20	234	0.515	3.23	1.77	341	13.4	56.8	21.3	0.171
13	52032	56.5	1.46	4.65	315	20	0.543	1.69	1.25	240	0.596	3.95	1.82	351	29.8	63.7	27.8	0.343
14	52026	61.0	2.18	5.28	315	20	0.616	1.83	1.22	235	0.676	4.27	1.77	342	44.7	71.3	32.8	0.514
15	52038	65.9	2.91	5.97	315	20	0.697	1.98	1.19	230	0.765	4.61	1.74	334	59.6	79.4	38.3	0.686
16	52035	88.9	4.68	6.33	505	20	1.19	2.67	1.06	204	1.3	6.22	1.55	298	71.5	138.0	59.0	0.686
17	52055	116.0	6.81	6.76	737	20	1.85	3.48	0.970	187	2.0	8.12	1.42	273	83.4	220.0	86.4	0.686
18	52012	110.0	9.35	8.88	505	20	1.66	3.30	0.996	192	1.82	7.70	1.45	280	143.0	235.0	87.4	1.371
19	52017	179.0	12.5	7.51	698	17	0.97	5.37	1.66	160	1.065	12.5	2.33	274	107.0	455.0	163.0	0.686
20	52031	256.0	19.8	8.23	1114	17	1.70	7.68	1.50	145	1.86	17.9	2.19	211	131.0	800.0	272.0	0.686
21	52103	220.0	24.5	8.77	688	17	1.12	6.60	1.72	165	1.23	15.4	2.51	241	238.0	503.0	212.0	1.371
22	52128	304.0	39.4	9.49	1104	17	1.94	9.12	1.53	147	2.13	21.3	2.24	215	286.0	896.0	341.0	1.371
23	52022	256.0	49.1	11.3	688	17	1.44	7.68	1.63	157	1.58	17.9	2.38	229	477.0	629.0	291.0	2.742
24	52042	347.0	78.7	12.0	1104	17	2.45	10.4	1.45	140	2.69	24.3	2.12	204	572.0	1109.0	453.0	2.742
25	52100	422.0	145.0	15.4	1089	17	3.11	12.7	1.43	138	3.41	29.5	2.08	200	1117.0	1342.0	633.0	5.142
26	52112	878.0	510.0	20.3	2871	17	10.8	26.3	1.1	106	11.8	61.5	1.61	155	2205.0	4895.0	1891.0	6.855
27	52426	1014.0	813.0	22.2	2856	17	11.7	24.4	1.02	98.1	12.9	71.0	1.66	159	3814.0	5077.0	2299.0	10.968

copper loss = iron loss

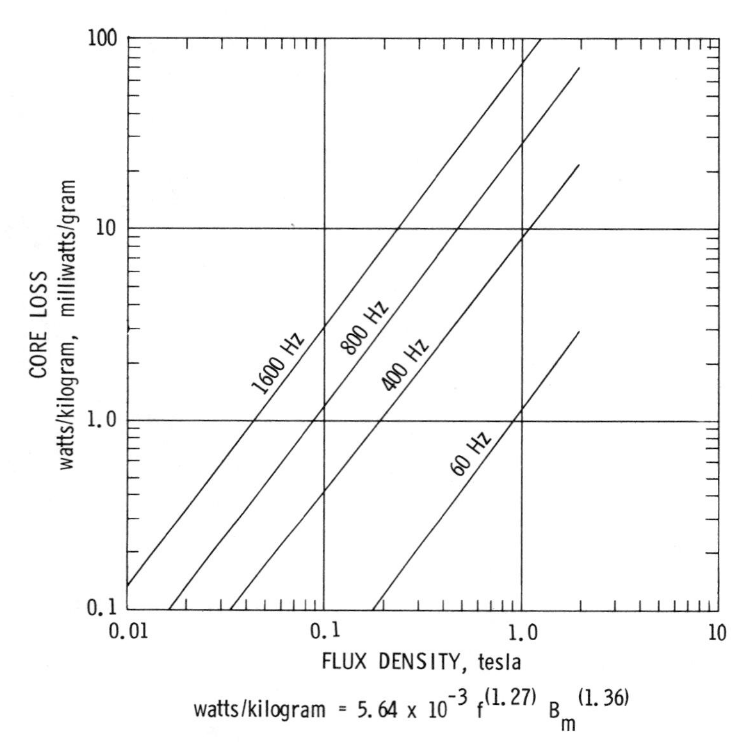

FIG. 3.11 Magnetics, Inc., Supermendur 4 mil.

The current density is given by Eq. (3.17):

$$J = K_j A_p^{-0.14} \text{ A/cm}^2$$

$$= 366 \times 205.82^{-0.14} \text{ A/cm}^2 = 173.62 \text{ A/cm}^2$$

The primary current I_p is calculated by

$$I_p = \frac{P_t - P_o}{V_i} \text{ A}$$

$$= \frac{539 - 262.5}{115}$$

$$= 2.4 \text{ A}$$

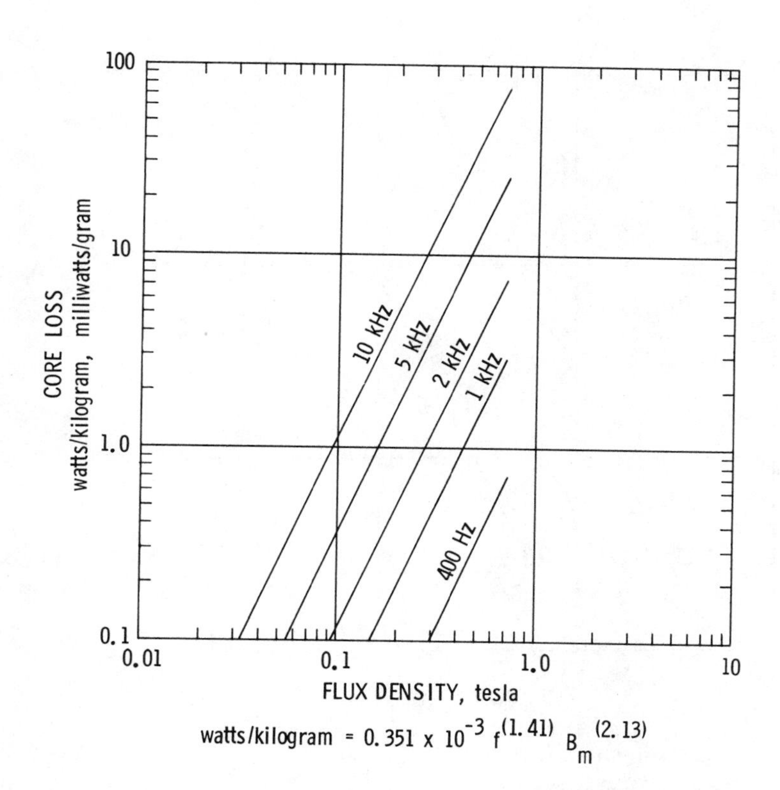

FIG. 3.12 Arnold Engineering Permalloy 2 mil cut C cores.

The bare wire size, $A_{w(B)}$, for the primary winding is

$$A_{w(B)} = \frac{I_p}{J} = \frac{2.40}{173.62} = 13.82 \times 10^{-3}\ \text{cm}^2$$

From Table 3.8, the value of $A_{w(B)}$ falls between sizes AWG 15 and AWG 16.

Select AWG 16 for the primary winding.

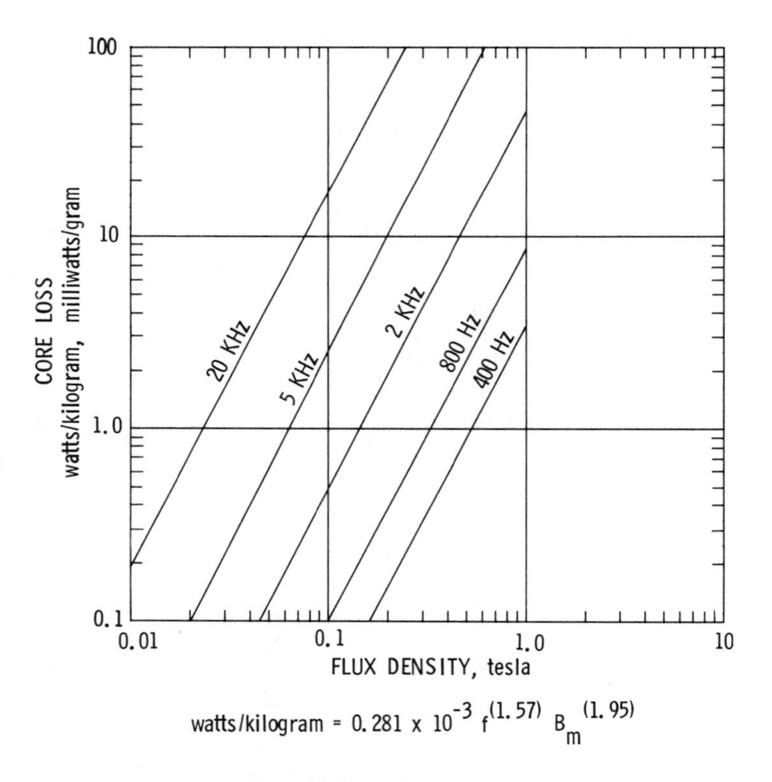

FIG. 3.13 Magnetics, Inc., alloy 48, 4 mil.

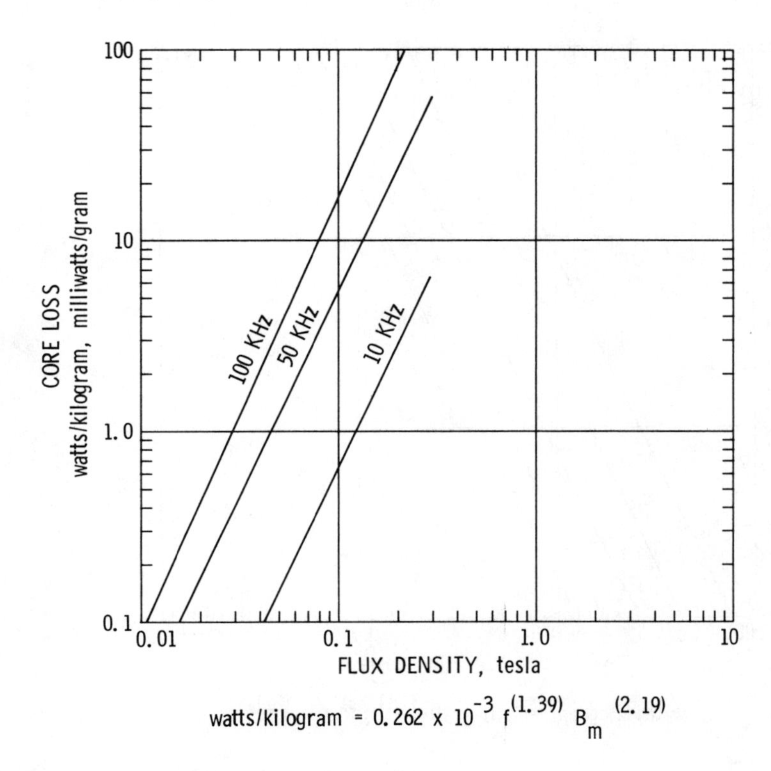

FIG. 3.14 Siemens Siferrit N27.

The primary winding resistance from column C of Table 3.8 is

$$R_p = MLT \times N \times \mu\Omega/cm \times \zeta \times 10^{-6} \tag{3.30}$$

$$= 27.7 \times 130 \times 131.8 \times 1.098 \times 10^{-6}$$

$$= 0.521\ \Omega$$

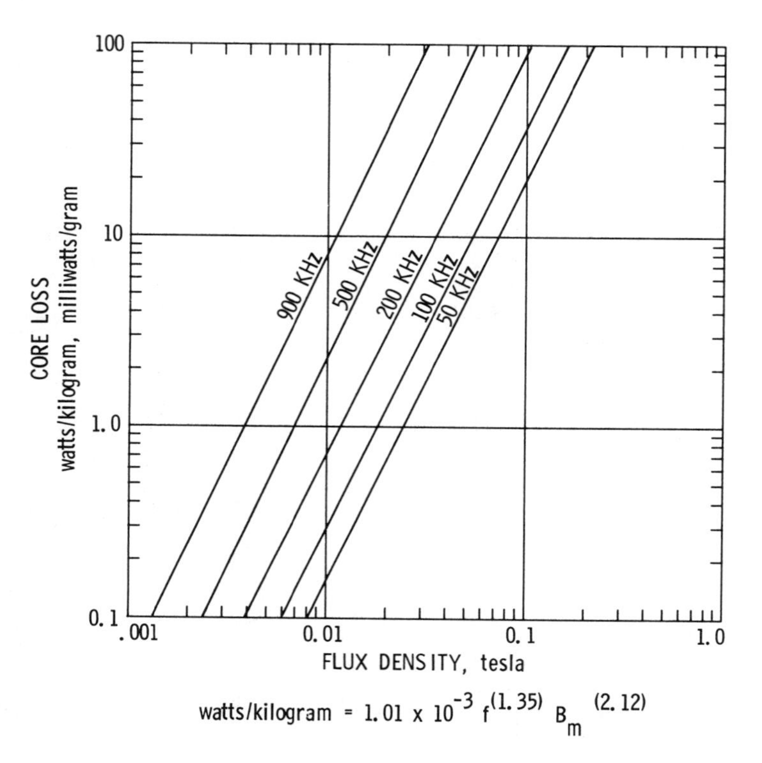

FIG. 3.15 Ferroxcube, 3C8 material.

The primary copper loss is therefore

$$P_{cu} = I_p^2 R_p = 2.40^2 \times 0.521 = 3.00 \text{ W}$$

The bare wire size for the secondary winding is

$$A_{w(B)} = \frac{I_o}{J} = \frac{7.5}{173.62} = 43.2 \times 10^{-3} \text{ cm}^2$$

Select AWG 11 for the secondary winding.
The number of secondary turns is

$$N_s = \frac{N_p}{V_i} V_o = \frac{130}{115} \times 35 = 39.56, \quad \text{say 40 turns}$$

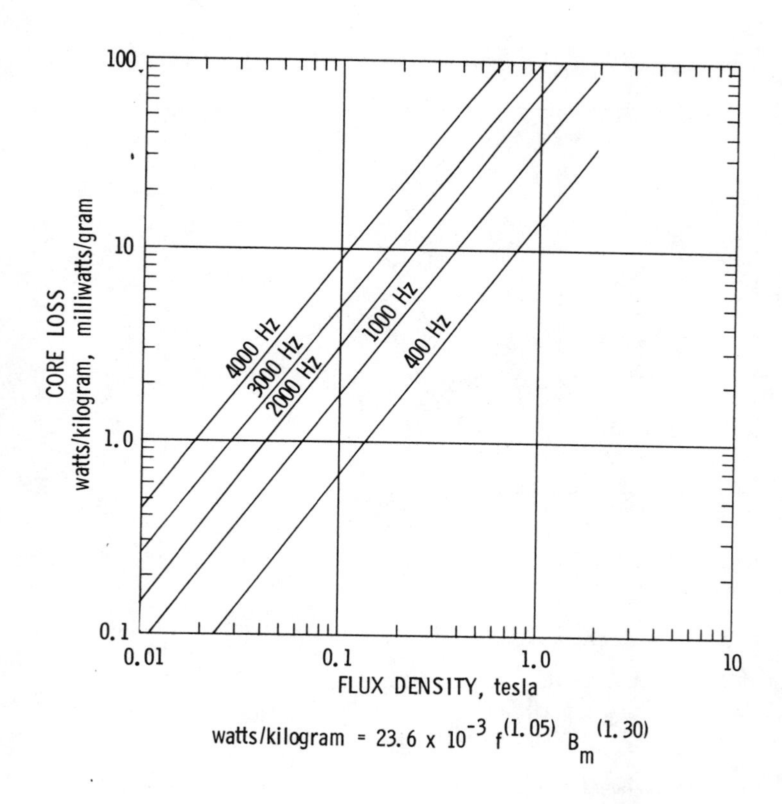

FIG. 3.16 Magnetics, Inc., Supermendur 2 mil.

The secondary winding resistance is

$$R_s = \text{MLT} \times N \times \mu\Omega/\text{cm} \times \zeta \times 10^{-6}\ \Omega$$

$$= 27.7 \times 40 \times 41.37 \times 1.098 \times 10^{-6}\ \Omega$$

$$= 0.05\ \Omega$$

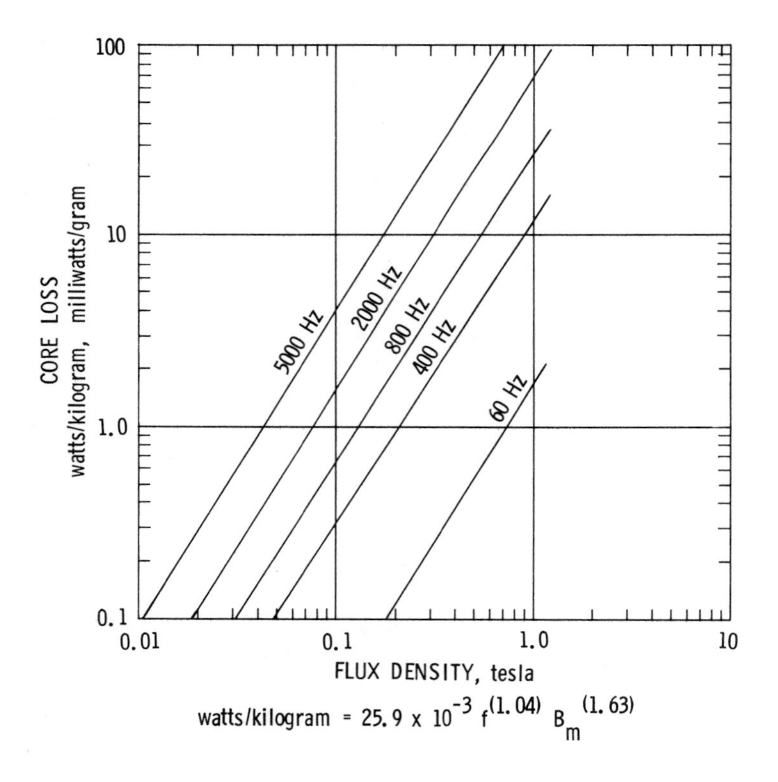

FIG. 3.17 Magnetics, Inc., Magnesil 2 mil.

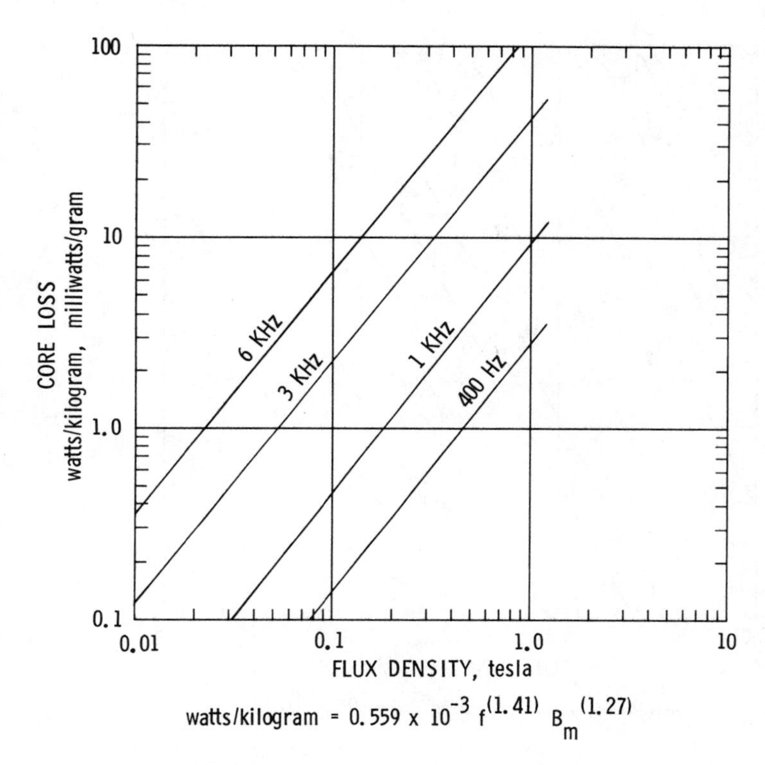

FIG. 3.18 Magnetics, Inc., Orthonol 2 mil.

The secondary loss is

$$P_{cu} = I_s^2 R_s = 7.5^2 \times 0.05 = 2.83 \text{ W}$$

Total loss:

$$P_\Sigma = \text{primary } P_{cu} + \text{secondary } P_{cu} + P_{fe} \text{ W}$$

$$= 3 + 2.83 + 3.946 \text{ W}$$

$$= 9.78 \text{ W}$$

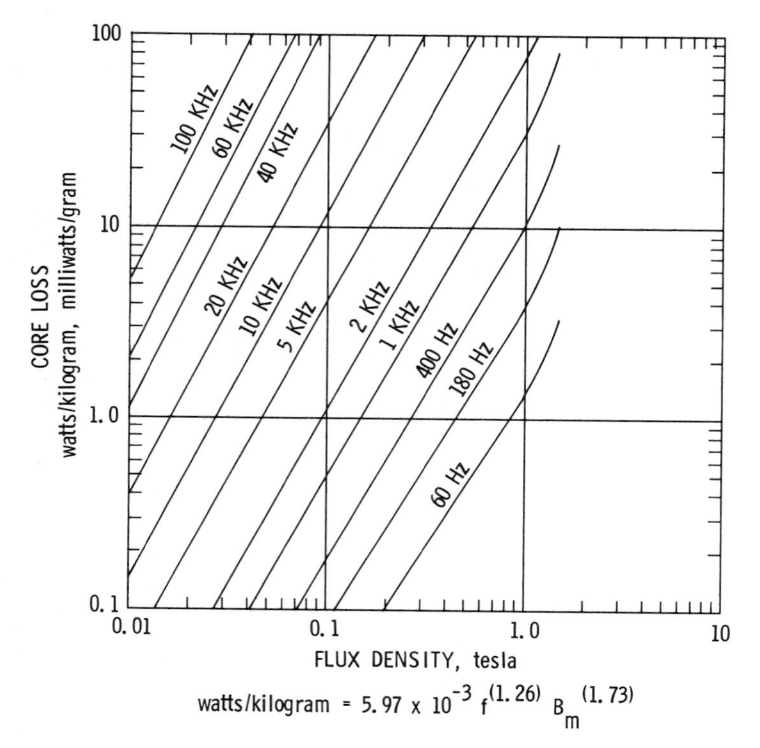

FIG. 3.19 Arnold Silectron 2 mil.

EXAMPLE 3.2 Design of 400-Hz power transformer using the core geometry approach (complete details and derivations of formulas used in this method can be found in McLyman [120, 122]). In this approach, the regulation and power capability of the transformer are related to the core geometry constant K_g, the electrical constant K_e, and the regulation constant α:

$$P_t = 2K_g K_e \alpha = P_o \frac{1}{\eta} + \sqrt{2} = 214.62 \qquad (3.31)$$

The core geometry constant K_g is given by [120]

$$K_g = \frac{W_a A_c^2 K_u}{MLT} \tag{3.32}$$

The constant K_e is given by

$$K_e = 0.145 K^2 f^2 B_m^2 \times 10^{-4} \tag{3.33}$$

TABLE 3.8 Wire Table

AWG Wire Size	Bare Area cm² 10⁻³ (footnote b)	Bare Area Cir-Mil[a]	Resistance $\frac{10^{-6}\,\Omega}{\text{cm at }20°C}$	Heavy Synthetics Area cm² 10⁻³	Heavy Synthetics Area Cir-Mil[a]	Diameter cm	Diameter Inch[a]	Turns-Per: cm	Turns-Per: Inch[a]	Turns-Per: cm²	Turns-Per: Inch²	Weight gm/cm
10	52.61	10384	32.70	55.9	11046	0.267	0.1051	3.87	9.5	10.73	69.20	0.468
11	41.68	8226	41.37	44.5	8798	0.238	0.0938	4.36	10.7	13.48	89.95	0.3750
12	33.08	6529	52.09	35.64	7022	0.213	0.0838	4.85	11.9	16.81	108.4	0.2977
13	26.26	5184	65.64	28.36	5610	0.190	0.0749	5.47	13.4	21.15	136.4	0.2367
14	20.82	4109	82.80	22.95	4556	0.171	0.0675	6.04	14.8	26.14	168.6	0.1879
15	16.51	3260	104.3	18.37	3624	0.153	0.0602	6.77	16.6	32.66	210.6	0.1492
16	13.07	2581	131.8	14.73	2905	0.137	0.0539	7.32	18.6	40.73	262.7	0.1184
17	10.39	2052	165.8	11.68	2323	0.122	0.0482	8.18	20.8	51.36	331.2	0.0943
18	8.228	1624	209.5	9.326	1857	0.109	0.0431	9.13	23.2	64.33	414.9	0.07472
19	6.531	1289	263.9	7.539	1490	0.0980	0.0386	10.19	25.9	79.85	515.0	0.05940
20	5.188	1024	332.3	6.065	1197	0.0879	0.0346	11.37	28.9	98.93	638.1	0.04726
21	4.116	812.3	418.9	4.837	954.8	0.0785	0.0309	12.75	32.4	124.0	799.8	0.03757
22	3.243	640.1	531.4	3.857	761.7	0.0701	0.0276	14.25	36.2	155.5	1003	0.02965
23	2.588	510.8	666.0	3.135	620.0	0.0632	0.0249	15.82	40.2	191.3	1234	0.02372
24	2.047	404.0	842.1	2.514	497.3	0.0566	0.0223	17.63	44.8	238.6	1539	0.01884
25	1.623	320.4	1062.0	2.002	396.0	0.0505	0.0199	19.80	50.3	299.7	1933	0.01498
26	1.280	252.8	1345.0	1.603	316.8	0.0452	0.0178	22.12	56.2	374.2	2414	0.01185
27	1.021	201.6	1687.6	1.313	259.2	0.0409	0.0161	24.44	62.1	456.9	2947	0.00945
28	0.8046	158.8	2142.7	1.0515	207.3	0.0366	0.0144	27.32	69.4	570.6	3680	0.00747
29	0.6470	127.7	2664.3	0.8548	169.0	0.0330	0.0130	30.27	76.9	701.9	4527	0.00602
30	0.5067	100.0	3402.2	0.6785	134.5	0.0294	0.0116	33.93	86.2	884.3	5703	0.00472
31	0.4013	79.21	4294.6	0.5596	110.2	0.0267	0.0105	37.48	95.2	1072	6914	0.00372
32	0.3242	64.00	5314.9	0.4559	90.25	0.0241	0.0095	41.45	105.3	1316	8488	0.00305
33	0.2554	50.41	6748.6	0.3662	72.25	0.0216	0.0085	46.33	117.7	1638	10565	0.00241
34	0.2011	39.69	8572.8	0.2863	56.25	0.0191	0.0075	52.48	133.3	2095	13512	0.00189
35	0.1589	31.36	10849	0.2268	44.89	0.0170	0.0067	58.77	149.3	2645	17060	0.00150
36	0.1266	25.00	13608	0.1813	36.00	0.0152	0.0060	65.62	166.7	3309	21343	0.00119
37	0.1026	20.25	16801	0.1538	30.25	0.0140	0.0055	71.57	181.8	3901	25161	0.000977
38	0.08107	16.00	21266	0.1207	24.01	0.0124	0.0049	80.35	204.1	4971	32062	0.000773
39	0.06207	12.25	27775	0.0932	18.49	0.0109	0.0043	91.57	232.6	6437	41518	0.000593
40	0.04869	9.61	35400	0.0723	14.44	0.0096	0.0038	103.6	263.2	8298	53522	0.000464
41	0.03972	7.84	43405	0.0584	11.56	0.00863	0.0034	115.7	294.1	10273	66260	0.000379
42	0.03166	6.25	54429	0.04558	9.00	0.00762	0.0030	131.2	333.3	13163	84901	0.000299
43	0.02452	4.84	70308	0.03683	7.29	0.00685	0.0027	145.8	370.4	16291	105076	0.000233
44	0.0202	4.00	85072	0.03165	6.25	0.00635	0.0025	157.4	400.0	18957	122272	0.000195
	A	B	C	D	E	F	G	H	I	J	K	L

[a] *This data from REA Magnetic Wire Datalator*

[b] *This notation means the entry in the column must be multiplied by* 10^{-3}

Source: Courtesy of Arnold Engineering Co., Marengo, Illinois.

TABLE 3.9 Coefficient K_g for Laminations[a]

Core[b]	10^{-3} K_g	W_a, cm^2	A_c, cm^2	MLT, cm	G, cm	D, cm
EE 3031	0.103	0.176	0.0502	1.72	0.714	0.239
EE 2829	0.356	0.252	0.0907	2.33	0.792	0.318
EI 187	2.75	0.530	0.204	3.20	1.113	0.478
EE 2425	8.37	0.807	0.363	5.08	1.27	0.635
EE 2627	51.1	1.11	0.816	5.79	1.748	0.953
EI 375	63.8	1.51	0.816	6.30	1.905	0.953
EI 50	144	1.21	1.45	7.09	1.91	1.27
EI 21	181	1.63	1.45	7.57	2.06	1.27
EI 625	441	1.89	2.27	8.84	2.38	1.59
EI 75	1100	2.72	3.27	10.6	2.86	1.91
EI 87	2390	3.71	4.45	12.3	3.33	2.22
EI 100	4500	4.83	5.81	14.5	3.81	2.54
EI 112	8240	6.12	7.34	16.0	4.28	2.86
EI 125	14100	7.57	9.07	17.7	4.76	3.18
EI 138	25400	9.20	11.6	19.5	5.24	3.49
EI 150	35300	10.9	13.1	21.2	5.72	3.81
EI 175	75900	14.8	17.8	24.7	6.67	4.45
EI 36	74900	21.2	15.3	26.5	6.67	4.13
EI 19	135000	33.8	17.8	31.7	7.62	4.45

[a]Where $K_u = 0.4$. [b]Magnetic Metals.

The specification of the example transformer is as follows: The circuit configuration is shown in Fig. 3.7b.

$$V_o = 28 + V_f$$

where V_f is diode forward voltage drop (assume $V_f = 1$ V), I_o = 3.0 A, V_i = 115-V ac sinusoid, f = 400 Hz, η = 95% (target efficiency), and α = 2% (target regulation).

Using Eq. (3.33),

$$K_e = 0.145 \times 4.44^2 \times 400^2 \times 0.9^2 \times 10^{-4} = 37$$

TABLE 3.10 Coefficient K_g for Powder Core[a]

Core[b]	10^{-3} K_g	W_a, cm^2	A_c, cm^2	MLT, cm
55051	0.901	0.381	0.113	2.16
55121	4.00	0.713	0.196	2.74
55848	8.26	1.14	0.232	2.97
55059	17.4	1.407	0.327	3.45
55894	55.3	1.561	0.639	4.61
55586	77.7	4.00	0.458	4.32
55071	108	2.93	0.666	4.80
55076	134	3.64	0.670	4.88
55083	316	4.27	1.060	6.07
55090	639	6.11	1.32	6.66
55439	852	4.27	1.95	7.62
55716	712	7.52	1.24	6.50
55110	1123	9.48	1.44	7.00

[a] Where K_u = 0.4.

[b] Magnetics, Inc.

TABLE 3.11 Coefficient K_g for Pot Cores[a]

Core[b]	10^{-3} K_g	W_a, cm^2	A_c, cm^2	MLT, cm
9 × 5	0.109	0.065	0.10	1.85
11 × 7	0.343	0.095	0.16	2.2
14 × 8	1.09	0.157	0.25	2.8
18 × 11	4.28	0.266	0.43	3.56
22 × 13	10.9	0.390	0.63	4.4
26 × 16	27.9	0.530	0.94	5.2
30 × 19	71.6	0.747	1.36	6.0
36 × 22	171	1.00	2.01	7.3
47 × 28	584	1.80	3.12	9.3
59 × 36	1683	2.77	4.85	12.0

[a]Where K_u = 0.31. [b]Siemens.

The core geometry constant K_g is calculated using Eq. (3.31) (see Tables 3.9 to 3.11):

$$K_g = \frac{P_t}{2K_e \alpha} = 1.45$$

Select from Table 3.12 a C-core, AL-19 with K_g = 1.60. The number of primary turns is calculated using Eq. (3.29):

$$N_p = \frac{115 \times 10^4}{4.44 \times 2.87 \times 0.9 \times 400} = 250 \text{ turns}$$

The effective window area $W_{a(eff)}$ is

$$W_{a(eff)} = W_a S_3$$

TABLE 3.12 Coefficient K_g for C Cores[a]

Core[b]	10^{-3} K_g	W_a, cm^2	A_c, cm^2	MLT, cm	G, cm	D, cm
AL-2	6.27	1.006	0.264	4.47	1.587	0.635
AL-3	14.4	1.006	0.406	5.10	1.587	0.952
AL-5	30.5	1.423	0.539	5.42	2.22	0.952
AL-6	47.8	1.413	0.716	6.06	2.22	1.27
AL-124	63.1	2.02	0.716	6.56	2.54	1.27
AL-8	106	2.87	0.806	7.06	3.015	0.952
AL-9	173	2.87	1.077	7.69	3.015	1.27
AL-10	248	2.87	1.342	8.33	3.015	1.587
AL-12	256	3.63	1.260	9.00	2.857	1.27
AL-135	273	4.083	1.260	9.50	2.857	1.27
AL-78	399	4.53	1.340	8.15	5.715	1.91
AL-18	530	6.30	1.257	7.51	3.927	1.27
AL-15	648	5.037	1.80	10.08	3.967	1.587
AL-16	869	5.037	2.15	10.72	3.967	1.905
AL-17	1380	5.037	2.87	11.99	3.967	2.54
AL-19	1600	6.30	2.87	12.98	3.967	2.54
AL-20	2370	6.30	3.58	13.62	3.967	2.54
AL-22	2940	7.804	3.58	13.62	4.92	2.54
AL-23	4210	7.804	4.48	14.98	4.92	3.175
AL-24	3910	11.16	3.58	14.62	5.875	2.54

[a]Where $K_u = 0.4$. [b]Arnold Engineering Co.

TABLE 3.13 Coefficient K_g for Tape-Wound Toroids[a]

Core[b]	10^3 Kg	W_a, cm^2	A_c, cm^2	MLT, cm
52402	0.0472	0.502	0.022	2.06
52153	0.254	0.502	0.053	2.22
52107	0.0860	0.982	0.022	2.21
52403	0.107	1.28	0.022	2.30
52057	0.456	1.56	0.043	2.53
52000	1.07	0.982	0.086	2.70
52063	1.62	1.56	0.086	2.85
52002	1.81	1.76	0.086	2.88
52007	10.6	1.56	0.257	3.87
52167	17.4	1.56	0.343	4.23
52094	20.8	1.56	0.386	4.47
52004	12.7	4.38	0.171	4.02
52032	44.3	4.38	0.343	4.65
52026	87.7	4.38	0.514	5.28
52038	138	4.38	0.686	5.97
52035	203	6.816	0.686	6.33
52055	276	9.93	0.686	6.76
52012	587	6.94	1.371	8.88
52017	459	18.3	0.686	7.51
52031	668	29.2	0.686	8.23
52103	1570	18.3	1.371	8.77
52128	2220	28.0	1.371	9.49
52022	4870	18.3	2.742	11.30
52042	6790	27.1	2.742	12.0
52100	18600	27.1	5.142	15.4
52112	68100	73.6	6.855	20.3
52426	159000	73.6	10.968	22.2

[a]Where K_u = 0.4. [b]Magnetics, Inc.

TABLE 3.14 Magnetic Core Material Characteristics

Trade names	Composition	Saturated flux density,[a] (tesla)	DC coercive force, amp-turn/cm	Squareness ratio	Material density, g/cm[b]	Loss factor at 3 kHz and 0.5 T, W/kg
Magnesil Silectron Microsil Supersil	3% Si 97% Fe	1.5-1.8	0.5-0.75	0.85-1.0	7.63	33.1
Deltamax Orthonol 49 Sq. Mu	50% Ni 50% Fe	1.4-1.6	0.125-0.25	0.94-1.0	8.24	17.66
Allegheny 4750 48 Alloy Carpenter 49	48% Ni 52% Fe	1.15-1.4	0.062-0.187	0.80-0.92	8.19	11.03
4-79 Permalloy Sq. Permalloy 80 Sq. Mu 79	79% Ni 17% Fe 4% Mo	0.66-0.82	0.025-0.05	0.80-1.0	8.73	5.51
Supermalloy	78% Ni 17% Fe 5% Mo	0.65-0.82	0.0037-0.01	0.40-0.70	8.76	3.75

[a] 1 T = 10^4 Gauss.

[b] 1 g/cm^3 = 0.036 $lb/in.^3$

TABLE 3.15 DC Inductor Examples

TOROIDAL CORES: SINGLE LAYER WINDINGS								
DC CURRENT / WIRE* SIZE / PART #	1.0 amp #28 AWG	2.5 amps #24 AWG	5.0 amps #20 AWG	7.5 amps #18 AWG	10 amps #16 AWG	15 amps #14 AWG	20 amps #12 AWG	30 amps #10 AWG
T50-26	92.0 μh 63 TURNS	25.6 μh 37 TURNS	7.8 μh 21 TURNS	4.2 μh 16 TURNS	2.1 μh 11 TURNS	1.1 μh 8 TURNS	0.6 μh 6 TURNS	0.4 μh 5 TURNS
T68-26A	260 μh 79 TURNS	76.8 μh 47 TURNS	24.4 μh 28 TURNS	12.8 μh 21 TURNS	6.8 μh 15 TURNS	3.4 μh 11 TURNS	1.9 μh 8 TURNS	1.0 μh 6 TURNS
T90-26	700 μh 120 TURNS	214 μh 74 TURNS	67.2 μh 44 TURNS	36.6 μh 34 TURNS	21.4 μh 26 TURNS	10.7 μh 19 TURNS	5.9 μh 14 TURNS	2.8 μh 10 TURNS
T106-26	1040 μh 125 TURNS	320 μh 77 TURNS	102 μh 46 TURNS	56.9 μh 36 TURNS	32.0 μh 27 TURNS	16.4 μh 20 TURNS	8.4 μh 14 TURNS	4.1 μh 10 TURNS
T131-26	1700 μh 141 TURNS	528 μh 87 TURNS	176 μh 53 TURNS	96.0 μh 41 TURNS	54.0 μh 31 TURNS	29.3 μh 24 TURNS	16.5 μh 18 TURNS	8.1 μh 13 TURNS
T157-26	3140 μh 213 TURNS	960 μh 132 TURNS	324 μh 82 TURNS	178 μh 64 TURNS	105 μh 50 TURNS	55.1 μh 38 TURNS	31.5 μh 29 TURNS	16.2 μh 22 TURNS
T184-26	5500 μh 213 TURNS	1730 μh 132 TURNS	580 μh 82 TURNS	320 μh 64 TURNS	190 μh 50 TURNS	99.6 μh 38 TURNS	58.0 μh 29 TURNS	29.3 μh 22 TURNS
T300-26D	21,400 μh 435 TURNS	6560 μh 272 TURNS	2240 μh 169 TURNS	1240 μh 135 TURNS	740 μh 105 TURNS	400 μh 82 TURNS	230 μh 63 TURNS	124 μh 49 TURNS
T400-26D	149,000 μh 507 TURNS	15,000 μh 317 TURNS	5120 μh 197 TURNS	2920 μh 157 TURNS	1700 μh 122 TURNS	907 μh 95 TURNS	525 μh 73 TURNS	284 μh 57 TURNS

E-CORES: FULL BOBBIN WINDINGS								
DC CURRENT / WIRE* SIZE / PART #	2.5 amp #20 AWG	5.0 amps #18 AWG	10 amps #14 AWG	15 amps #12 AWG	20 amps #11 AWG	30 amps FOIL	50 amps FOIL	100 amps FOIL
E137-26 .007 BUTT GAP	1090 μh 111 TURNS	408 μh 70 TURNS	70 μh 28 TURNS	29.3 μh 18 TURNS	17.5 μh 14 TURNS	10.2 μh 11 TURNS	4.1 μh 7 TURNS	0.8 μh 3 TURNS
E168-26 .015 BUTT GAP	5440 μh 215 TURNS	2080 μh 139 TURNS	370 μh 56 TURNS	147 μh 35 TURNS	92.5 μh 28 TURNS	45.6 μh 20 TURNS	12.2 μh 10 TURNS	4.1 μh 6 TURNS
E220-26 .020 BUTT GAP	15,000 μh 303 TURNS	5440 μh 194 TURNS	1000 μh 79 TURNS	418 μh 50 TURNS	258 μh 40 TURNS	116 μh 27 TURNS	41.2 μh 16 TURNS	10.3 μh 8 TURNS

*Based on maximum temperature rise of 40°C due to copper loss.

Note: This table assumes <1% ripple current. The presence of significant ripple current will result in both greater inductance and higher operating temperature.

Source: Courtesy of Micrometals, Inc., Anaheim, California.

For $S_3 = 0.75$,

$$W_{a(eff)} = 6.30 \times 0.75 = 4.725 \text{ cm}^2$$

$$\text{Primary winding area} = 0.4W_{a(eff)} = 0.4 \times 4.725$$
$$= 1.89 \text{ cm}^2$$

The wire area A_w, with insulation, using a fill factor S_2 of 0.6 is

$$A_w = \frac{W_a S_2}{N_p} = \frac{1.89 \times 0.6}{250} = 0.004536 \text{ cm}^2$$

Select the wire size from Table 3.8, column D:

$$\text{AWG } 22 = 0.003857 \text{ cm}^2$$

The primary winding resistance is

$$R_p = \text{MLT} \times N_p \times \mu\Omega/\text{cm} \times 10^{-6}$$
$$= 12.98 \times 250 \times 531.4 \times 10^{-6} = 1.724 \ \Omega$$

The primary current

$$I_p = \frac{VA}{V_i} = \frac{87}{115} = 0.7565 \text{ A}$$

The primary copper loss is

$$P_{cu} = I_p^2 R_p = 0.7565^2 \times 1.724 = 0.987 \text{ W}$$

The number of secondary turns

$$N_s = \frac{N_p}{V_i} V_o = \frac{250}{115} \times 29 = 63 \text{ turns}$$

The secondary wire area

$$A_w = \frac{W_a S_2}{2N_s} = \frac{0.6 \times 4.725 \times 0.6}{2 \times 63} = 0.0135 \text{ cm}^2$$

Select the wire size from Table 3.8, column D:

$$\text{AWG } 17 = 0.01168 \text{ cm}^2$$

The secondary winding resistance is

$$R_s = \text{MLT} \times N_s \times \mu\Omega/\text{cm} \times 10^{-6}$$

$$= 12.98 \times 63 \times 165.8 \times 10^{-6}$$

$$= 0.136 \ \Omega$$

The secondary loss is $I_s^2 R_s = 3.5^2 \times 0.136 = 1.22$ W.

$$P_\Sigma = \frac{P_o}{\eta} - P_o = \frac{87}{0.95} - 87 = 4.579 \text{ W}$$

The optimum efficiency, $P_{cu} = P_{fe} = (1/2) \times 4.579 = 2.289$ W. Therefore, the core loss is

$$\frac{P_{fe}}{W_t} \times 10^3 \text{ mW/g} = 2.289 \text{ W}$$

From Table 3.5, column 14, AL-19 has a weight of 328 g. Therefore,

$$\frac{2.289}{328} \times 10^3 = 6.98 \text{ mW/g}$$

With reference to Fig. 3.19, for a flux density of 0.9 T, the core loss at 400 Hz is quite close to 8.5 mW/g. Therefore, the total loss is

$$P_{\Sigma} = P_{cu} + P_{fe}$$

$$= 0.987 + 1.22 + 2.289 = 4.496\ \text{W}$$

which will have an efficiency of about 94%. The regulation factor α is given by

$$\alpha = \frac{P_{cu\ pri} + P_{cu\ sec}}{P_o + P_{cu\ pri} + P_{cu\ sec}} \times 100\%$$

$$= \frac{0.987 + 1.22}{87 + 0.987 + 1.22} \times 100\%$$

$$= 2.47\%$$

3.2 DESIGN OF CURRENT TRANSFORMER

The current transformer is normally used for sampling a time-varying current within a circuit. The primary purpose of using the transformer is to obtain isolation from the actual circuit where the sample is to be taken. The ideal current transformer provides a current step-down at a designated ratio for efficiency and convenience. Ideally, a transformer with a 1:1000 turns ratio should measure a current of 1 A by providing a current of 1 mA in the secondary circuit.

From Eq. (3.12), referring to Fig. 3.20,

$$v_2 = KfN_2B_mA_c \tag{3.34}$$

The induced voltage of a current transformer is determined by three parameters: the secondary load R, the secondary winding resistance r_2, and the secondary current I_2:

$$v_2 = I_2(r_2 + R) \tag{3.35}$$

Substituting Eq. (3.35) into Eq. (3.34) for v_2 and rearranging, a rationalized expression is obtained for

$$A_c = \frac{I_2(r_2 + R) \times 10^4}{KN_2fB_m} \tag{3.36}$$

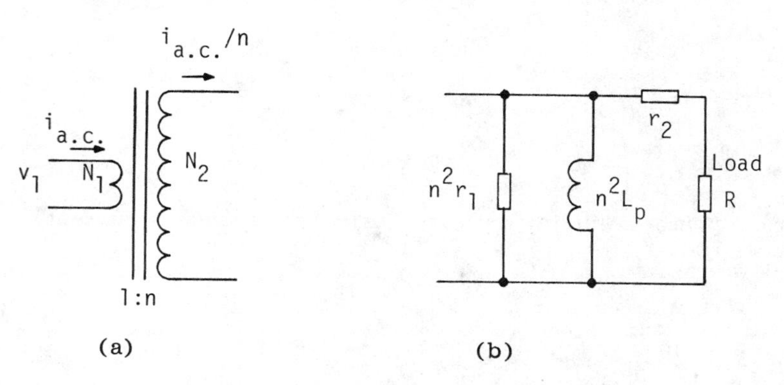

FIG. 3.20 Current transformer. (a) Ideal current transformer. (b) Simplified equivalent circuit, all values referred to secondary.

With reference to Fig. 3.18b, a correct choice of flux density B_m would result in appropriate values of inductance L_s and r_1 (where r_1 represents the shunt resistive loss of the primary winding).

To obtain a close ratio approximation, r_1 and L_s must be large, i.e.,

$$n^2 r_1 >> r_2 + R$$

$$n^2 \omega L_s >> r_2 + R$$

It follows that $\omega L_s > R$ is a basic requirement, since $R >> r_2$.

The primary inductance is given by

$$L_p = \frac{0.4\pi N_1^2 A_c \mu_\Delta}{l_m} \times 10^{-8} \text{ H} \tag{3.37}$$

EXAMPLE 3.3 Design a current transformer to meet the following specifications:

$I_i = 4.5$ A max.

$V_o = 5.0$ V max.

$$R_o = 1\ k\Omega$$

$$f = 35\ kHz$$

$$B_m = 0.2\ T$$

Assume a primary winding with one turn; the secondary current is

$$I_s = \frac{V_o}{R_o} = \frac{5}{1000} = 0.005\ A$$

The secondary turns:

$$N_s = \frac{I_i N_p}{I_s} = \frac{4.5 \times 1}{.005} = 900 \text{ turns}$$

Since f = 35 kHz, ferrite is selected for the transformer core. The minimum core area is calculated using Eq. (3.36), assuming a 1-V diode forward voltage drop:

$$A_c = \frac{(5 + 1) \times 10^4}{4 \times 0.2 \times 35 \times 10^3 \times 900}$$

$$= 2.38 \times 10^{-3}\ cm^2$$

Note that the small core requirement is due to the low volt-second requirement of the primary winding.

Since R_o is selected as 1 kΩ, it is desirable to ensure r_2 is very much smaller than R_o so that the voltage drop across R_o is very much greater than that across r_2.

Because of this requirement, a core should be selected for adequate space for windings to reduce wire loss. Therefore, in this case, an oversize core is chosen for ease of implementation as well as low secondary loss.

From Table 3.4, a gapless pot core of N27 material, size 14 × 8, is chosen for evaluation: From columns 3 and 16, the window area is calculated:

$$\frac{0.0393\ \text{cm}^4}{0.25\ \text{cm}^2} = 0.1572\ \text{cm}^2 = W_a$$

Since the primary is only one turn, in this case, it is safe to allot 75% of the available area for the secondary winding. By assuming a winding efficiency of 75%, the effective window area $W_{a(eff)}$ is

$$W_{a(eff)} = 0.75 \times 0.75 \times 0.1572\ \text{cm}^2 = 0.088\ \text{cm}^2$$

The wire area A_w is

$$A_w = \frac{0.088}{900} = 98.25 \times 10^{-6}\ \text{cm}^2$$

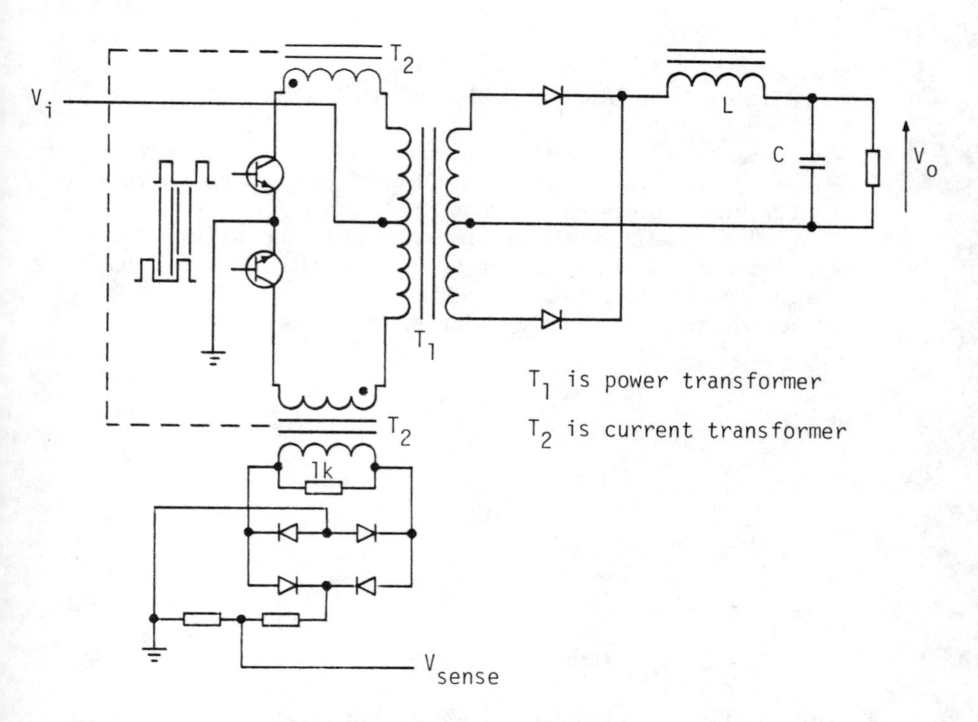

FIG. 3.21 Current transformer application in push-pull power conversion.

From Table 3.8, 0.09825×10^{-3} cm^2 falls between sizes of AWG 37 and AWG 38.

Select AWG 38.

The secondary winding resistance is

$$r_2 = MLT \times N_2 \times \mu\Omega/cm \times 10^{-6}$$

$$= 2.8 \times 900 \times 21266 \times 10^{-6}$$

$$= 53.59\ \Omega$$

This is approximately 5% of R_o.

A typical application of this transformer in a push-pull power converter is illustrated in Fig. 3.21, where two primary windings in opposite polarities are required to sense two switch currents as shown.

3.3 DESIGN OF INDUCTORS WITH DC BIAS

The dc biased inductor is the most essential magnetic component in a power converter. It exists in all basic power converters and is used in input filters as well as output filters.

The basic equations will be developed in a step-by-step but concise manner. The application of these equations will be demonstrated in an example.

In accordance with Faraday's law of induction, the induced emf E is

$$E = N\frac{d\varphi}{dt} = L\frac{di}{dt} \tag{3.38}$$

$$L\ di = N\ d\varphi \tag{3.39}$$

$$L \int di = N \int d\phi \tag{3.40}$$

$$Li = N\varphi \tag{3.41}$$

$$Li = NBA_c, \quad \text{since } \varphi = BA_c \tag{3.42}$$

Multiplying both sides of Eq. (3.42) by I/2 gives

$$(1/2)LI^2 = (1/2)NIBA_c \tag{3.43}$$

$$B = \mu_\Delta \mu_o H \tag{3.44}$$

But by Ampere's law, magnetomotive force

$$\text{mmf} = \oint H dl_m \tag{3.45}$$

$$NI = Hl_m \tag{3.46}$$

$$H = \frac{NI}{l_m} \tag{3.47}$$

Substituting Hl_m for NI in Eq. (3.43) gives

$$(1/2)LI^2 = (1/2)Hl_m BA_c \tag{3.48}$$

But $l_m A_c$ is the core volume Vol_c; therefore,

$$(1/2)LI^2 = (1/2)HBVol_c \tag{3.49}$$

Equation (3.49) shows that the energy storage capability of an inductor is directly proportional to its core volume [95, 105].

From Eq. (3.42),

$$I = \frac{NBA_c}{L} \tag{3.50}$$

Multiplying both sides by N gives

$$NI = \frac{N^2BA_c}{L} \tag{3.51}$$

Substituting Eq. (3.46) for NI gives

$$Hl_m = \frac{N^2BA_c}{L} \tag{3.52}$$

$$L = \frac{N^2 B A_c}{H l_m} \tag{3.53}$$

But

$$\mu_\Delta \mu_o = \frac{B}{H} \tag{3.54}$$

Therefore,

$$L = \frac{\mu_\Delta \mu_o N^2 A_c}{l_m} \tag{3.55}$$

Equations (3.38) to (3.55) show the qualitative relations of magnetic and electric circuits, assume basically linear relationships, and were given without concern with systems of units. The following expressions will be rationalized with the correct constants to reflect the units given in the list of symbols provided at the end of this volume.

The number of turns is given by Eq. (3.42) as

$$N = \frac{LI}{BA_c} \times 10^4 \tag{3.56}$$

The energy storage capability of an inductor with distributed gap is given by Eq. (3.48) and is rationalized as

$$(1/2)LI^2 = (1/2)HBVol_c \times 10^{-4} \tag{3.57}$$

and for the inductor with air gap,

$$(1/2)LI^2 = (1.2)HBVol_g \times 10^{-4} \tag{3.58}$$

Equation (3.55) relates the A_L value given by the core manufacturer to inductance and the square of the number of turns for distributed gap cores by

$$L = \frac{0.4\pi \mu_\Delta N^2 A_c}{l_m} \times 10^{-8} \tag{3.59}$$

$$L = \frac{0.4\pi\mu_\Delta N^2 A_c}{l_g + (l_m/\mu_\Delta)} \times 10^{-8} \tag{3.60}$$

Multiplying Eq. (3.56) by I gives

$$NI = \frac{LI^2}{BA_c} \times 10^4 \tag{3.61}$$

The number of amp-turns NI is allocated into the window area W_a with a window utilization factor K_u for a given current density J according to the following relation:

$$JK_uW_a = NI \tag{3.62}$$

Equations (3.61) and (3.62) are equal, or

$$JK_uW_a = \frac{LI^2}{BA_c} \times 10^4 \tag{3.63}$$

$$A_cW_a = A_p = \frac{LI^2}{BJK_u} \times 10^4 \tag{3.64}$$

Inspection of Eq. (3.64) indicates that for a chosen core material, B is determined. For a desired temperature rise, J is determined. K_u is determined by core geometry and insulation requirements. With B, J, and K_u values assigned, the area product A_p is seen to relate directly to the energy storage capability of the inductor.

From Appendix D, Eq. (D.28),

$$J = K_jA_p^{-0.125} \tag{3.65}$$

Substituting Eq. (3.65) into Eq. (3.64) for J gives

$$A_p = \frac{LI^2 \times 10^4}{BK_uK_jA_p^{-0.125}} \tag{3.66}$$

$$A_p^{0.875} = \frac{LI^2 \times 10^4}{BK_uK_j} \tag{3.67}$$

$$A_p = \left(\frac{LI^2 \times 10^4}{BK_uK_j}\right)^{1.14} \tag{3.68}$$

From Eq. (3.44),

$$B = \mu_\Delta \mu_o H$$

Multiplying both sides of Eq. (3.42) by I/2 gives

$$(1/2)LI^2 = (1/2)NIBA_c \tag{3.69}$$

$$(1/2)LI^2 = (1/2)NI\mu_\Delta \mu_o HA_c \tag{3.70}$$

Substituting Eq. (3.47) into Eq. (3.69) for H gives

$$(1/2)LI^2 = \frac{\mu_\Delta \mu_o N^2 I^2 A_c}{2l_m} \tag{3.71}$$

Substituting Eq. (3.47) into Eq. (3.44) for H gives

$$B = \mu_\Delta \mu_o \frac{NI}{l_m} \tag{3.72}$$

Therefore,

$$I = \frac{Bl_m}{\mu_\Delta \mu_o N} \tag{3.73}$$

But, from Eq. (3.62),

$$I = \frac{K_u W_a J}{N} \tag{3.74}$$

Equating Eqs. (3.73) and (3.74) gives

$$\frac{K_u W_a J}{N} = \frac{B l_m}{\mu_\Delta \mu_o N} \tag{3.75}$$

or, rationalized,

$$\mu_\Delta = \frac{B l_m \times 10^{-2}}{\mu_o W_a J K_u} \tag{3.76}$$

Substituting $4\pi \times 10^{-7}$ for μ_o in Eq. (3.76) gives

$$\mu_\Delta = \frac{B l_m \times 10^{4}}{0.4\pi W_a J K_u} \tag{3.77}$$

EXAMPLE 3.4 Design a dc biased inductor of 420 μH with a bias of 7 A. [Refer to the buck-boost converter example in Chapter 1; $I = I_{max}$ is calculated using Eq. (1.8).] Allowing margin for short circuit protection, design for 8 A. The energy stored in the inductor is

$$(1/2)LI^2 = (1/2)(420 \times 10^{-6} \times 8^2) = 13.44 \text{ mJ}$$

The area product A_p from Eq. (3.68), is

$$A_p = \left(\frac{LI^2 \times 10^4}{B_m K_u K_j}\right)^{1.14}$$

$$= \left(\frac{26.88 \times 10^{-3} \times 10^4}{0.3 \times 0.4 \times 403}\right)^{1.14}$$

$$= 7.067 \text{ cm}^4$$

From Table 3.3, column 3, select core 55090-A2 (Magnetics, Inc.) or A-090086-2 (Arnold Engineering) with an A_p of 8.06 cm^4. From Eq. (D.28) (Appendix D), the current density is

$$J = K_j A_p^{-0.12}$$

$$= 403 \times 7.067^{-0.12}$$

$$= 318.71 \text{ A/cm}^2$$

The required core permeability, according to Eq. (3.77), is

$$\mu_\Delta = \frac{B_m l_m \times 10^4}{0.4\pi W_a J K_u}$$

$$= \frac{0.3 \times 11.62 \times 10^4}{0.4\pi \times 6.11 \times 318.71 \times 0.4} = 35.614$$

Note that $B_m = B_{ac} + B_{dc}$:

$$B_m = \frac{0.4\pi N(\Delta I/2 + I) \times 10^{-4}}{l_m/\mu_\Delta} \tag{3.78}$$

See Fig. 3.22 and Eqs. (1.5) and (1.8). Select a μ of 60 material from the manufacturer's data (see, for example, Magnetics, Inc., catalog MPP-303T); the inductance per 1000 turns is 86 mH, or

$$N = 1000 \sqrt{\frac{L}{86}}$$

$$= 1000 \sqrt{\frac{0.42}{86}}$$

$$= 70 \text{ turns}$$

The bare wire size $A_{w(B)}$ is

$$A_{w(B)} = \frac{I}{J} = \frac{8}{318.71} = 0.0251 \text{ cm}^2$$

Select AWG 14 with an area of 0.02082 cm^2. The winding resistance is

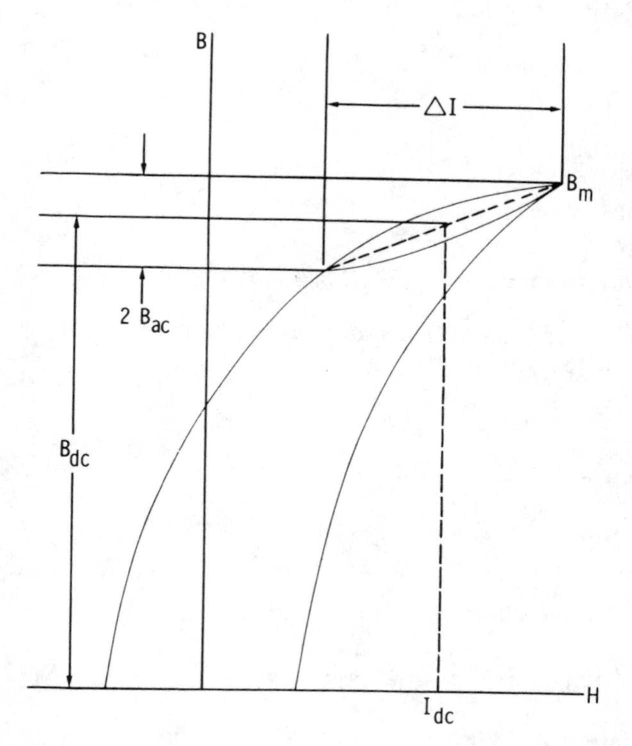

FIG. 3.22 Flux density versus $I_{dc} + \Delta I$.

$$R = MLT \times N \times \mu\Omega/cm \times \zeta \times 10^{-6}$$

$$= 6.66 \times 70 \times 82.80 \times 1.098 \times 10^{-6}$$

$$= 0.0424\ \Omega$$

The winding loss is

$$P_{cu} = I^2R = 7^2 \times 0.0424 = 2.078\ W$$

The current of 7 A is used here for operating condition calculation, whereas the 8-A rating is designed for a margin used for short-circuit protection purposes—not an operating condition. From Table 3.3, column 2,

$$A_t = 81.58 = \frac{P_\Sigma}{\psi} = \frac{P_{cu}}{\psi} = \frac{2.078}{\psi}$$

Therefore,

$$\psi = \frac{2.078}{81.58} = 0.0255 \text{ W/cm}^2$$

which will produce a temperature rise of 25°C [120].

EXAMPLE 3.5 Design of a dc biased inductor with a discrete gap, using the core geometry approach:

Inductance = 420 μH.
Direct current = 7.7 A.
Alternating current = 0.3 A.
Output power = 56 W.
Frequency = 33 kHz.
Regulation = 3%.

The energy handling capability of the inductor is

$$(1/2)LI^2 = 1/2 \times 420 \times 10^{-6} \times 8^2 = 13.44 \text{ mJ}$$

The electrical constant K_e is given by [120]

$$K_e = 0.145 P_o B_m^2 \times 10^{-4} \tag{3.79}$$

For a ferrite pot core, use a flux density B_m of 0.25 T; therefore,

$$K_e = 0.145 \times 56 \times 0.25^2 \times 10^{-4}$$

$$= 50.75 \times 10^{-6}$$

The core geometry constant K_g is given by [120]

$$K_g = \frac{(\text{energy})^2}{K_e \alpha} = \frac{(13.44 \times 10^{-3})^2}{50.75 \times 10^{-6} \times 3} \tag{3.80}$$

$$= 1.186 \text{ cm}^5$$

From Table 3.11, select a pot core 59 × 36 with a K_g of 1.683 cm^5. The current density is

$$J = \frac{LI^2 \times 10^4}{B_m K_u A_p} \text{ A/cm}^2$$

$$= \frac{26.88 \times 10^{-3} \times 10^4}{0.25 \times 0.4 \times 13.43}$$

$$= 200.15 \text{ A/cm}^2$$

The bare wire size is

$$A_{w(B)} = \frac{I}{J} = \frac{8}{200.15} = 0.03997 \text{ cm}^2$$

Select AWG 11 with an area of 0.04168 cm^2. The effective window area is

$$W_{a(eff)} = W_a S_3$$

For $S_3 = 0.75$,

$$W_{a(eff)} = 2.77 \times 0.75 = 2.0775 \text{ cm}^2$$

The number of turns N is

$$N = \frac{W_{a(eff)} S_2}{A_w} = \frac{2.0775 \times 0.6}{0.04168} = 30 \text{ turns}$$

The air gap is given by

$$l_g = \frac{0.4\pi N^2 A_c \times 10^{-8}}{L} \text{ cm} \qquad (3.81)$$

$$= \frac{0.4\pi \times 30^2 \times 4.85 \times 10^{-8}}{420 \times 10^{-6}}$$

$$= 0.13 \text{ cm}$$

To convert 0.13 cm into mils, multiply by 393.7:

$$0.13 \times 393.7 = 51.5 \text{ mils}$$

Rounding this figure off to 52 mils requires a spacer of 26 mils to be inserted between the two halves of the pot core to contribute a total of 52 mils in the magnetic path.

The fringing flux factor is given by

$$F = 1 + \frac{l_g}{\sqrt{A_c}} \log_e \frac{2G}{l_g} \tag{3.82}$$

where G is related to the length of coil, obtainable from manufacturer's data:

$$F = 1 + \frac{0.13}{\sqrt{4.85}} \log_e \frac{2 \times 2.36}{0.13}$$

$$= 1.212$$

Adjust the number of turns by using the equation

$$N = \frac{l_g L}{0.4\pi A_c F \times 10^{-8}} \text{ turns} \tag{3.83}$$

$$= \frac{0.13 \times 420 \times 10^{-6}}{0.4 \times 4.85 \times 1.212 \times 10^{-8}} = 27 \text{ turns}$$

The winding resistance is

$$R = \text{MLT} \times N \times \mu\Omega/\text{cm} \times 10^{-6}$$

$$= 12 \times 27 \times 41.37 \times 10^{-6} = 0.0134 \ \Omega$$

The copper loss is

$$P_{cu} = I^2R = 8^2 \times 0.0134 = 0.858 \text{ W}$$

The regulation α is

$$\alpha = \frac{P_{cu}}{P_o + P_{cu}} \times 100\%$$

$$= \frac{0.858}{56.858} \times 100\% = 1.5\%$$

From Eq. (3.78),

$$B_{ac} = \frac{0.4\pi N(\Delta I/2) \times 10^{-4}}{l_g + (l_m/\mu_\Delta)}$$

$$= \frac{0.4\pi \times 27 \times (0.3/2) \times 10^{-4}}{0.13}$$

$$= 0.2088 \text{ T, below } 0.25 \text{ T, is acceptable}$$

The core loss factor due to ac flux at 33 kHz is calculated using the formula in Fig. 3.14:

$$\text{Loss in W/kg} = 0.262 \times 10^{-3} f^{1.39} B^{2.19}$$

$$= 0.262 \times 10^{-3} \times (33 \times 10^3)^{1.39} \times 0.2088^{2.19}$$

$$= 16.18995 \text{ W/kg}$$

or

$$\text{Loss in mW/gm} = 16.18995 \text{ mW/gm}$$

Therefore

$$Pfe = (\text{mW/gm}) \times Wt \times 10^{-3}$$

$$= 16.18995 \times 270 \times 10^{-3}$$

$$= 4.371 \text{ mW} = 0.00437 \text{ W}$$

Tht total loss is

$$P_\Sigma = P_{cu} + P_{fe} = 0.858 + 0.00437 = 0.86237 \text{ W}$$

$$\psi = \frac{P_\Sigma}{A_t} = \frac{0.86237}{121.56} = 0.007094 \text{ W/cm}^2$$

indicating a temperature rise of less than 25°C.

4

Stability Considerations

4.1 LOOP GAIN ASSESSMENT

The purpose of applying feedback is to modify or improve the response of the open-loop system. A proper evaluation of the loop gain of the system serves to provide the essential information for tailoring the final loop gain response for stable system operation. Figure 4.1 shows the block diagram of an amplifier with negative feedback. The closed-loop gain of this arrangement is

$$\frac{V_o}{V_i} = \frac{A}{1 + A\beta} \tag{4.1}$$

where A, in this case, is the open-loop gain of the amplifier. The product $A\beta$ is commonly referred to as the loop gain T(s) of the system. This concept is simply visualized as the loop being broken at point P_1, and $A\beta$ is the gain around this loop. Figure 4.2 shows a similar arrangement but with an extra block α. This block α could assume the form of a passive network, an active network, an amplifier, etc. The purpose of inserting this block is to demonstrate the effect of its presence on the overall loop gain. If the loop is now broken at point P_2, the loop gain is seen to be

$$T(s) = A\alpha\beta \tag{4.2}$$

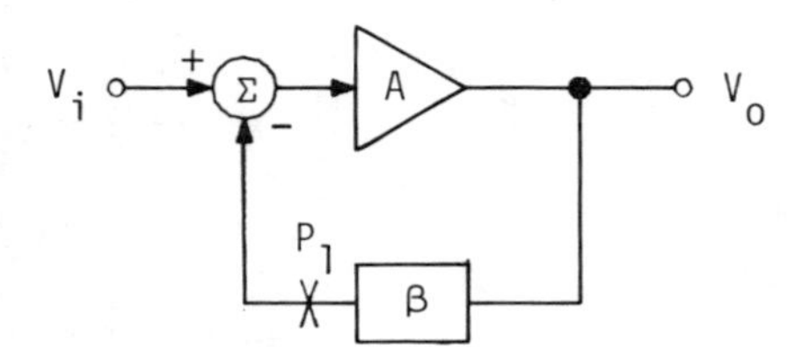

FIG. 4.1 Amplifier with feedback network β.

For a switch mode power converter, the typical system block diagram would appear as shown in Fig. 4.3. By breaking the feedback loop at point P_3, the loop gain of the system is

$$T(s) = M_c H_e(s) A \beta$$

where M_c is the control modulation function of the power stage.

It is appropriate to point out here that in Chapter 2 the programs provided for converter analysis essentially calculate the small signal control to output transfer functions, $M_c H_e(s)$. The block β normally consists of the voltage divider network for sampling the output voltage, and the error amplifier A is usually gain limited and is at a location convenient for the application of frequency compensation, if required.

By means of the calculator programs, the task of obtaining the overall loop gain is reduced to the process of assessing the responses of the blocks A and β and combining the results graphically. For buck and buck-derived converters with output filter damping, the calculated $M_c H_e(s)$ response is usually adequate for assessing the overall loop gain characteristics. The attenuating network β could be arranged in combination with the error amplifier

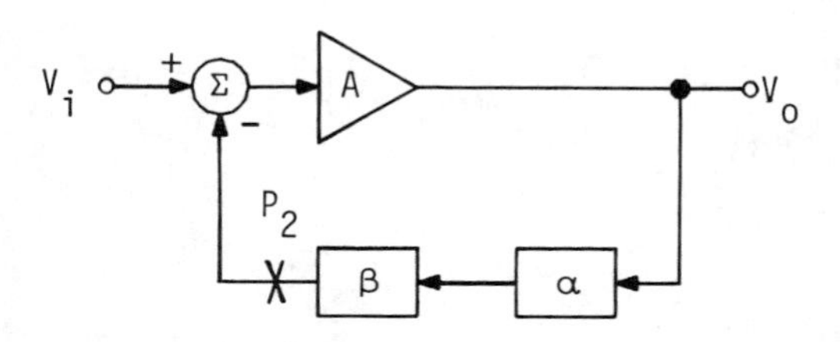

FIG. 4.2 Amplifier with feedback networks α and β.

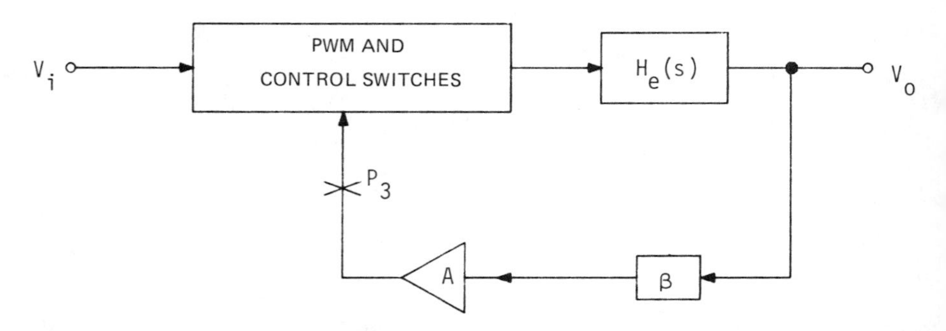

FIG. 4.3 Block diagram of switch mode power converter.

closed-loop gain to result in a 0 dB gain, so that $A\beta = 1$. Then the loop gain $M_cH_e(s)$ provides a true representation of the overall loop gain. This method of description is given to help the reader to visualize the comprehensive results obtainable from calculator analysis and is not a recommendation to set the loop gain this way. If gain adjustment is required to tailor the loop gain for better phase or gain margin, it can then be accomplished by adjusting the closed-loop gain of the error amplifier. Figure 4.4 shows a simple buck converter arrangement. Analogous to the amplifier in Fig. 4.1, the switch mode power converter of Fig. 4.3 has the following characteristics:

The loop gain is

$$T(s) = M_cH_e(s)A\beta \tag{4.4}$$

The open-loop gain is

$$M_iH_e(s) \tag{4.5}$$

The closed-loop gain is

$$\frac{V_o}{V_i} = \frac{M_iH_e(s)}{1 + M_cH_e(s)A\beta} \tag{4.6}$$

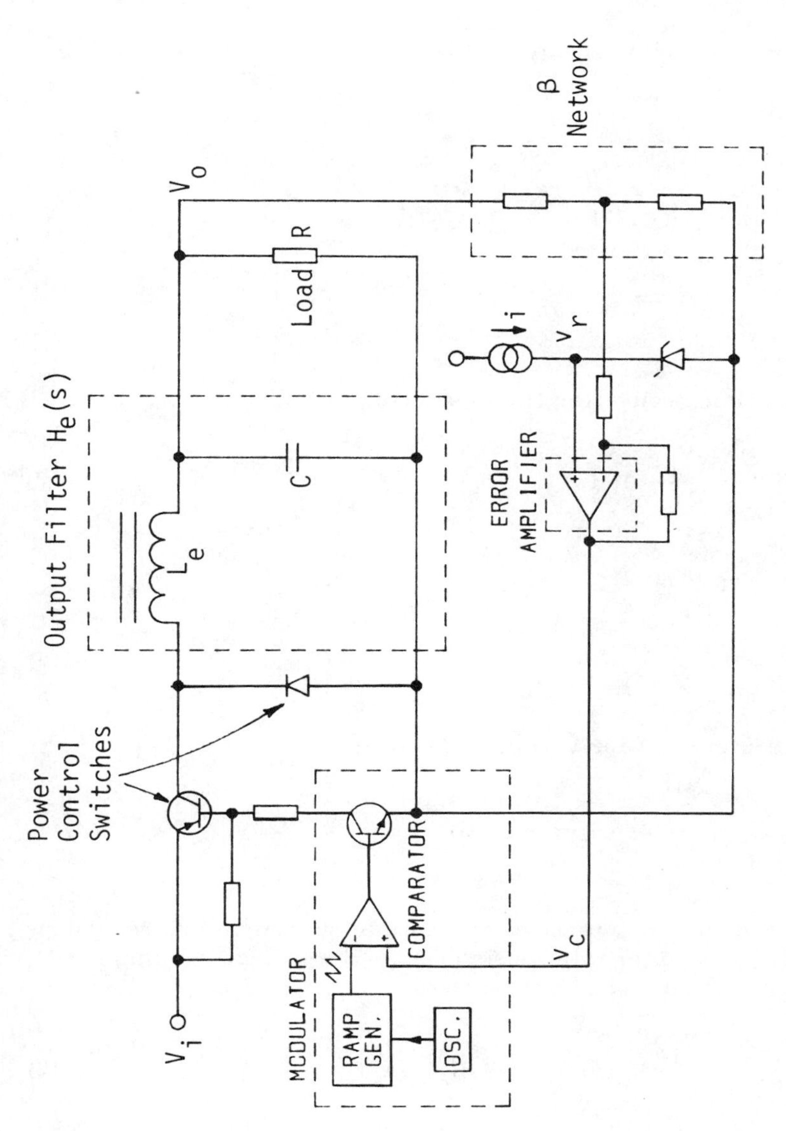

FIG. 4.4 Buck converter example of Fig. 4.3.

From Appendix A, Eq. (A.57),

$$\frac{\hat{v}_o(s)}{\hat{d}(s)} = \frac{V_o}{D} H_e(s) \tag{4.7}$$

and from Eq. (A.72),

$$\frac{\hat{v}_o(s)}{\hat{v}_c(s)} = \frac{\hat{v}_o(s)}{\hat{d}(s)} \frac{1}{V_m} \tag{4.8}$$

For the buck converter [Chapter 2, Eq. (2.1)],

$$\frac{\hat{v}_o(s)}{\hat{v}_c(s)} = \frac{V_o}{DV_m} H_e(s) = M_c H_e(s) \tag{4.9}$$

From Chapter 1, Eq. (1.13),

$$\frac{V_o}{V_i} = D \tag{4.10}$$

Substituting $V_i D$ for V_o in Eq. (4.9) gives

$$\frac{\hat{v}_o(s)}{\hat{v}_c(s)} = \frac{V_i}{V_m} H_e(s) \tag{4.11}$$

To investigate the effect of error amplifier gain on line regulation, insert an amplifier with a gain of A_r and a voltage sampling network β so that Eq. (4.11) becomes

$$\frac{\hat{v}_o(s)}{\hat{v}_c(s)} = \frac{V_i}{V_m} H_e(s) A_r \beta = M_c H_e(s) A_r \beta \tag{4.12}$$

Closing the feedback loop will provide the following relation:

$$\frac{\Delta v_o}{\Delta v_i} = \frac{M_i H_e(s)}{1 + M_c H_e(s) A_r \beta} \tag{4.13}$$

For the case of loop gain very much greater than unity, i.e., $M_c H_e(s) A_r \beta >> 1$,

$$\frac{\Delta v_o}{\Delta v_i} \cong \frac{DH_e(s)}{M_c H_e(s) A_r \beta} = \frac{D}{M_c A_r \beta} \tag{4.14}$$

Substituting Eq. (4.10) into Eq. (4.14) for D gives

$$\frac{\Delta v_o}{\Delta v_i} = \frac{V_o/V_i}{M_c A_r \beta} \tag{4.15}$$

Therefore,

$$\frac{\Delta v_o}{V_o} = \frac{v_i/V_i}{(V_{i_{min}}/V_m) A_r \beta} \quad \textit{for line regulation} \tag{4.16}$$

or

$$\frac{\Delta v_o/V_o}{\Delta v_i/V_i} = \frac{1}{(V_{i_{min}}/V_m) A_r \beta} \tag{4.17}$$

EXAMPLE If it is desired to maintain a 0.5% change in output voltage for a change of 20% in input voltage, then, from Eq. (4.17),

$$\frac{0.5}{20} = \frac{1}{(V_{i_{min}}/V_m) A_r \beta}$$

Therefore, for $V_{i_{min}} = 12$ V, $V_m = 2.5$ V, and $\beta = 0.5$. A_r must have a gain of 16.67, since

$$\frac{20}{0.5} = \frac{V_{i_{min}}}{V_m} A_r \beta = 40$$

Table 4.1 shows the relationships of the input to output and control to output ports of the modulator for the three basic converter configurations.

TABLE 4.1 Input and Control to Output Transfer Relationships of the Three Basic Configurations

	Buck	Boost	Buck-boost
M_c	$\frac{V_o}{DV_m}$	$\frac{V_o}{D_o V_M}\left(1 - \frac{sL}{D_o^2 R}\right)$	$\frac{V_o}{DD_o V_m}\left(1 - \frac{sDL}{D_o^2 R}\right)$
M_i	D	$\frac{1}{D_o}$	$\frac{D}{D_o}$

4.2 FREQUENCY RESPONSE CONCEPTS

In a power converter, negative feedback is employed around a loop containing amplifier(s), comparator, modulator, filter components, and so on. At zero or very low frequencies, the loop gain (including the error amplifier stage) exhibits a phase lag of approximately 180° by virtue of the negative polarity of the feedback signal. As the frequency increases, the components within this loop will begin to contribute a certain amount of phase shift (usually a lag due to the presence of a low-pass filter at the output). If at some particular frequency this phase lag is equal to 180°, then a total of 360° lag is reached. Under this condition, the feedback signal will be *in phase* with the input signal. If at this particular frequency the loop gain is unity (0 dB) or greater, the system will oscillate.

Note that if an analysis is made excluding the error amplifier, which may contribute an initial lag of 180°, then the phase range of interest will be within 180°. This is true of calculator programs where the error amplifier has been left out, because the major problem with switch mode power converters has been with the modulator part of the system, for which the calculator programs are written as an aid to loop analysis and design.

To understand how the reactive elements contribute to the additional phase shift, it would be appropriate to investigate the characteristics of some commonly encountered networks. Figure 4.5 shows a low-pass network comprising one resistor and one capacitor. The transfer function of this network is

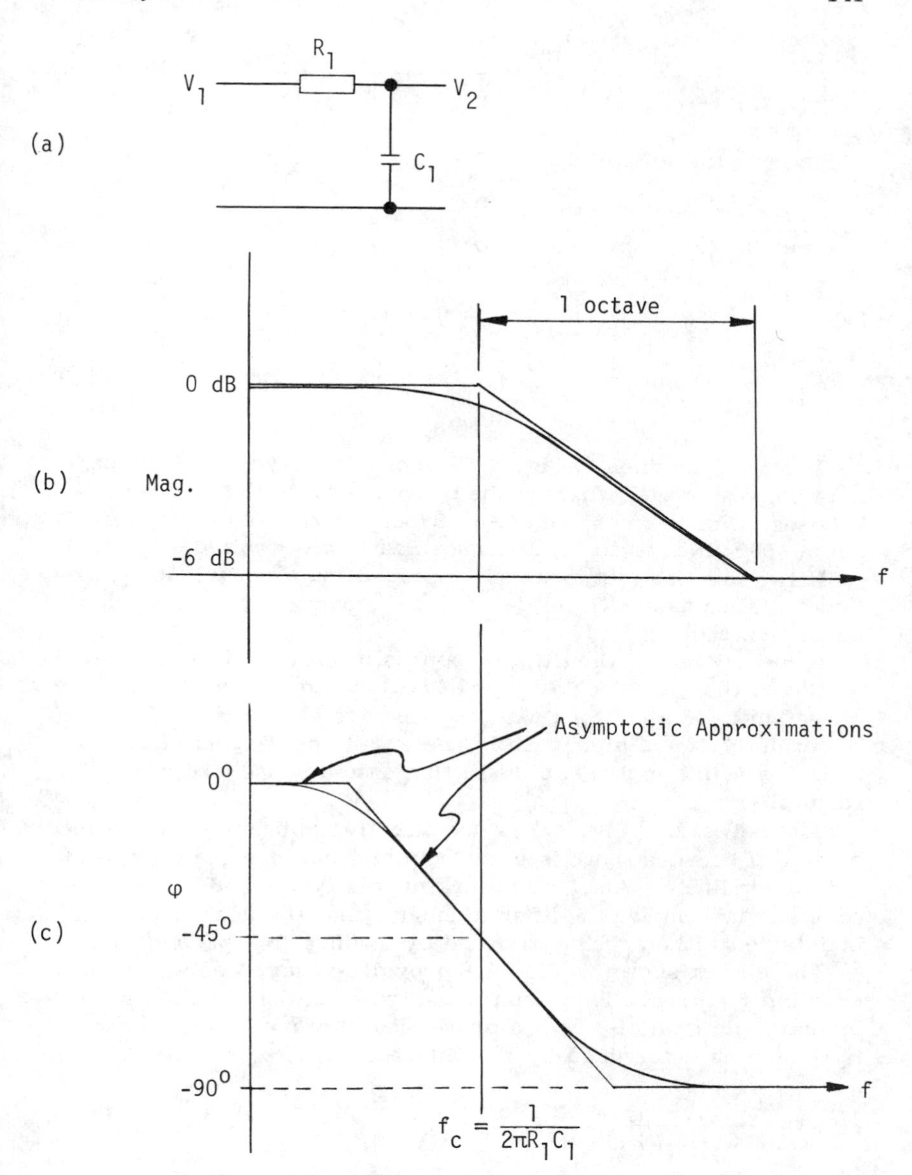

FIG. 4.5 RC low-pass section.

$$\frac{V_2}{V_1} = \frac{X_{cl}}{R_1 + X_{cl}} = \frac{1/j\omega C_1}{R_1 + (1/j\omega C_1)} = \frac{1}{1 + j\omega C_1 R_1} \tag{4.18}$$

which is in the form of

$$\frac{1}{1 + j(\omega/\omega_1)}$$

where

$$\omega_1 = \frac{1}{C_1 R_1} \tag{4.19}$$

The straight lines in Fig. 4.5b depict the asymptotes of the network responses, whereas the curves provide the actual responses. Notice the slope of -6 dB/octave (or -20 dB/decade) and a -90° phase shift for a network with one reactive element.

If two identical networks are cascaded via a unity gain isolation amplifier as shown in Fig. 4.6a, the overall responses will be as shown in Figs. 4.6b and c.

If $R_1C_1 > R_2C_2$, then the responses in Fig. 4.6d will result.

Notice the -12 dB/octave (or -40 dB/decade) slope and a -180° phase shift for a network with two reactive elements.

Similarly, for a high-pass RC network, the responses shown in Fig. 4.7 will apply. Notice in this case a phase lead of 90° results.

The network of Fig. 4.7 is not directly applicable within a feedback loop because there is no dc coupling between the input and output terminals. Also, a network of this type should ideally be coupled via a buffer amplifier to ensure that the desired response is obtained without being modified by loading or impedance effects.

The circuit shown in Fig. 4.8 provides isolation as well as dc coupling features. This amplifier stage essentially provides a noninverting dc gain of 1 with a phase lead network, whose corner frequency is determined by the values of R_1, R_2, and C_1; that is,

$$f_c = \frac{1}{2\pi C_1 (R_1 // R_2)} \tag{4.20}$$

Figures 4.9 and 4.10 show the magnitude and phase characteristics of this stage.

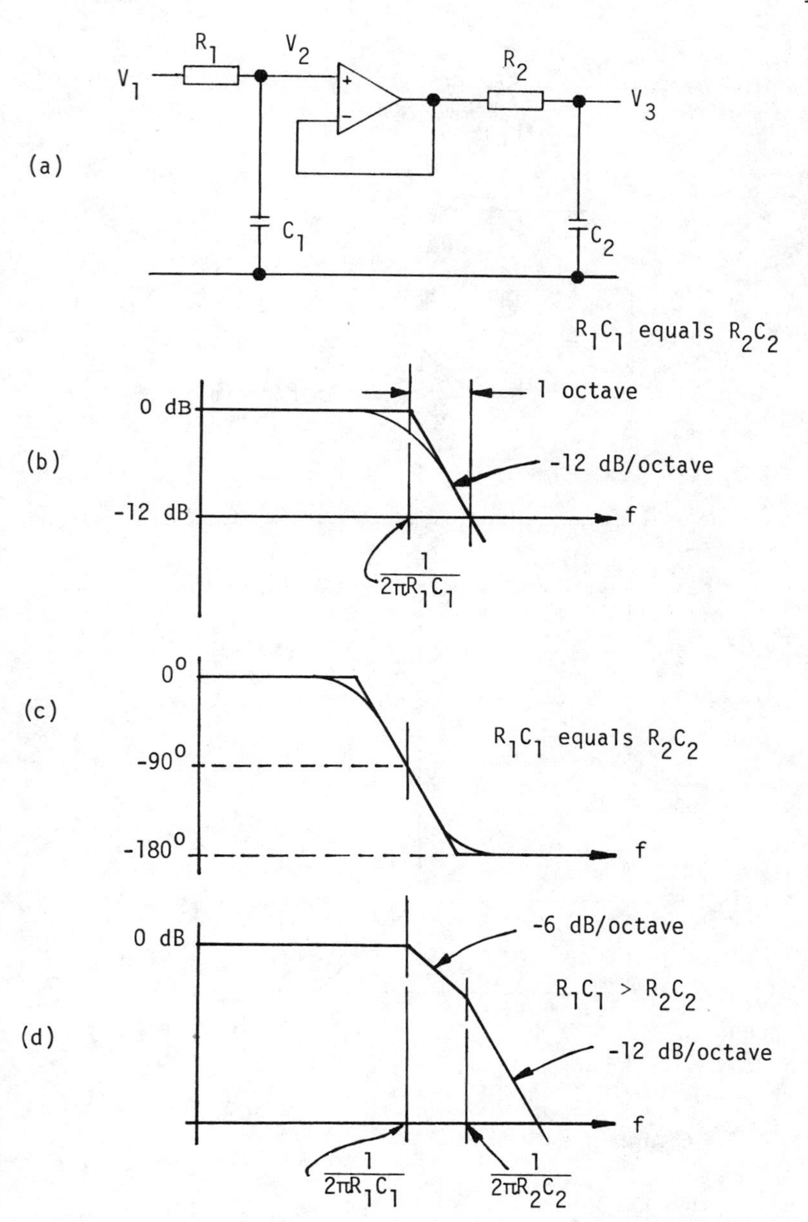

FIG. 4.6 Cascaded RC sections.

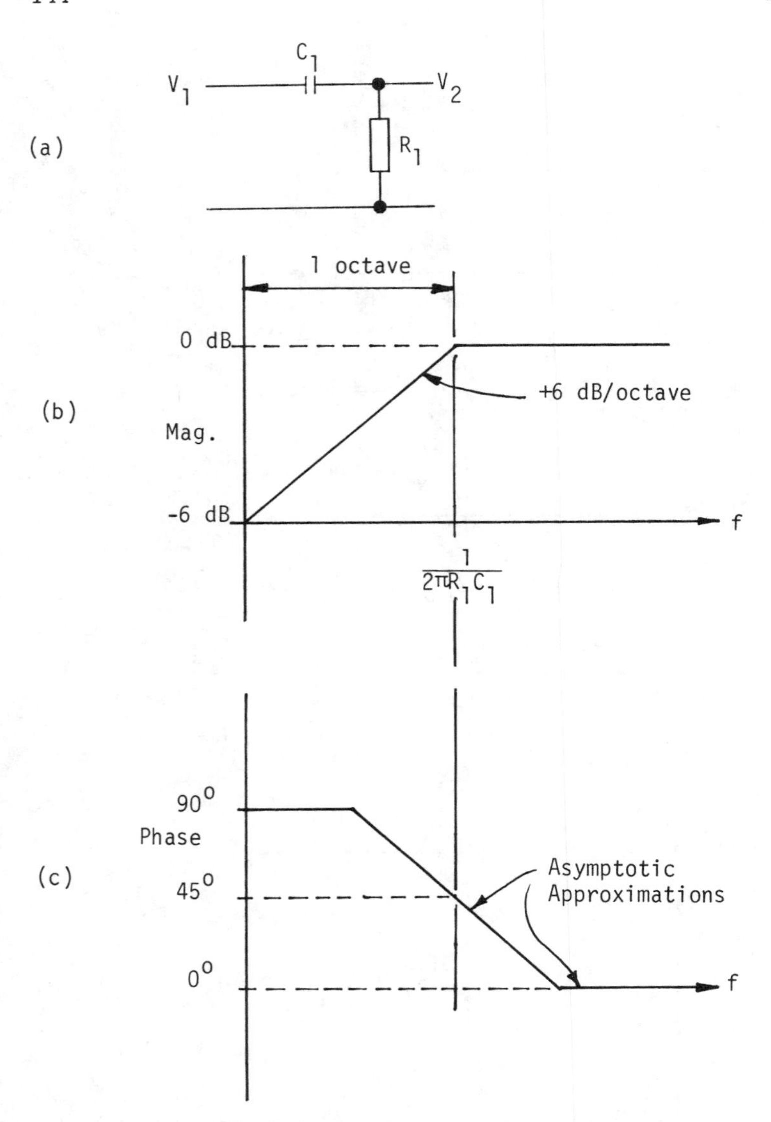

FIG. 4.7 RC high-pass section.

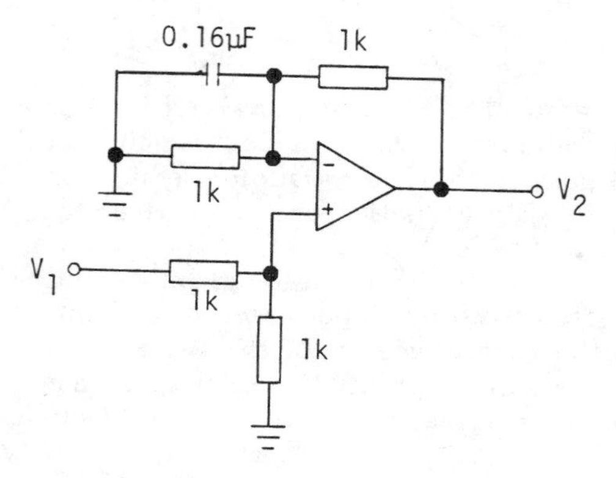

FIG. 4.8 Lead network with isolation amplifier.

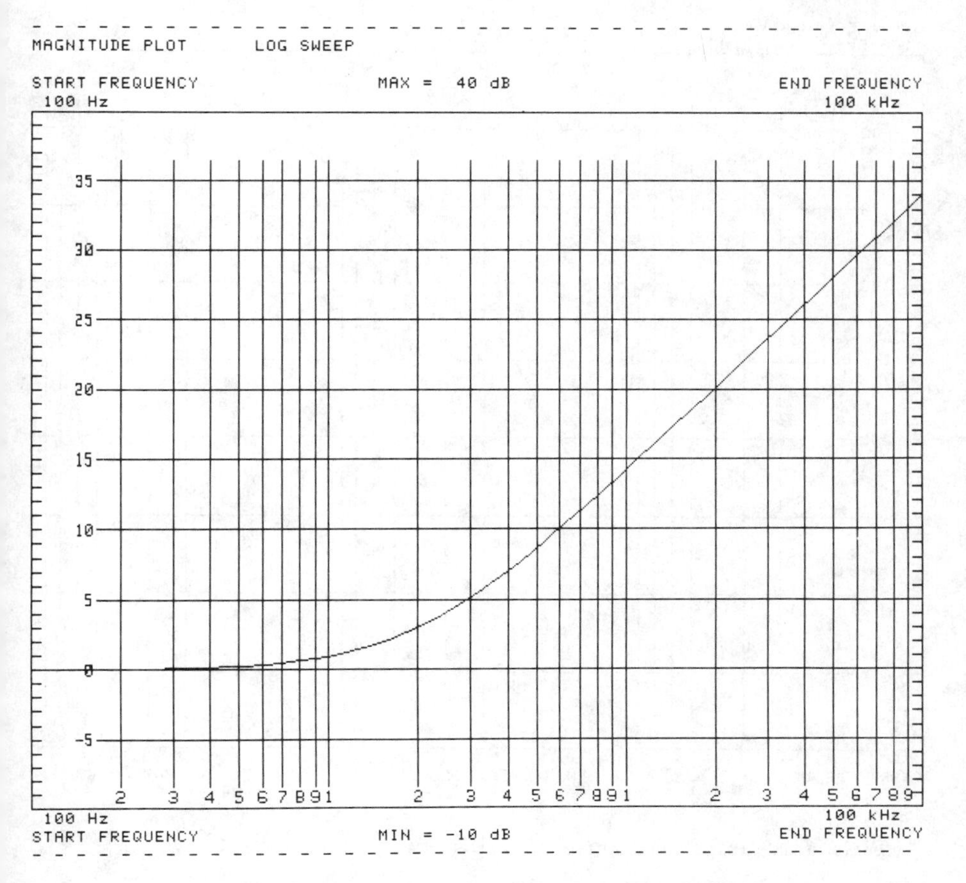

FIG. 4.9 Magnitude response of lead network with amplifier.

Another commonly encountered network in power converters is the integrator. It has a general characteristic as detailed in Fig. 4.11. Notice that a reduced gain is possible by adding a resistor R_2 in parallel with C_1, but this gain reduction does not affect the 0-dB crossover frequency of $1/(2\pi R_1 C_1)$.

In the power converter, the greatest contributor of phase shift is the output filter. It consists of two principal reactive elements (L_e and C) which determine the characteristics of the filter. Because there are two reactive elements connected in a low-pass configuration, the rate of attenuation is, asymptotically, -12 dB/octave,

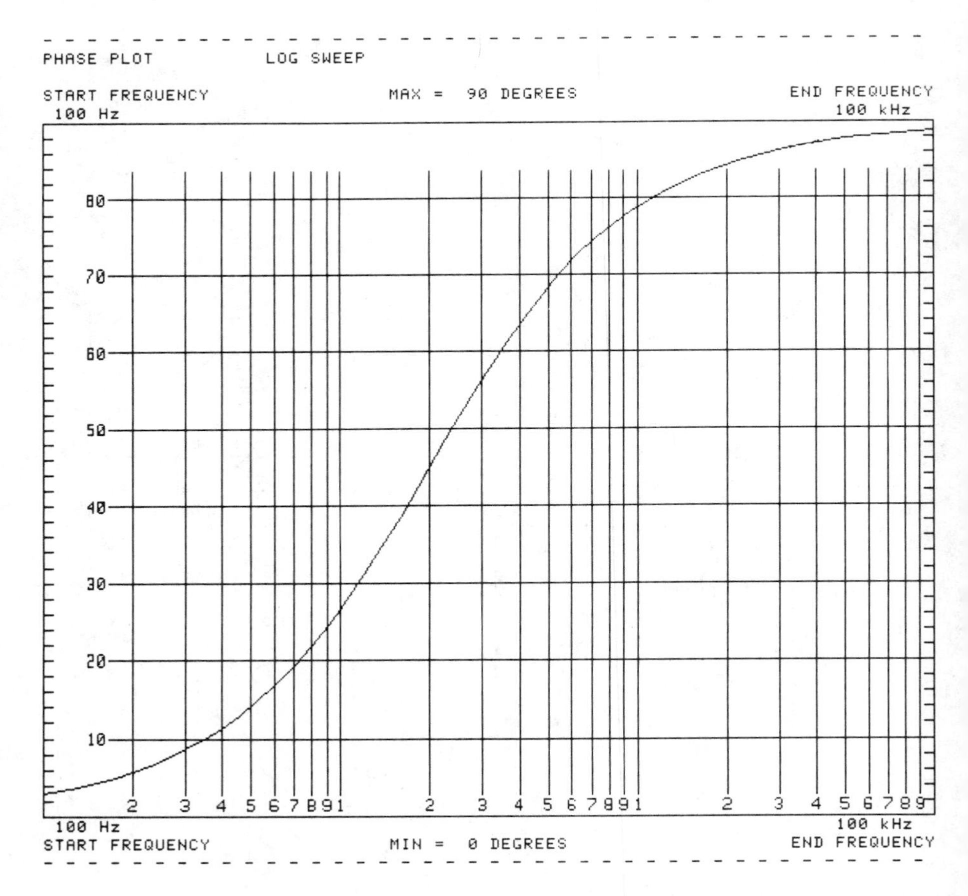

FIG. 4.10 Phase response of lead network with amplifier.

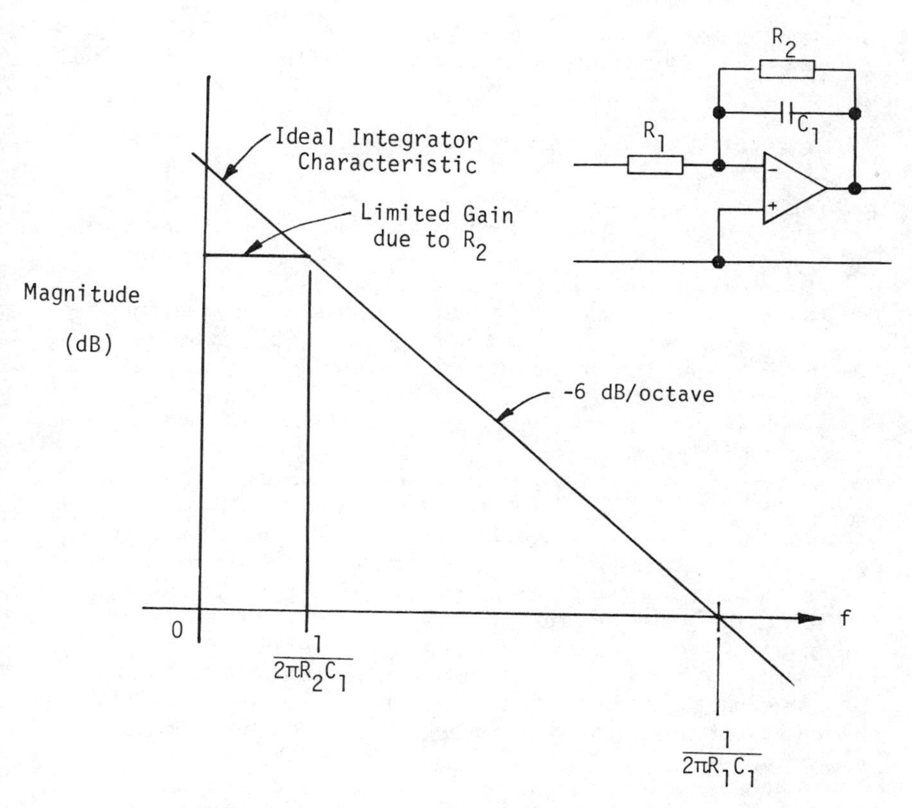

FIG. 4.11 Magnitude response of integrator.

and the phase lag is expected to reach 180° at some frequency beyond the pole (or tuned) frequency. The characteristics of a single-stage LC filter are given in Appendix B.

It is worthy of consideration that the phase and amplitude responses are interrelated [133]. *The phase response is very much dependent on the slope of the magnitude response.* The accuracy of this statement is affected only if there are other break frequencies in the immediate vicinity. In general, if the magnitude response is having a slope of 0 dB/decade at a given frequency f_1 and the nearest break frequency is at least a decade away, then the phase shift will be 0° at f_1. By the same reasoning, if the slope is ±6 dB/octave (or ±20 dB/decade) at f_1, the phase will approach ±90°. Similarly, for ±12 dB/octabe, the phase will approach ±180°; for ±18 dB/octave the phase will approach ±270°; etc.

4.3 GAIN AND PHASE MARGINS

A system can be conditionally stable, unconditionally stable, or unstable. Because of this, the relative stability of the system must be assessed.

The relative stability of a control system is assessed by the amount of gain and phase margins of the system. For an initially stable linear system, a further increase in gain is required to drive the system into instability; this gain change (in decibels) is defined as the gain margin G_m of the system. The gain margin is observed as the gain (decibels) at the frequency where the phase shift reaches -180°.

As the magnitude response of the system begins to decline toward higher frequencies, a frequency is reached at which the magnitude is at 0 dB (which is referred to as the 0-dB crossover frequency); the corresponding phase shift ϕ_o at this frequency provides a measure of the phase margin, $\varphi_m = 180° + \varphi_o$. For system stability, the phase margin φ_m should always be positive. The phase margin is observed as the phase (degree) at the frequency where the magnitude response has declined to the 0-dB level.

The accepted norm [133, 134] *for an unconditionally stable linear system is to have a gain margin G_m of 6 dB and a phase margin φ_m of 45°.*

To show the roles played by the gain margin and phase margin, the buck converter example given in Chapter 2 is recalled here for further discussion.

For simplicity, assume the combination of the error amplifier, the β network, and the comparator within the pulse width modulator contributes a gain of -1, i.e., a unity gain with 180° phase reversal. With this assumption, Figs. 4.12 to 4.14 represent the (control to output) loop gain of the buck converter, excluding the error amplifier combination. The phase range of interest is, therefore, from 0° to -180°. The figures are reproduced here for the convenience of the readers.

To determine the phase margin φ_m of the undamped converter,

1. Observe the frequency (Fig. 4.12) at which the gain curve crosses the 0-dB lines:
   a. For the D_H curve, this frequency is approximately 1200 Hz, and

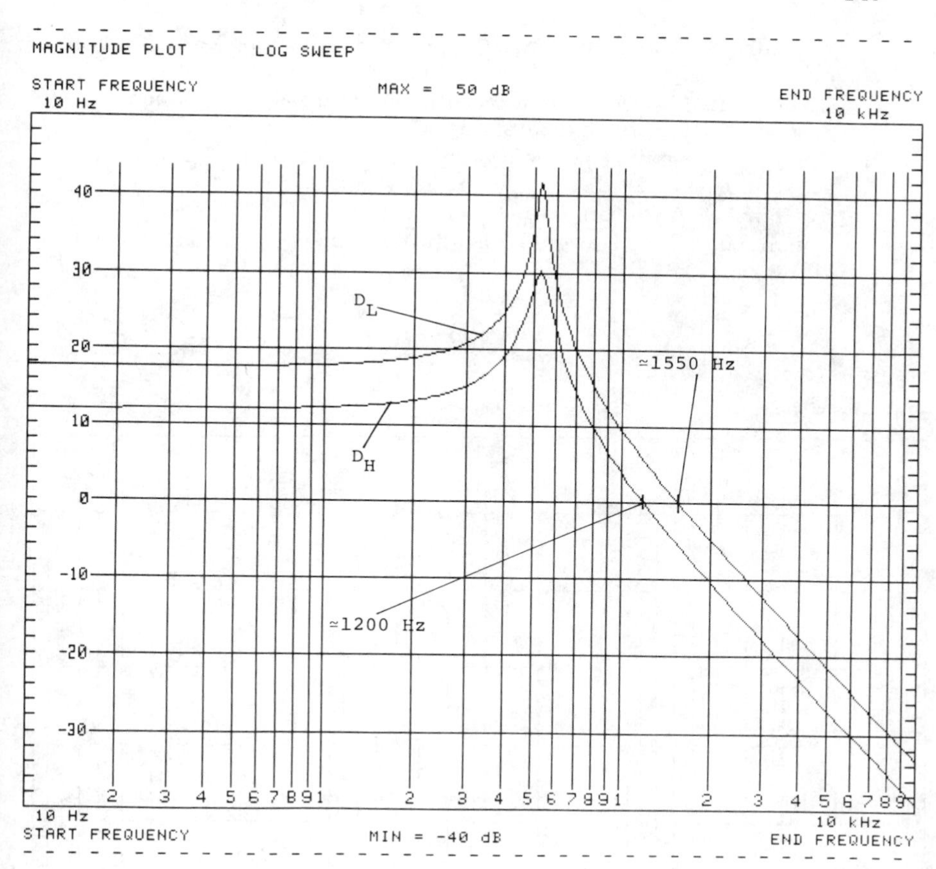

FIG. 4.12 Buck converter (no output filter damping).

b. For the D_L curve, this frequency is approximately 1550 Hz.

2. Observe in Fig. 4.14 the undamped phase responses at the frequencies of 1200 Hz (for D_H) and 1550 Hz (for D_L).
3. The two phase readings (for D_H and D_L) are within 0.5°, and the phase margins are not more than 3° in either case. Conclusion: φ_m is less than 3° for D_H or D_L.

To determine the gain margin G_m of the undamped converter,

1. Observe in Fig. 4.14 that the undamped phase responses reach -180° at approximately 10 kHz.
2. Read off the gain (decibels) from Fig. 4.12 at this frequency:
   a. For $D_H \cong -39$ dB, $G_{mH} = 39$ dB, and
   b. For $D_L \cong -34$ dB, $G_{mL} = 34$ dB.

   Conclusion: A large gain margin alone does not promise good stability.

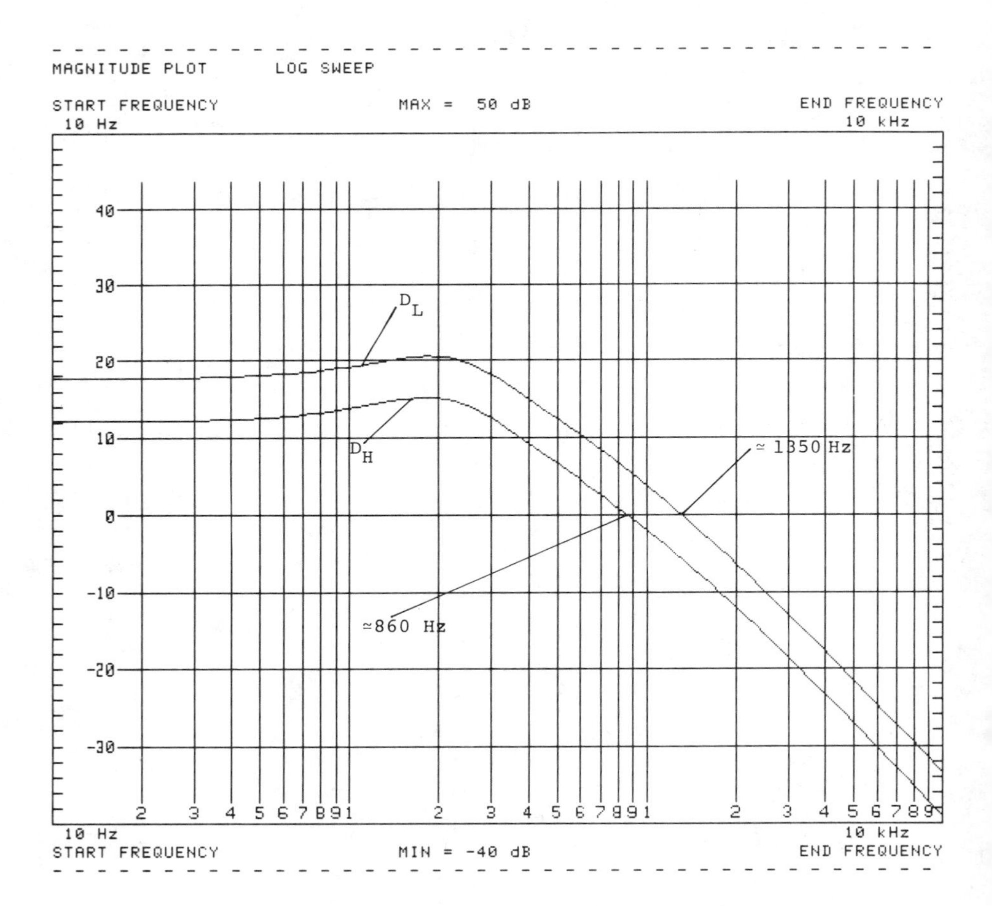

FIG. 4.13 Buck converter with output filter damping.

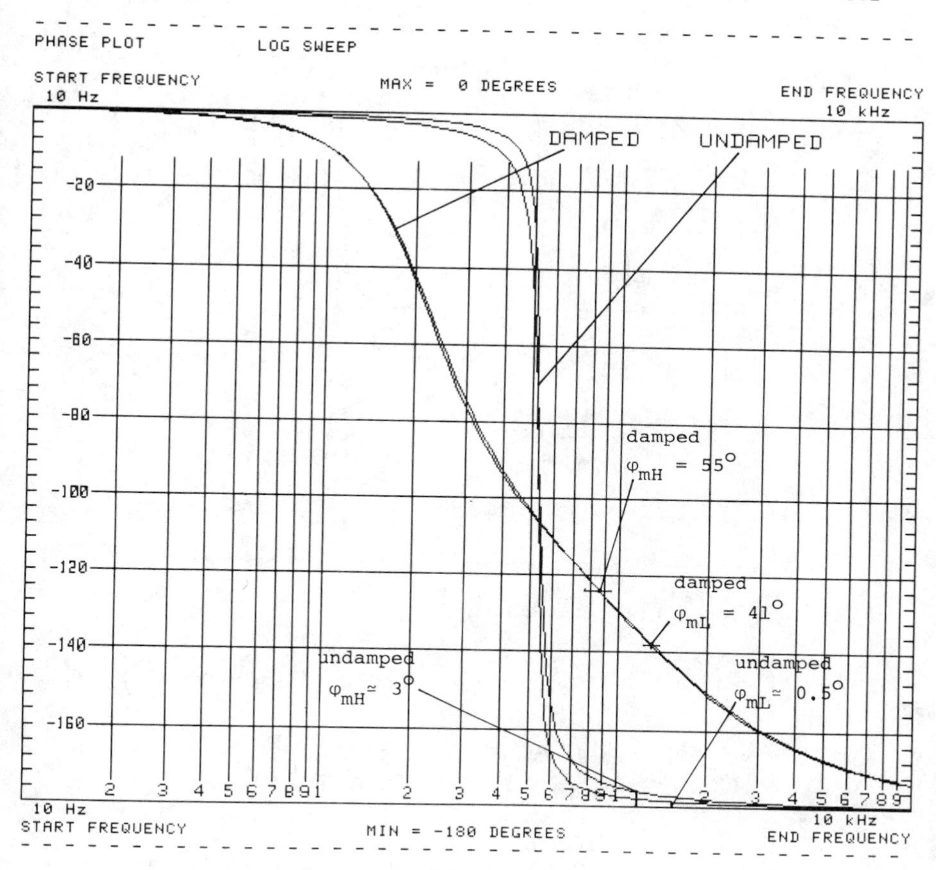

FIG. 4.14 Damped and undamped curves.

Now obtain the phase margins for the damped converter:

1. The frequency at which the gain curve crosses the 0-dB line:
   a. For the D_H curve, this frequency is approximately 860 Hz, and
   b. For the D_L curve, this frequency is approximately 1350 Hz.
2. Observe the damped phase responses in Fig. 4.14 at 860 Hz (for D_H) and 1350 Hz (for D_L).

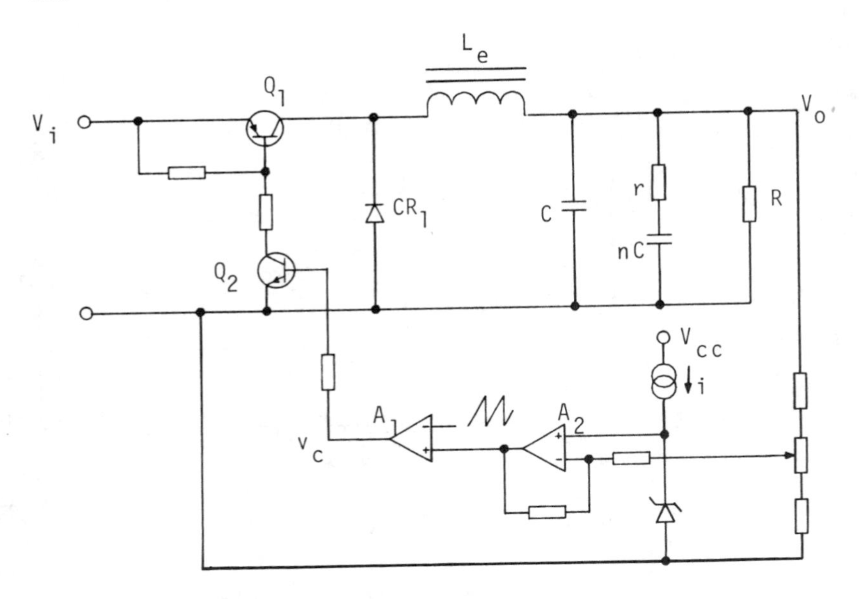

FIG. 4.15 Regulated buck converter.

3. The two phase readings are $-129°$ (for D_H) and $-139°$ (for D_L). Therefore, $\varphi_{mH} = 180° - 125° = 55°$, and $\varphi_{mL} = 180° - 139° = 41°$.

As for the gain margins, it is observed that the phase responses never quite reach $-180°$ even at 100 kHz; therefore, it will suffice to say that $G_m > 50$ dB for this circuit.

From the preceding exercise, it is clear that output filter damping provides extra phase margin to obtain better relative stability for a closed-loop system, which would appear as shown in Fig. 4.15. Note that for accurate control of output filter damping, the damping capacitor nC should be of the low-esr and low-esl type. The damping resistor r should be of sufficient wattage and mechanical bonding strength to withstand surge currents up to short-circuit current level. The damping capacitance value (nC), however, is much less critical.

The process of tailoring the loop gain will be dealt with in Section 4.7.

4.4 OUTPUT FILTER DAMPING

This section provides the theoretical basis for output filter damping.

From Eq. (A.55), Appendix A, the output filter transfer function $H_e(s)$ is a function with a quadratic form in the denominator, which can be put in the form of

$$H(j\omega) = \frac{1}{-\omega^2T^2 + 2\zeta j\omega T + 1} \tag{4.21}$$

where ζ is the damping factor.

At zero frequency or very low frequencies, the first two terms in the denominator are very small compared to unity; therefore,

$$H(0) = 1, \quad \text{with } 0° \text{ phase shift} \tag{4.22}$$

and the gain is 0 dB. This is the first asymptote: the line joining the point of f = 0 Hz at 0 dB to the point at the break (or corner) frequency at 0 dB.

At the high end of the frequency scale,

$$|H(j\omega)| = \left|\frac{1}{-\omega^2T^2}\right| \tag{4.23}$$

since the square terms in the denominator dominate the function, and the phase shift is 180°.

The gain at high frequency is therefore

$$-20 \log_{10} \omega^2T^2 = -40 \log_{10} \omega T \tag{4.24}$$

This is the second asymptote, with a slope of -40 dB/decade.

At the resonant frequency, $\omega T = 1$,

$$H(j\omega) = \frac{1}{2j\zeta\omega T} = \frac{1}{2j\zeta} \tag{4.25}$$

The gain in decible at this frequency is $20 \log_{10}(1/2\zeta)$ with a phase lag of 90°.

Theoretically, a zero damping factor therefore suggests an infinite gain at the resonant frequency. Damping, therefore, is a means of reducing peaking at the resonant frequency and, in turn, reduces the "steepness" of the slope of the magnitude response, which has a direct effect on the phase response. See Appendix B.

A further example of output filter damping as applied to the buck converter example is shown in Figs. 4.16 and 4.17. For the same loop gain arrangement as shown in Fig. 4.15, it is observed that there is a substantial increase in phase margin. The only change in this damping arrangement is the use of r = 0.05 Ω instead of the 0.1 Ω used before. However, the lower r is, the more unrealistic this arrangement becomes, because r is usually not a very high-wattage resistor, but the accuracy is important

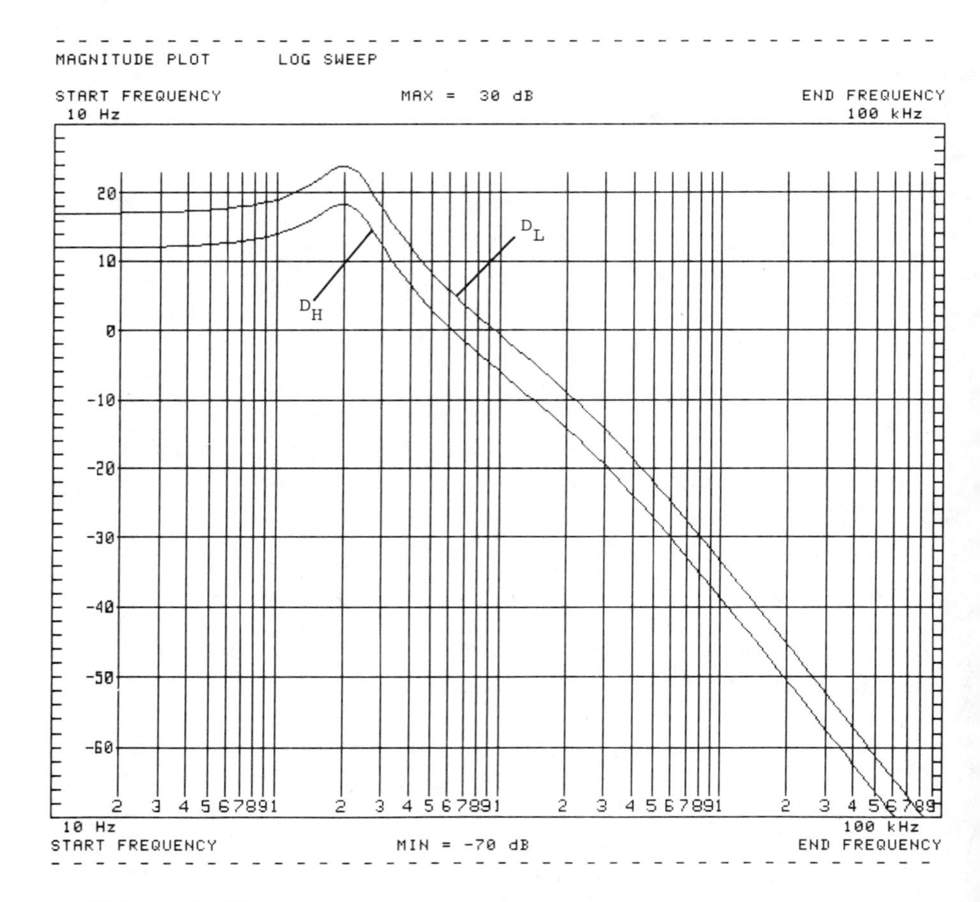

FIG. 4.16 Magnitude response for r = 0.05 Ω.

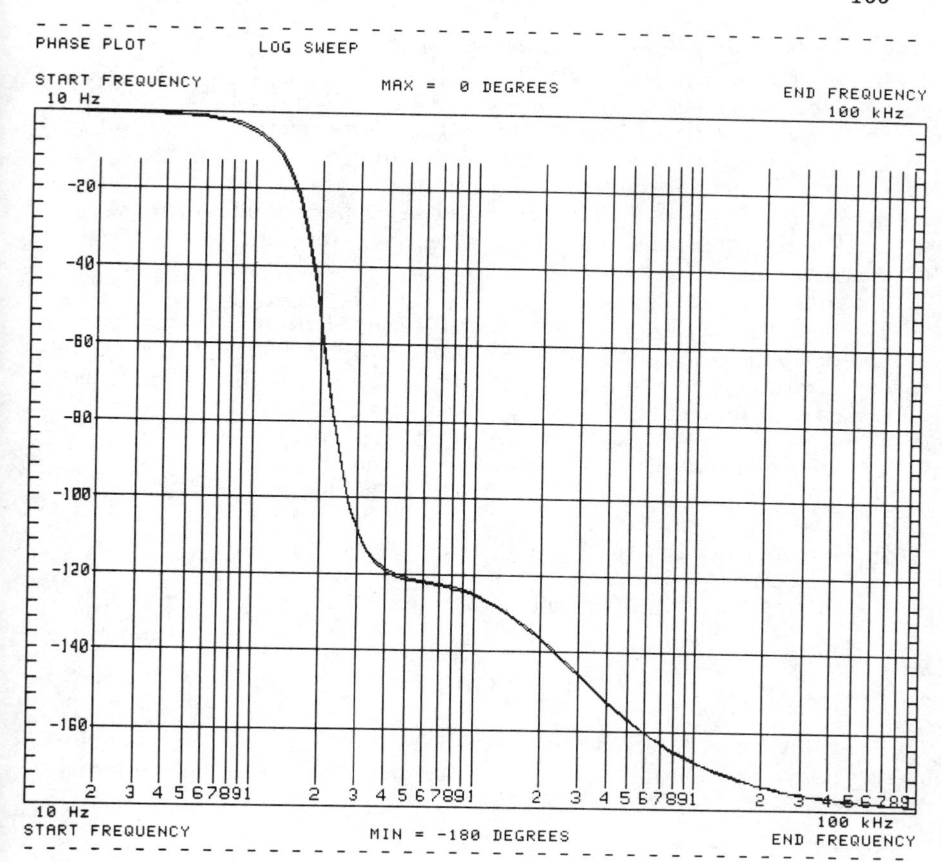

FIG. 4.17 Phase response for r = 0.05 Ω.

for the control of desirable damping characteristics. If the required value is very low, the circuit resistance as well as the esr of the capacitor will come into play. In general, it is undesirable to go below 0.1 Ω. Also, if there is inductance (esl) within the combination of r and nC, along with the low value of r, a high Q tuned circuit is formed. A parasitic circulating current could develop within the output filter circuit. This is another reason for the caution required when paralleling output filter capacitors for the reduction of esr effects.

A method of avoiding the use of very low values of r is to design the output filter with a ratio of L/C of at least 1. This is evident in the comparative study of Figs. B.5, B.6, B.18, B.19, B.32, and B.33 in Appendix B.

Another possible method of output filter damping is the use of the "trap" circuit. Since the LC output filter section is presenting a high impedance at the resonant frequency, it is theoretically feasible to use a circuit that has low impedance at its resonant frequency: the series tuned circuit. The responses shown in Figs. B.29 and B.30 assume no losses in the inductor, and this is therefore the worst possible case. Figure B.28 shows the circuit arrangement of this method.

This method is particularly useful in input filter damping, where the damping capacitor can become too large for practical purposes.

In conclusion, the process of output filter damping can be summarized as follows:

1. Determine the ratio of L/C of the given filter.
2. Select the set of curves with L/C ratio closest to the given ratio from Appendix B.
3. Chose the curve with the greatest phase margin.
4. Identify the r and nC values in the chosen curve.
5. Insert the identified values of r and nC in the given filter.

4.5 ASYMPTOTIC APPROXIMATIONS

Recalling Eq. (4.19) and Fig. 4.7, it is observed that for a single-reactance low-pass (lag) section, the form of the transfer function is

$$\frac{V_2(s)}{V_1(s)} = \frac{1}{1 + (s/\omega_1)} \tag{4.26}$$

and for a single-reactance high-pass (lead) section,

$$\frac{V_2(s)}{V_1(s)} = \frac{1}{1 + (\omega_1/s)} \tag{4.27}$$

For a two-reactance low-pass (lag) network, Eq. (4.21) shows the presence of a quadratic factor in the denominator of the transfer function in the following form:

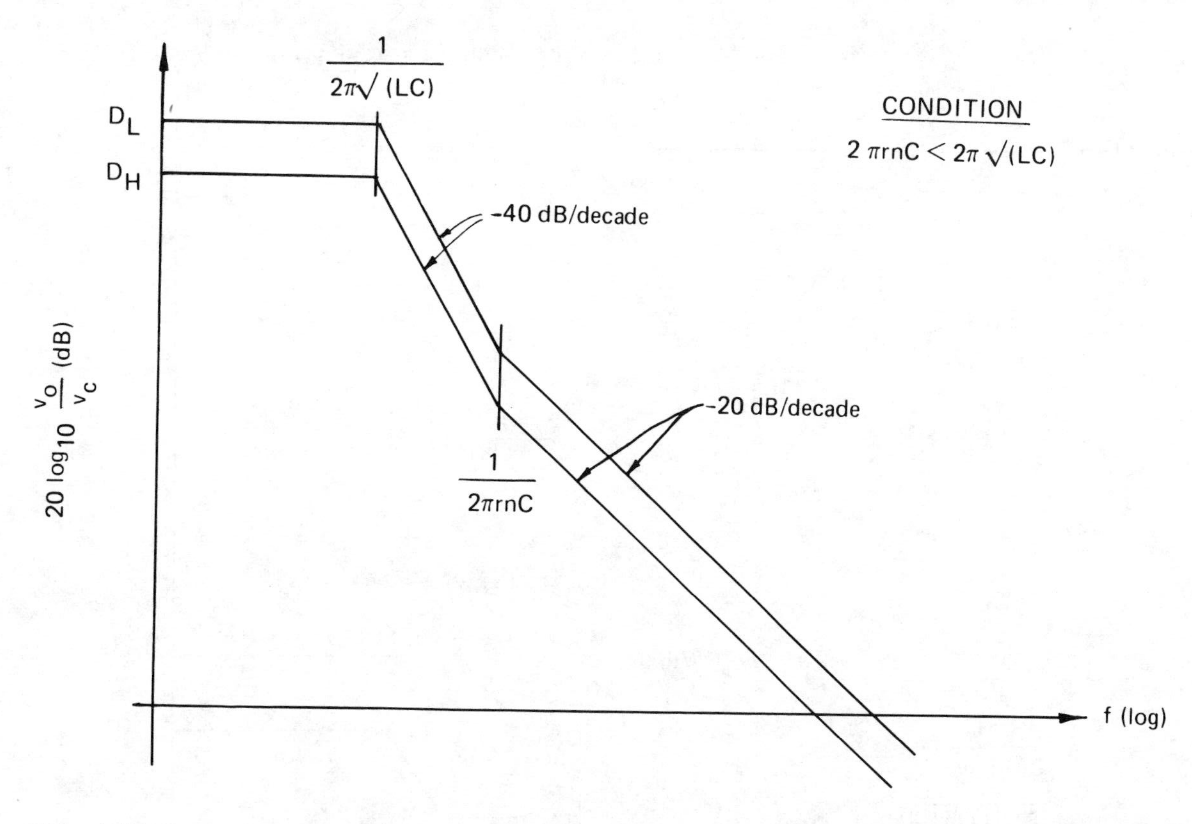

FIG. 4.18 Asymptotic approximations of the buck converter.

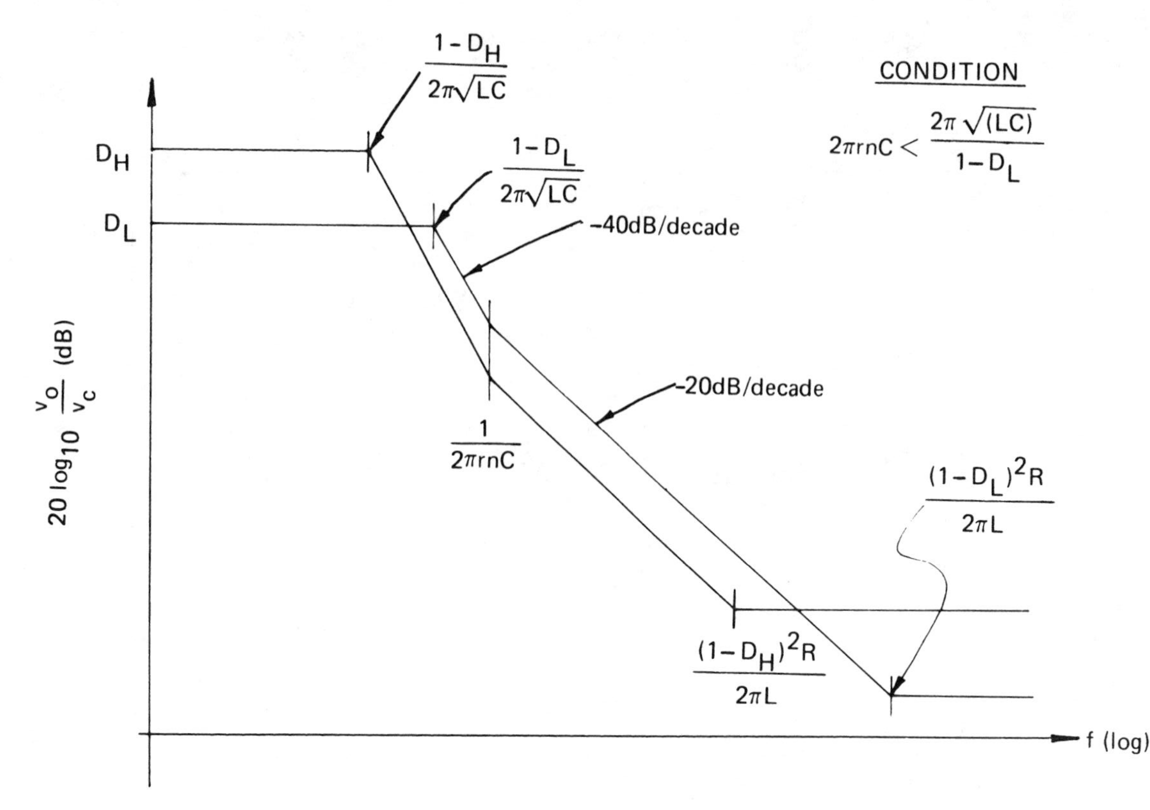

FIG. 4.19 Asymptotic approximations of the boost converter.

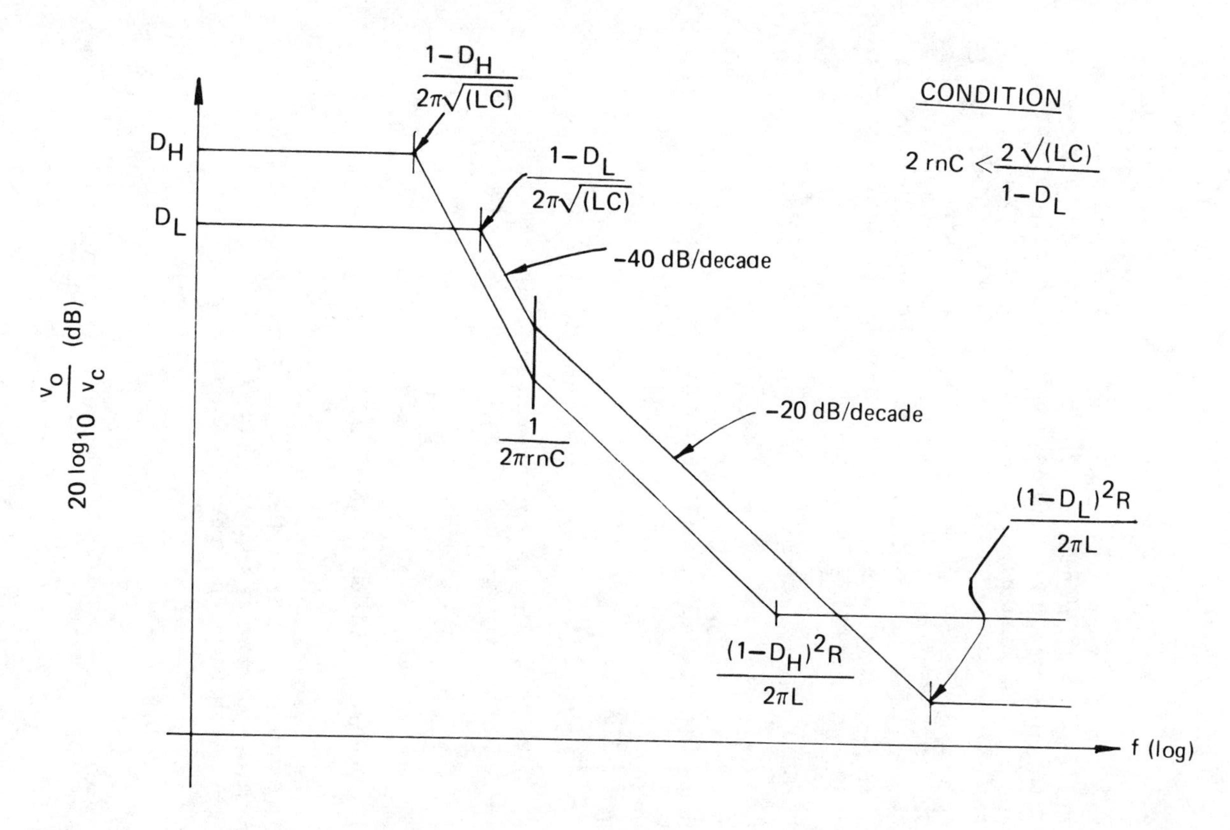

FIG. 4.20 Asymptotic approximations of the buck-boost converter.

$$H_e(s) = \frac{1}{s^2T^2 + 2\zeta s + 1} \tag{4.28}$$

The slope of the magnitude response is -40 dB/decade.

Equations (4.26) to (4.28) suggest that if the form of the transfer function is recognizable, the asymptotes for that transfer function can be constructed on a magnitude/frequency plot.

The asymptotes for Eq. (2.1) (Chapter 2) are shown in Fig. 4.18.

Similarly, the asymptotic approximations for boost and buck-boost converters are readily obtainable but must take into account the different effective output filter. See Fig. A.4 in Appendix A. It is noted that the effective inductance is given by

$$L_e = \frac{L}{(1 - D)^2} \tag{4.29}$$

Because of this, the resonant frequency of the output filter becomes

$$f = \frac{1 - D}{2\pi\sqrt{LC}} \tag{4.30}$$

Therefore, the asymptotic approximations for the boost and buck-boost converters are shown in Figs. 4.19 and 4.20, respectively.

4.6 COMPENSATION NETWORKS

A network suitable for use in lead compensation is shown in Fig. 4.21. This network is particularly useful for reducing the phase lag of the loop gain at a given frequency.

The parallel branch impedance due to C_1 and R_1 is

$$z_p = \frac{R_1}{1 + sC_1R_1} \tag{4.31}$$

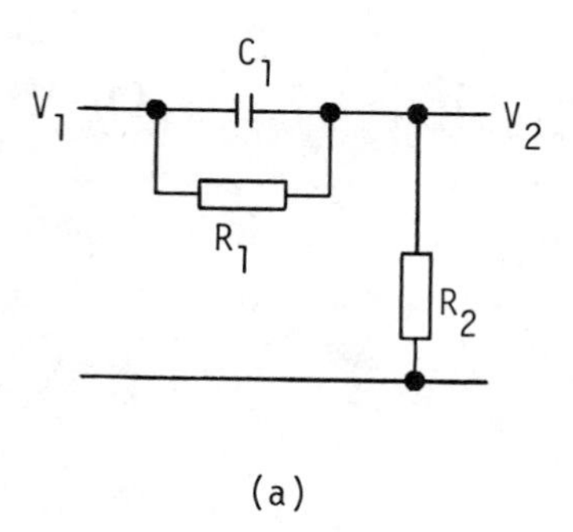

(a)

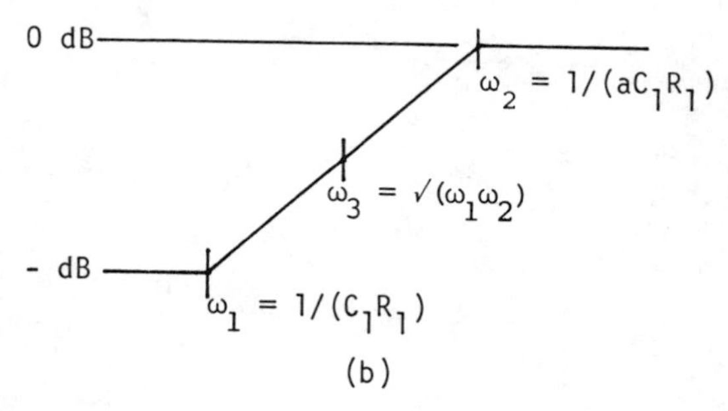

(b)

FIG. 4.21 Basic lead network.

The transfer function of this network is

$$\frac{V_2}{V_1} = \frac{R_2}{R_2 + z_p} = \frac{R_2}{R_2 + [R_1/(1 + sC_1R_1)]} \tag{4.32}$$

$$= \frac{R_2(1 + sC_1R_1)}{R_1 + R_2 + sC_1R_1R_2}$$

$$\frac{V_2}{V_1} = \frac{R_2}{R_1 + R_2} \frac{1 + sC_1R_1}{1 + sC_1[R_1R_2/(R_1 + R_2)]} \tag{4.33}$$

which is in the form of

$$\frac{V_2}{V_1} = a \frac{1 + (s/\omega_1)}{1 + (s/\omega_2)} \tag{4.34}$$

where

$$a = \frac{R_2}{R_1 + R_2}$$

$$\omega_1 = \frac{1}{C_1R_1}$$

$$\omega_2 = \frac{1}{aC_1R_1} = \frac{\omega_1}{a} \tag{4.35}$$

The phase lead introduced by this network is

$$\varphi = \tan^{-1} \frac{\omega}{\omega_1} - \tan^{-1} \frac{\omega}{\omega_2} \tag{4.36}$$

$$\tan \varphi = \tan(\varphi_1 - \varphi_2) \tag{4.37}$$

$$= \frac{\tan \varphi_1 - \tan \varphi_2}{1 + \tan \varphi_1 \tan \varphi_2} \tag{4.38a}$$

$$= \frac{(\omega/\omega_1) - (\omega/\omega_2)}{1 + (\omega/\omega_1)(\omega/\omega_2)} \tag{4.38b}$$

But

$$\omega_2 = \frac{\omega_1}{a} \tag{4.35}$$

Therefore,

$$\tan \varphi = \frac{(\omega/\omega_1) - (a\omega/\omega_1)}{1 + (\omega/\omega_1)(a\omega/\omega_1)} = \frac{(\omega/\omega_1)(1 - a)}{1 + (\omega/\omega_1)^2} \tag{4.39}$$

To find maximum phase lead in relation to a, differentiating $\tan \varphi$ with respect to ω/ω_1 gives

$$\frac{d(\tan \phi)}{d\left(\frac{\omega}{\omega_1}\right)} = \frac{\left[1 + a\left(\frac{\omega}{\omega_1}\right)^2\right](1 - a) - \frac{\omega}{\omega_1}(1 - a)\left(2a\frac{\omega}{\omega_1}\right)}{\left[1 + a\left(\frac{\omega}{\omega_1}\right)^2\right]^2} \tag{4.40}$$

This expression equals zero at maximum phase lead:

$$0 = \left[1 + a\left(\frac{\omega}{\omega_1}\right)^2\right](1 - a) - \frac{\omega}{\omega_1}(1 - a)\left(2a\frac{\omega}{\omega_1}\right) \tag{4.41}$$

$$1 + a\left(\frac{\omega}{\omega_1}\right)^2 = 2a\left(\frac{\omega}{\omega_1}\right)^2 \tag{4.42}$$

$$1 = a\left(\frac{\omega}{\omega_1}\right)^2 \tag{4.43}$$

$$\omega = \frac{\omega_1}{\sqrt{a}} \quad \text{for maximum phase lead} \tag{4.44}$$

Substituting $\omega_1/\sqrt{a}$ for ω in Eq. (4.39) gives

$$\tan o = \frac{(1/\sqrt{a})(1 - a)}{1 + a(1/\sqrt{a})^2} = \frac{1 - a}{2\sqrt{a}} \tag{4.45}$$

The maximum phase lead φ_d occurs at ω given by Eq. (4.44), and from Eq. (4.45),

$$\sin \varphi_d = \frac{1}{\sqrt{1 + \cot^2 \varphi_d}} = \frac{1}{\sqrt{1 + [41/(1 - a)^2]}} \tag{4.46}$$

Therefore,

$$\sin \varphi_d = \frac{1 - a}{1 + a} \tag{4.47}$$

To ensure accurate implementation, it is sometimes desirable to use an isolation amplifier. Figure 4.22 shows a noninverting phase lead network with an isolation amplifier. The series-parallel combination impedance due to R_1, R_2, and C_1 is

$$z_c = R_2 // \left(R_1 + \frac{1}{C_1}\right) = \frac{R_2[R_1 + (1/sC_1)]}{R_2 + R_1 + (1/sC_1)} \tag{4.48}$$

The transfer function of this network is

$$\frac{V_2}{V_1} = \frac{R_5}{R_4 + R_5}\left(1 + \frac{R_3}{z_c}\right) \tag{4.49}$$

$$= \frac{R_5}{R_4 + R_5}\left[1 + \frac{R_3}{R_2}\,\frac{R_2 + R_1 + (1/sC_1)}{R_1 + (1/sC_1)}\right] \tag{4.50}$$

$$= \frac{R_5}{R_4 + R_5}\,\frac{R_2 + R_3 + sC_1(R_1R_2 + R_2R_3 + R_1R_3)}{R_2(1 + sC_1R_1)} \tag{4.51}$$

$$= \frac{R_5}{R_4 + R_5}\,\frac{R_2 + R_3}{R_2}\,\frac{1 + \dfrac{sC_1(R_1R_2 + R_2R_3 + R_1R_3)}{R_2 + R_3}}{1 + sC_1R_1} \tag{4.52}$$

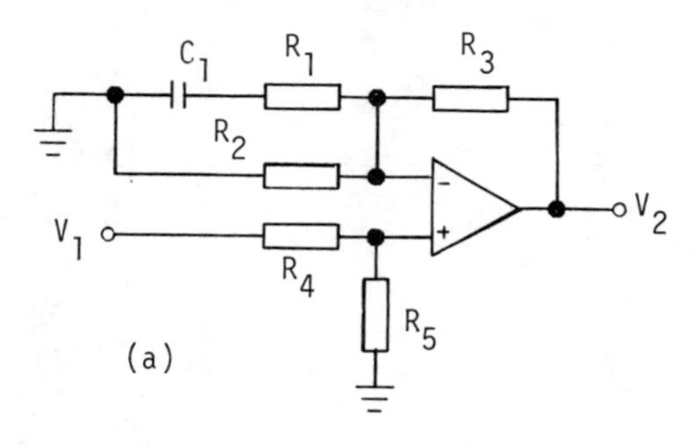

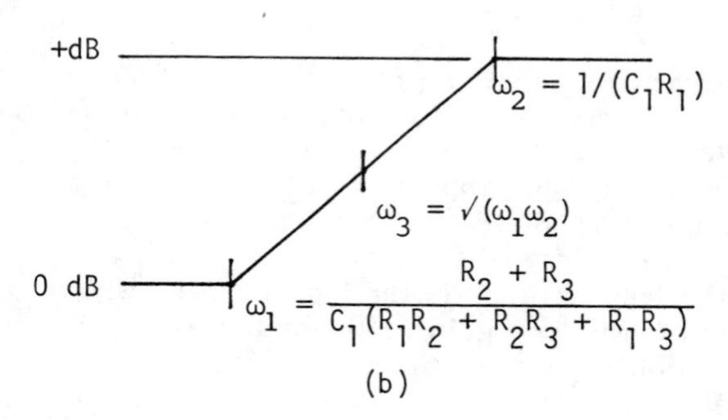

FIG. 4.22 Lead network with isolation amplifier.

which is in the form of

$$\frac{V_2}{V_1} = K\,\frac{1 + (s/\omega_1)}{1 + (s/\omega_2)} \tag{4.53}$$

where

$$K = \frac{R_5}{R_2}\,\frac{R_2 + R_3}{R_4 + R_5}$$

$$\omega_1 = \frac{R_2 + R_3}{C_1(R_1R_2 + R_2R_3 + R_1R_3)}$$

$$\omega_2 = \frac{1}{C_1R_1}$$

For $R_2 = R_3 = R_4 = R_5 = R_a$, K = 1,

$$\omega_1 = \frac{1}{C_1} \frac{1}{R_1 + (1/2R_a)} \tag{4.54}$$

or

$$R_a = \frac{1}{\pi f_1 C_1} - 2R_1 \tag{4.55}$$

If a gain other than unity is required, R_3 can be made larger or smaller than the other resistors. In that event, the relation of Eq. (4.52) should be used for designing the lead network. In summary, the lead network has the property of increasing the system bandwidth, reducing overshoot on the transient response, and improving the relative stability of the system.

The complement of the lead network is the lag network, which is commonly used to roll off the loop gain response before the phase lag reaches 180° to obtain a wider gain margin.

Figure 4.23 shows a basic lag network. The series branch impedance due to R_1 and C_1 is

$$z_s = R_1 + \frac{1}{sC_1} = \frac{sC_1R_1 + 1}{sC_1} \tag{4.56}$$

$$\frac{V_2}{V_1} = \frac{1 + sC_1R_1}{1 + sC_1(R_1 + R_2)}$$

$$= \frac{1 + sC_1R_1}{1 + sC_1R_1[(R_1 + R_2)/R_1]} \tag{4.57}$$

$$\frac{V_2}{V_1} = \frac{1 + sC_1R_1}{1 + saC_1R_1} \tag{4.58}$$

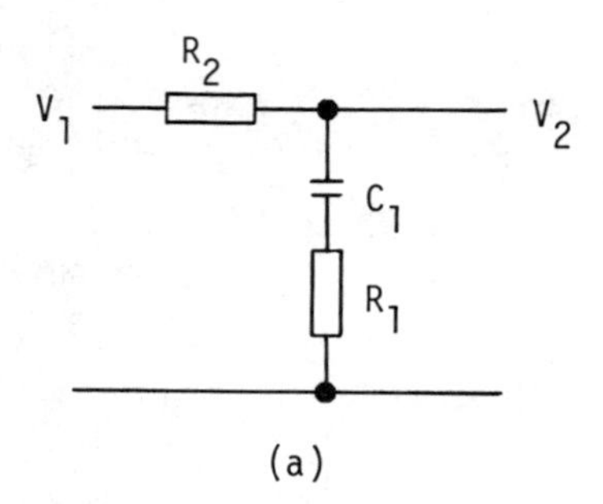

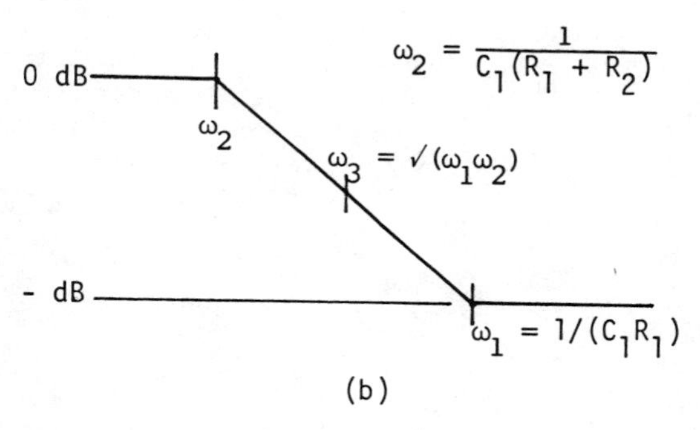

FIG. 4.23 Basic lag network.

where $a = (R_1 + R_2)/R_1$, and Eq. (4.58) is in the form of

$$\frac{V_2}{V_1} = \frac{1 + (s/\omega_1)}{1 + (s/\omega_2)} \tag{4.59}$$

where

$$\omega_1 = \frac{1}{C_1R_1}$$

$$\omega_2 = \frac{\omega_1}{a}$$

In general, one does not require an analytical approach to estimate the maximum phase point as derived in Eqs. (4.40) to (4.44); the maximum phase always lies within the geometric mean of ω_1 and ω_2 at ω_3 such that

$$\omega_3 = \sqrt{\omega_1 \omega_2} \tag{4.60}$$

The lag network is sometimes used with an isolation amplifier as shown in Fig. 4.24. The feedback impedance due to R_1, R_2, and C_1 is

$$Z_f = (R_1 + \frac{1}{sC_1})//R_2 = \frac{R_2(R_1 + \frac{1}{sC_1})}{R_2 + R_1 + \frac{1}{sC_1}}$$

$$Z_f = R_2 \frac{1 + sC_1R_1}{1 + sC_1(R_1 + R_2)} \tag{4.61}$$

The transfer function of the above circuit is:

$$\frac{V_2}{V_1} = \frac{R_5}{R_4 + R_5} \left[1 + \frac{Z_f}{R_3}\right]$$

$$= \frac{R_5}{R_4 + R_5} \left[1 + \frac{R_2}{R_3} \frac{1 + sC_1R_1}{1 + sC_1(R_1 + R_2)}\right]$$

$$= \frac{R_5}{R_4 + R_5} \frac{R_2 + R_3 + sC_1[R_3(R_1 + R_2)+R_1R_2]}{R_3[1 + sC_1(R_1 + R_2)]}$$

$$\frac{V_2}{V_1} = \frac{R_5}{R_4 + R_5} \frac{R_2 + R_3}{R_3} \frac{1 + sC_1(R_1 + R_2//R_3)}{1 + sC_1(R_1 + R_2)} \tag{4.62}$$

which is in the form of

$$\frac{V_2}{V_1} = K \frac{1 + \frac{s}{\omega_1}}{1 + \frac{s}{\omega_2}} \tag{4.63}$$

where

$$K = \frac{R_5}{R_4 + R_5} \frac{R_2 + R_3}{R_3} \tag{4.64}$$

$$\omega_1 = \frac{1}{C_1(R_1 + R_2//R_3)} \tag{4.65}$$

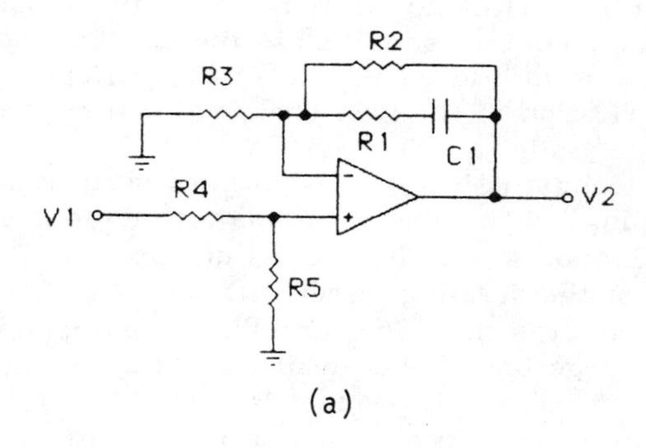

(a)

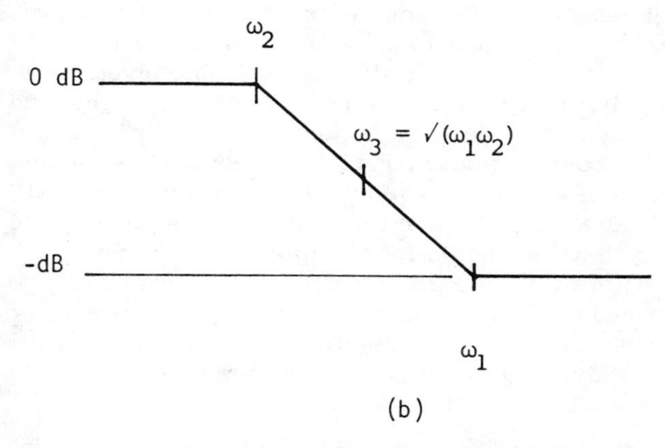

(b)

FIG. 4.24 A lag network with isolation amplifier.

$$\omega_2 = \frac{1}{C_1(R_1 + R_2)} \tag{4.66}$$

It will be observed that the networks shown in Figs. 4.21 to 4.24 can be combined to form lead-lag networks or lag-lead networks for tailoring the loop gain characteristic of any converter systems.

4.7 LOOP DESIGN

The purpose of loop design is to obtain a stable converter when operating in closed-loop feedback configuration.

First, it is necessary to point out that the change of gain within the loop gain transfer function does not alter the shape of the phase or magnitude characteristics unless the damping ratio is also altered (say, due to load change). The magnitude curve will be shifted in relation to the gain change, but the phase remains unchanged (see Fig. 4.25).

A very useful technique in loop design is the implementation of output filter damping. The concept of this technique is outlined in some detail in Section 4.4. The graphs in Appendix B provide a clear indication of the damping effect with various LC ratios. The general implication is that, for continuous conduction converter designs, a large L_e value is desirable. But for the bandwidth requirement, a small L_eC product is desirable. It is therefore sometimes useful to use two stages of output filter to obtain the desired overall response. The idea is to have a high enough pole frequency for the first stage to provide the bandwidth and another LC stage to attenuate the ripple to a desirable level. However, it is usually advisable to have the pole frequency of the second stage at least a couple of decades higher than the first pole frequency to avoid excessive phase lag within a narrow frequency band. Obviously, the switching frequency of the converter must also be considered here. The higher the switching frequency, the wider the frequency band available for manipulating the loop response. An advantage of output filter damping is its insensitivity to component tolerance. This method of obtaining phase margin is quite reliable and has the additional benefit of reducing noise within the loop. The disadvantage of this method, however, is the re-

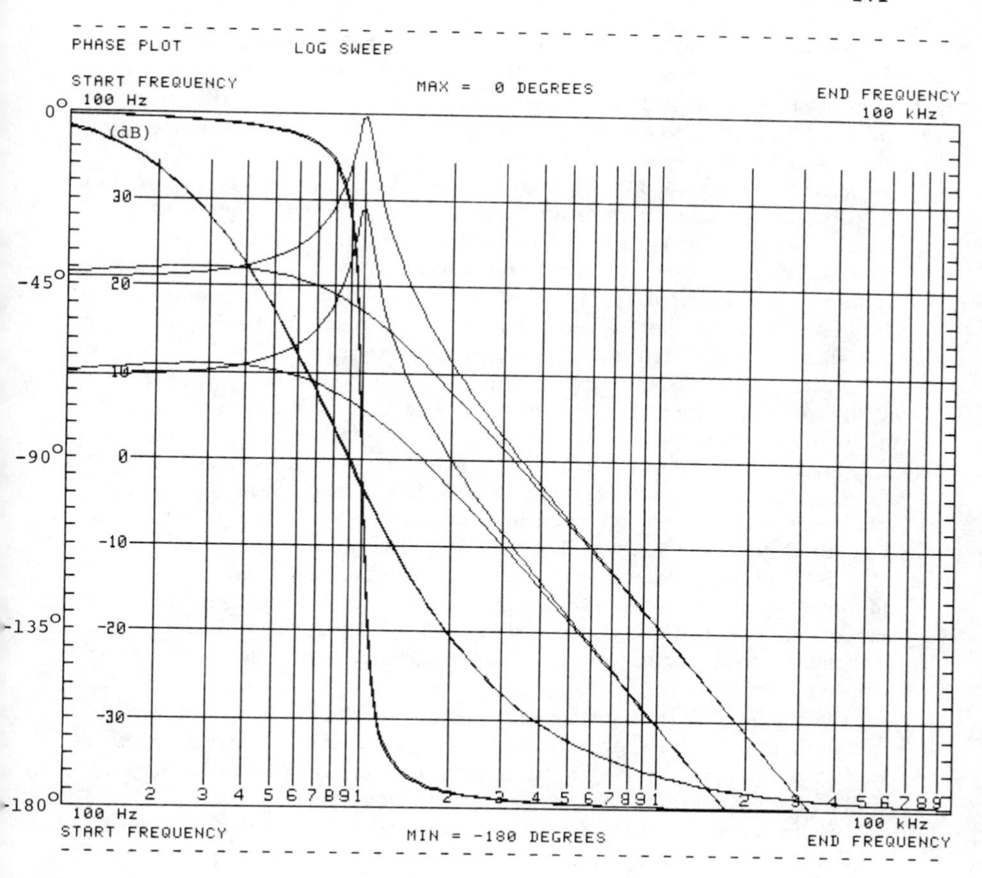

FIG. 4.25 Effect of gain variation on circuit response.

quirement of a bulky power component: the damping capacitor nC, which is several times of the output filter capacitor C.

A pole can be nulled by lag compensation aimed at the *dominant pole*. This is done by rolling off the output filter response well before the filter resonant frequency at a slope of -20 dB/decade, as shown in Fig. 4.26. In Fig. 4.26, a fictitious converter with 15-dB gain at zero frequency is assumed. To compensate this converter which has a filter resonant frequency at f, a new response with a slope of -20 dB/decade as shown is desirable, because of the gain/phase relationship mentioned: It is desirable to have the gain response crossing the 0-dB level at a slope of -20 dB/decade to ensure a moderate phase margin.

Notice that the roll-off corner frequency starts at a point below 0.02f. This means that the response of the loop gain starts to decrease from this point onward toward the higher frequencies. If f is at 1 kHz, the corner frequency is seen to be below 20 Hz, suggesting a drastic reduction of bandwidth in the loop gain transfer characteristic. This means that the converter so compensated is expected to react rather sluggishly under step load or line change. The advantage of dominant pole compensation is that it is easy to implement. The disadvantage is very conspicuous when line ripple rejection is required. The addition of an input filter is usually necessary when dominant pole compensation is employed in an off-line converter.

Compensation is sometimes achieved by the insertion of a lead network within the feedback loop to increase the phase margin within a small frequency band. The application of the basic lead network has been given in some detail by Pressman [48].

If the network of Fig. 4.22 is used, then the phase lead is deduced from Eq. (4.52), so that the lead angle at any given frequency is calculated as follows:

$$\varphi = \tan^{-1}\frac{\omega}{\omega_1} - \tan^{-1}\frac{\omega}{\omega_2} \tag{4.67}$$

$$\varphi = \tan^{-1}\frac{\omega C_1(R_1R_2 + R_2R_3 + R_1R_3)}{R_2 + R_3} - \tan^{-1}\omega C_1R_1 \tag{4.68}$$

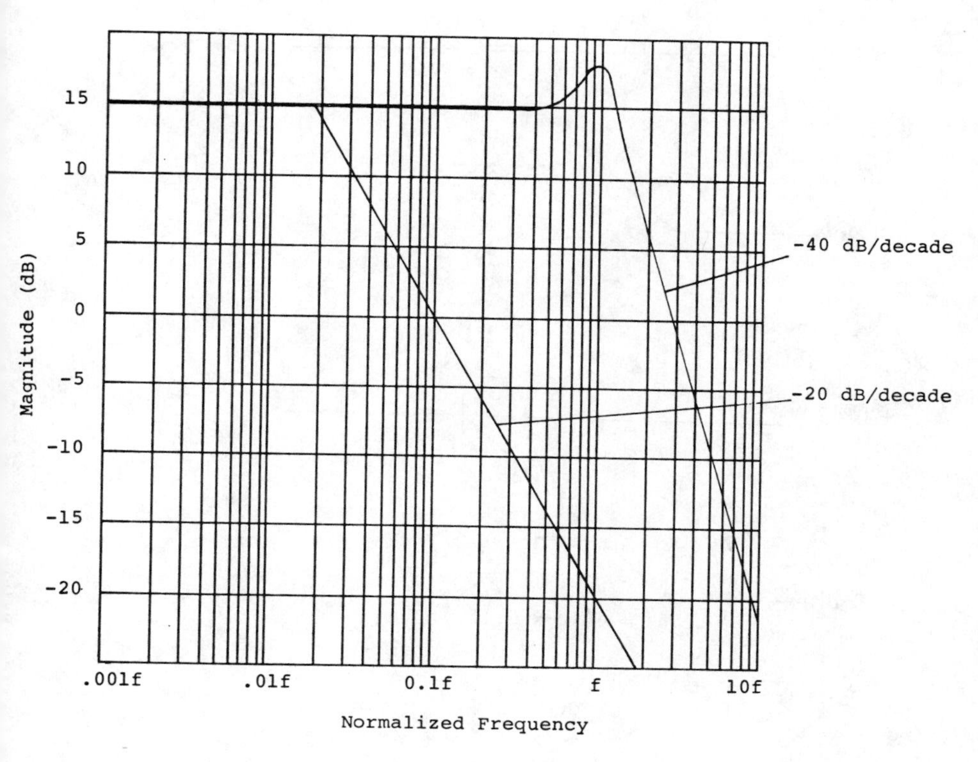

FIG. 4.26 Dominant pole compensation.

The frequencies f_1 and f_2 can be designed to span from one to two decades; see Fig. 4.27a. f_3 is the frequency at which the maximum phase lead is to be introduced. If f_1 = 1 kHz and f_1 to f_2 is to be contained within one decade, then f_2 will be 10 kHz and f_3 will be 3162 Hz. The design procedure of the lead network with isolation amplifier to span one decade can be summarized as follows:

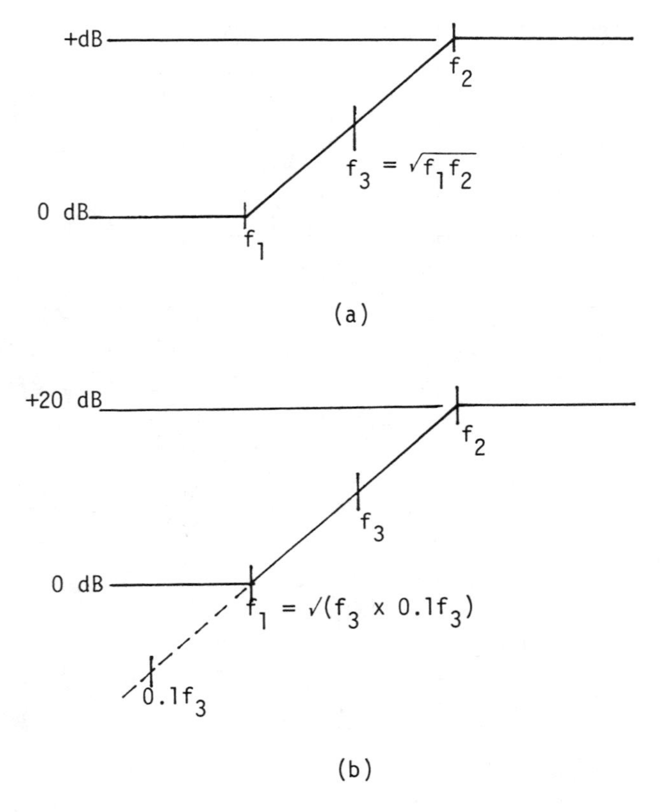

FIG. 4.27 Lead network design. (a) Frequency span. (b) Designing f_1.

1. A decision is made to produce additional phase lead at f_3.
2. The frequency span from f_1 to f_2 is to be designed to within one decade.
3. Calculate f_1 as outlined in Fig. 4.27b.
4. $f_2 = 10f_1$.
5. Assign a convenient value to R_1 (Fig. 4.22).
6. Calculate C_1 using the relation $\omega_2 = 1/(C_1R_1)$.
7. Calculate R_a using Eq. (4.55).

For a lead network to span two decades,

$$f_1 = 0.1f_3$$

Figure 4.28 shows the responses of a lead network with f_3 = 1 kHz spanning two decades. Figure 4.29 is an example of a lead network contained within one decade, with f_3 = 1 kHz.

As an illustration on the design details of the lead network, the buck converter example of Chapter 2 is recalled here. Figure 4.30 shows the undamped responses of this converter. Inspection of the curves in this figure shows that the phase margin is negative, suggesting an unstable system, if the loop is closed. Inspection of Figs. 4.28 and 4.30 indicates that the resonant frequency of the output filter is at 554 Hz and that a lead network centering at about 1 kHz would provide sufficient phase lead at 554 Hz for a reasonable phase margin. A lead network is introduced in the loop with f_3 = 1 kHz spanning two decades as shown in Fig. 4.31:

$$f_2 = \frac{1}{2\pi C_1 R_1}$$

Assign 470 Ω for R_1; then

$$C_1 = \frac{1}{2\pi f_1 R_1} = \frac{1}{2\pi \times 10k \times 470} = 0.0338\ \mu F$$

<u>Use 0.033 μF</u>

From Eq. (4.55),

$$R_a = \frac{1}{\pi f_1 C_1} - 2R_1$$

$$= \frac{1}{\pi \times 100 \times 0.033 \times 10^{-6}} - 940$$

$$= 95.5\ \Omega$$

<u>Use 95.3 Ω, standard 1% value</u>

The compensated responses are shown in Fig. 4.32. It can be seen that a phase margin of more than 30° is obtained for both the D_H and D_L curves. Note that, in the case of the buck converter, a lighter load gives a lower damping ratio. That is, if sufficient phase margin is obtained for the case of D_L, then it is

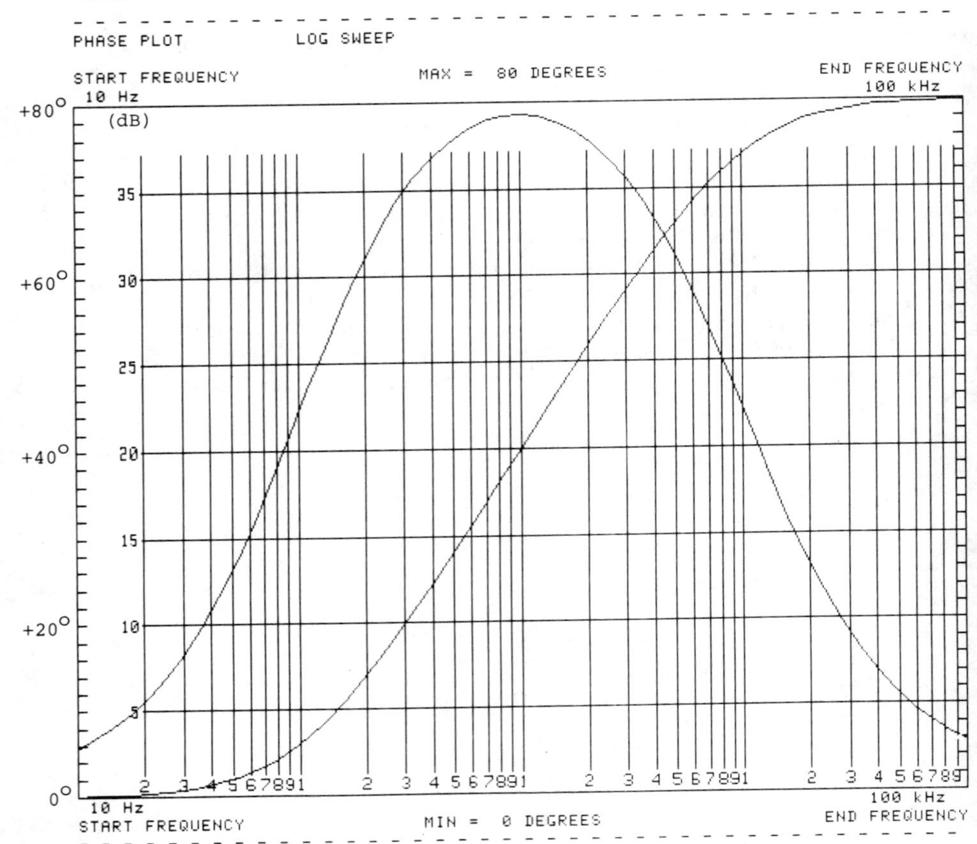

FIG. 4.28 Phase and magnitude responses for a two-decade span lead network.

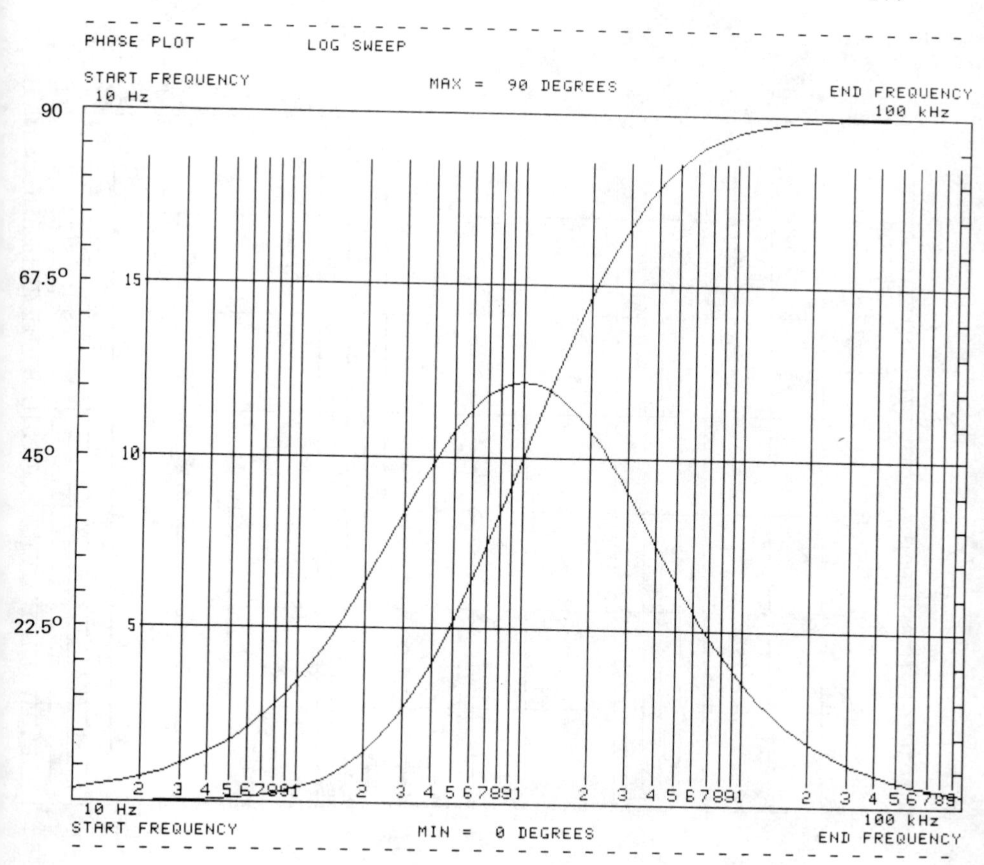

FIG. 4.29 Phase and magnitude responses for a one-decade span lead network.

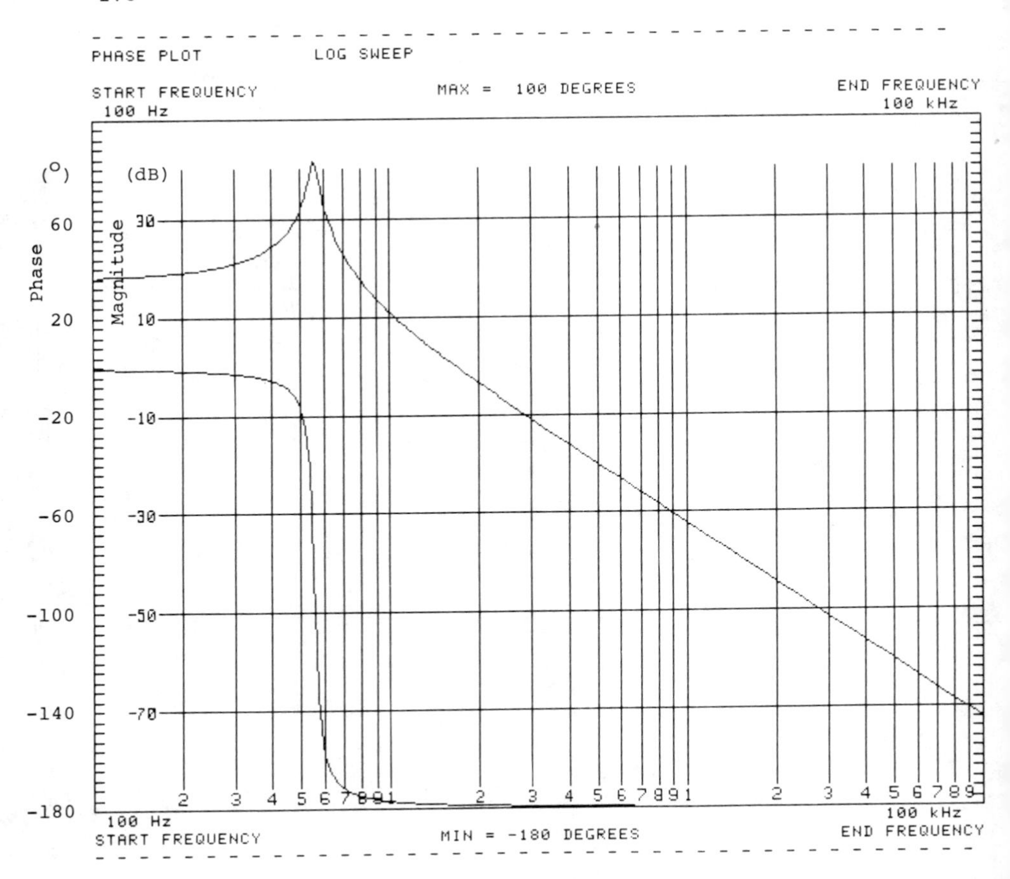

FIG. 4.30 Buck converter—no output filter damping, no compensation.

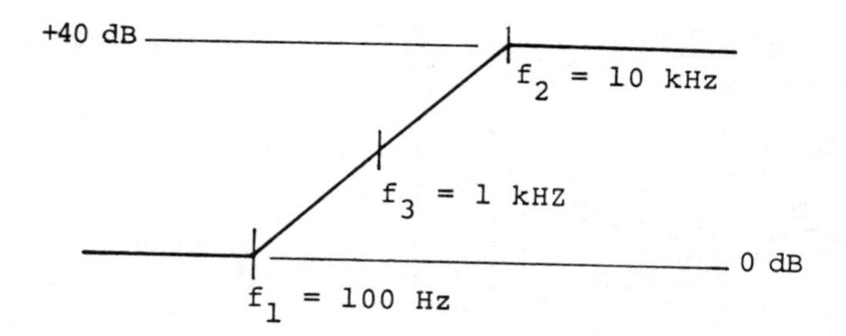

FIG. 4.31 Two-decade frequency span lead network.

likely that the D_H case will look after itself. Because the loop gain assessment up to this point has not been taking the error amplifier gain into account, the calculated response is only part of the complete loop gain. If the responses as shown in Fig. 4.32 are considered acceptable in every respect, then the attenuation of the sampling voltage divider network must be compensated for with the error amplifier gain by designing for

$$\frac{R_2 + R_3}{R_3} = \frac{R_5}{R_4}$$

as depicted in Fig. 4.33. It follows that, by adjustment of the ratio of R_5/R_4, the overall loop gain of the system can be varied to adjust the gain and phase margins.

In consideration of the lag network of Fig. 4.17, the phase lag is given by

$$\varphi = \tan^{-1} \omega C_1 R_1 - \tan^{-1} a\omega C_1 R_1 \tag{4.69}$$

The maximum phase lag is

$$\varphi_g = \tan^{-1} \frac{1 - a}{2\sqrt{a}} \tag{4.70}$$

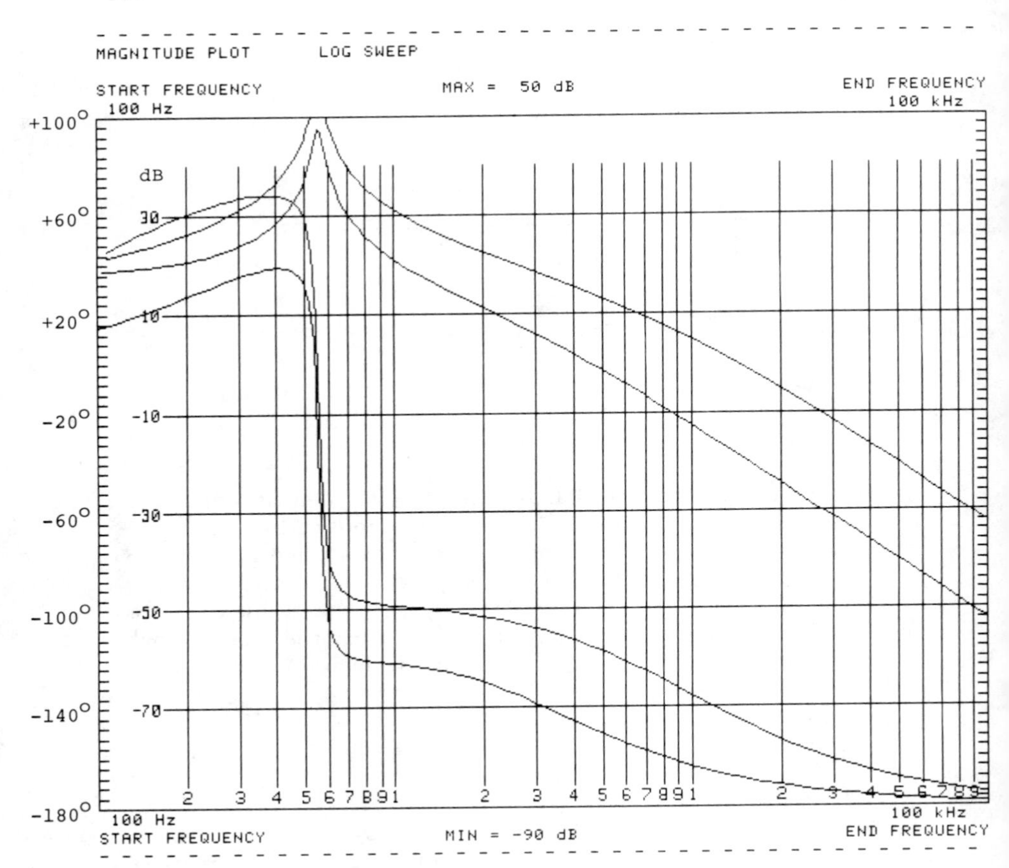

FIG. 4.32 Buck converter with lead compensation.

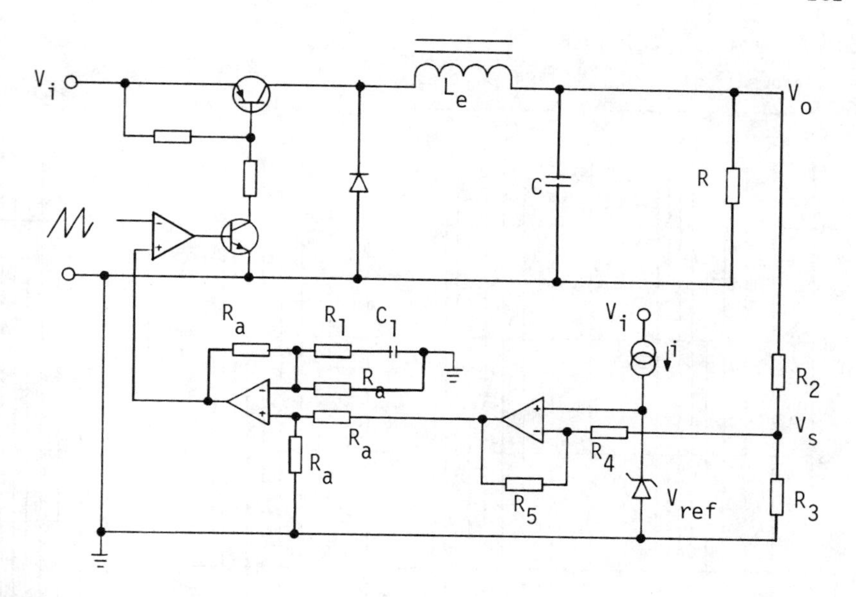

FIG. 4.33 Lead-compensated buck converter.

Similarly, the lag network with isolation amplifier has a phase lag of

$$\varphi = \tan^{-1} \frac{\omega C_1 R_1 R_2 R_3}{R_1 R_2 + R_1 R_3 + R_2 R_3} \tag{4.71}$$

$$- \tan^{-1} \frac{\omega C_1 R_1 R_2 R_3}{R_2 R_3 + R_1 R_3}$$

Figures 4.34 and 4.35 show, respectively, the magnitude and phase of the basic lag network of Fig. 4.23. The design procedure of this network and of the lag network with the isolation amplifier is very similar to that given for the lead network and will not be elaborated here.

The problem of loop design, however, is not always a straightforward and clear-cut affair. The buck-boost converter is a

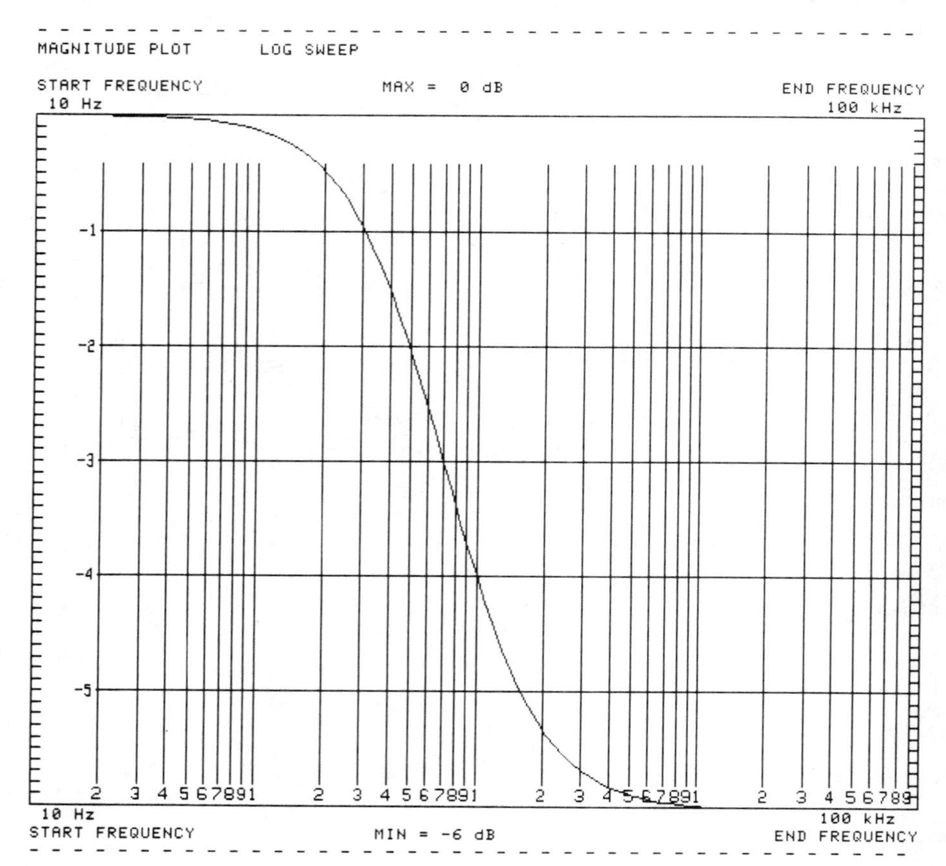

FIG. 4.34 Magnitude response of basic lag network of Fig. 4.23.

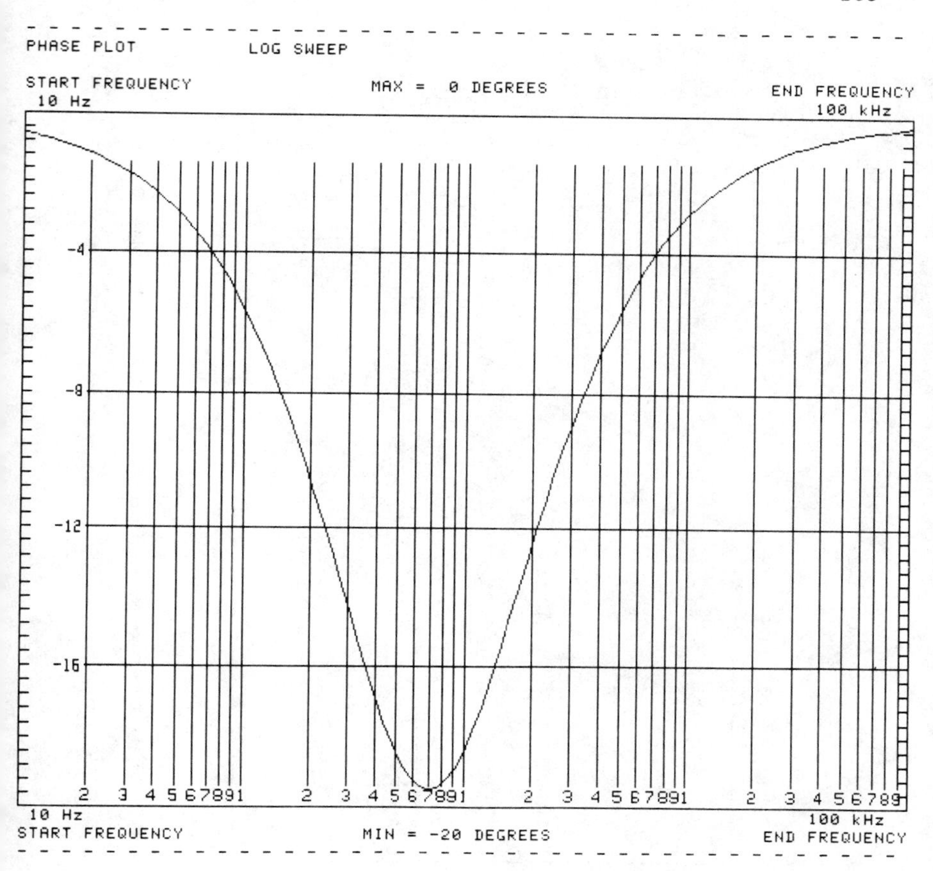

FIG. 4.35 Phase response of basic lag network of Fig. 4.23.

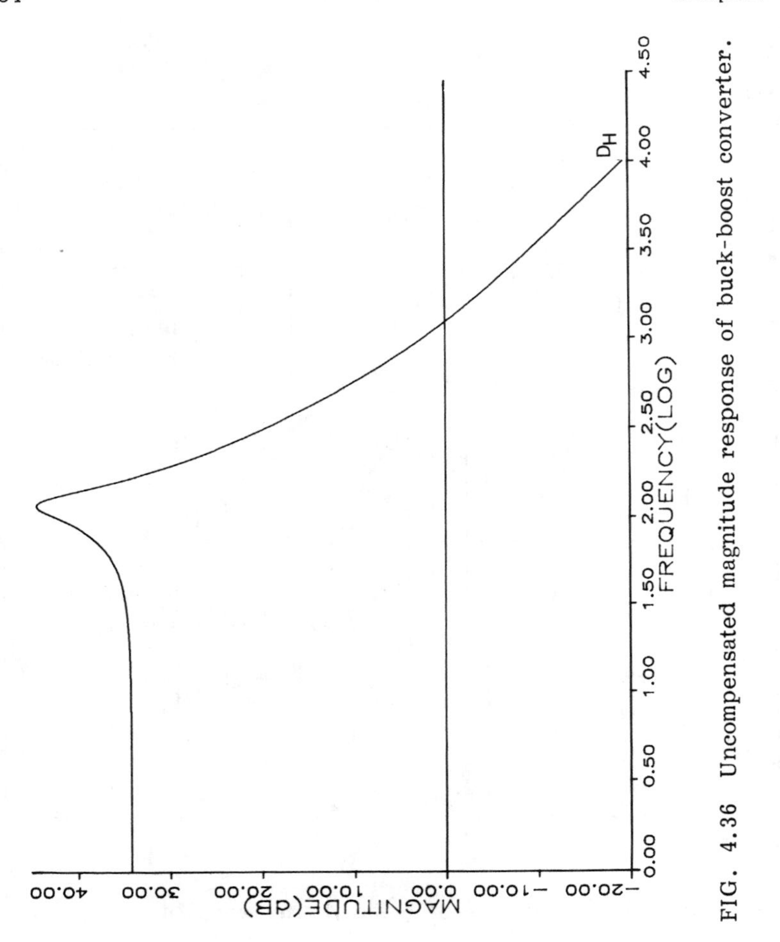

FIG. 4.36 Uncompensated magnitude response of buck-boost converter.

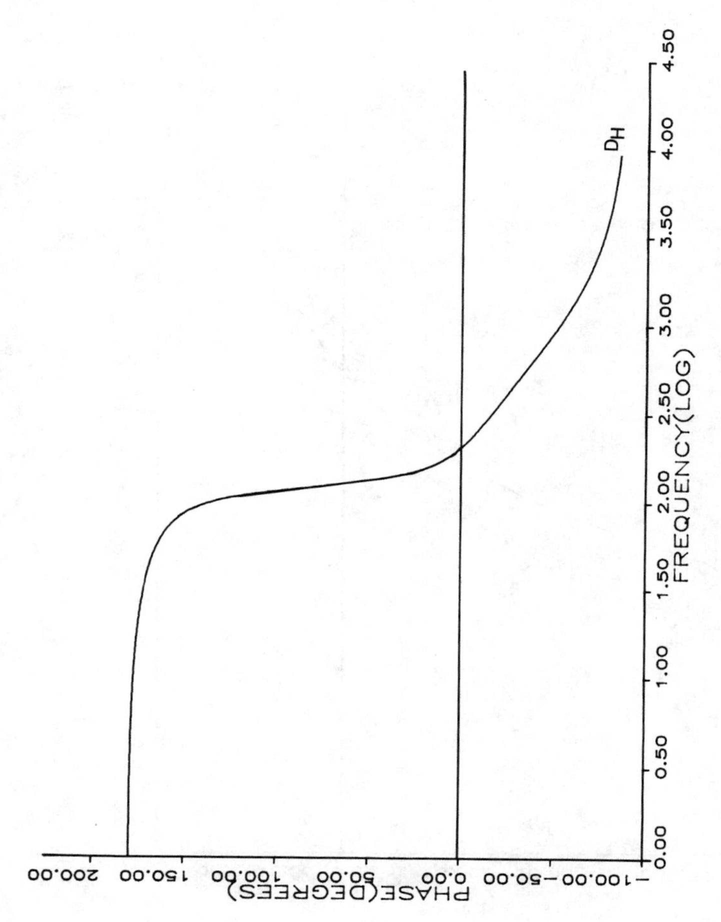

FIG. 4.37 Uncompensated phase response of buck-boost converter.

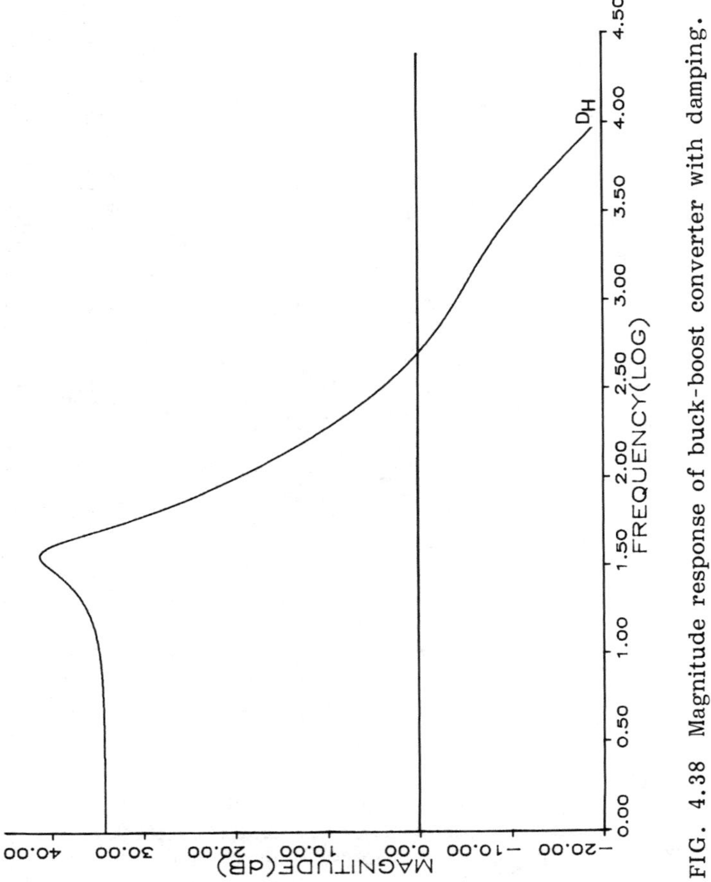

FIG. 4.38 Magnitude response of buck-boost converter with damping.

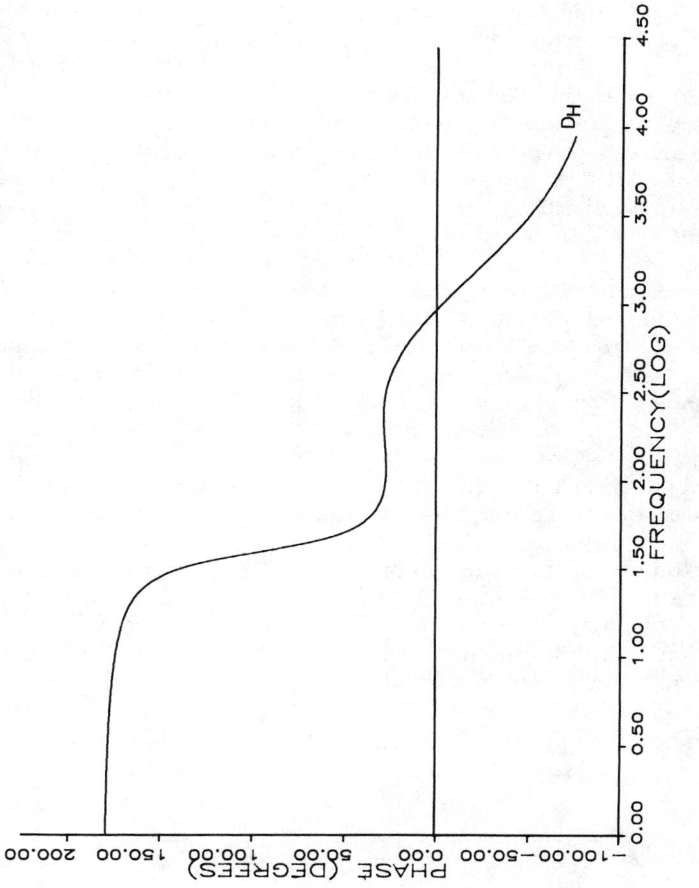

FIG. 4.39 Phase response of buck-boost converter with damping.

typical case in question. The reader will find that the boost and buck-boost converters are not easily amenable to lead compensation because of the inherent nonlinearity of the circuit topology. Recalling now the buck-boost converter example in Chapter 1, the loop gain of this converter is analyzed using SPICE models developed for this purpose [182]. The models used with the Mitel version of SPICE 2F are listed in Appendix E. Figures 4.36 and 4.37 show, respectively, the uncompensated magnitude and phase responses of the buck-boost converter. Since the L_e/C ratio works out to be approximately 12 times, Fig. B.33 is used to estimate the output filter component values. The selection of a damping capacitor nC of 3300 μF and a damping resistor r of 0.25 Ω provided the responses shown in Figs. 4.38 and 4.39. The phase margin is estimated, graphically, at approximately 30°. Note that all this analysis is only an engineering prediction of a tentative design. The final circuit should always be checked with actual measurements. This example provided a gain of about 34 dB at low frequencies. This gain may be adjusted down to a 28- to 30-dB area to ensure that the phase margin stays within known bounds.

If output filter damping is not used in this circuit, one other possible alternative would be to employ dominant pole compensation. If an input filter is added to suppress interference, the interaction of input filter with the feedback loop should also be considered. See Chapter 5 for details.

5

Input Filter Interactions

5.1 SINGLE-STAGE LC FILTER RESPONSE

For power conversion equipment application, the single-stage LC filter is usually designed with a prime consideration for low insertion loss. The inductor is usually designed for a high Q factor and the capacitor chosen with a low esr and high-energy density characteristics. Since the source feeding this filter is usually of low output impedance (a battery bank or other low-output impedance power sources), the view from the output port of the unloaded filter toward the input port is essentially that of a high Q parallel LC tuned circuit. See Fig. 5.1.

The behavior of this circuit is such that the impedance across the LC combination is at a maximum at the resonent frequency of

$$f = \frac{1}{2\pi\sqrt{LC}}$$

The typical responses of a loaded filter of this type are shown in Figs. B.2 and B.3 in Appendix B.

5.2 NATURE OF THE NEGATIVE RESISTANCE OSCILLATOR

Consider the circuit in Fig. 5.2. The dc source responsible for energizing this circuit is omitted, since, in this analysis, only the ac conditions are of interest. Writing equations for i_1 and i_2:

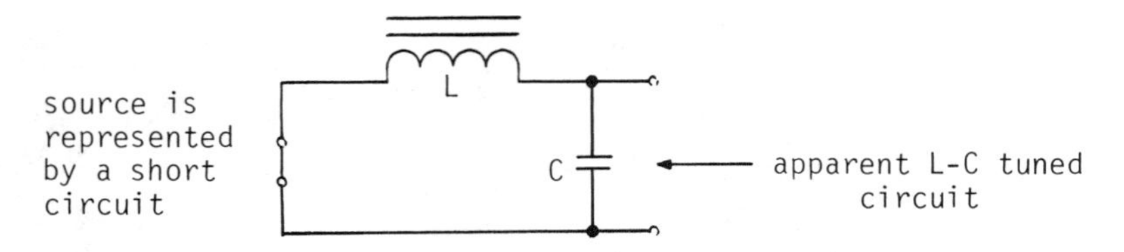

FIG. 5.1 View of the input filter from the output port.

$$(i_1 + i_2)R = \frac{i_2 j}{\omega C} \tag{5.1}$$

$$i_1 R = -i_2 (R - \frac{j}{\omega C}) \tag{5.2}$$

$$0 = R(i_1 + i_2) + i_1(r_L + j\omega L) \tag{5.3}$$

$$Ri_2 = -i_1(R + r_L + j\omega L) \tag{5.4}$$

$$i_1 i_2 R^2 = i_1 i_2 (R - \frac{j}{\omega C})(R + r_L + j\omega L) \tag{5.5}$$

$$i_1 i_2 (Rr_L + j\omega LR - \frac{jR}{\omega C} - \frac{jr_L}{\omega C} + \frac{L}{C}) = 0 \tag{5.6}$$

If the circuit is oscillatory, $i_1 i_2 \neq 0$, and the condition for oscillation is

$$Rr_L + \frac{L}{C} + j\left(\omega LR - \frac{R + r_L}{\omega C}\right) = 0 \tag{5.7}$$

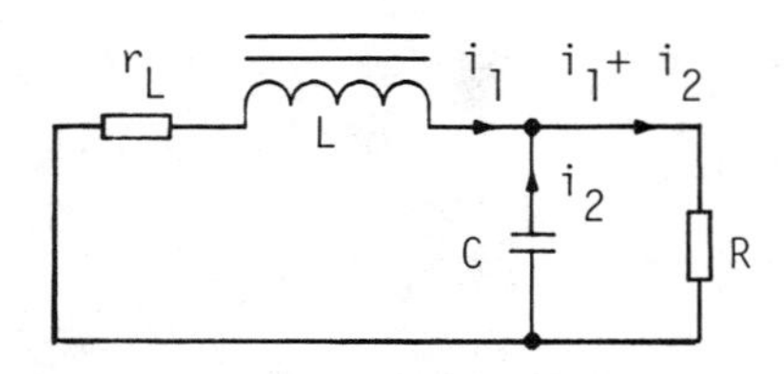

FIG. 5.2 Circuit for analysis.

To maintain oscillation, the real part

$$Rr_L + \frac{L}{C} = 0$$

or

$$R = -\frac{L}{Cr_L} \qquad (5.8)$$

The frequency of oscillation is given by the imaginery part:

$$\omega LR - \frac{R + r_L}{\omega C} = 0$$

$$\omega^2 = \frac{1 + (r_L/R)}{LC} \qquad (5.9)$$

By the substitution of Eq. (5.8) into Eq. (5.9) gives

$$\omega^2 = \frac{1 - (r_L^2 C/L)}{LC}$$

$$= \frac{1}{LC} - \frac{r_L^2}{L}$$

$$f = \frac{1}{2\pi}\sqrt{\frac{1}{LC} - \frac{r_L^2}{L^2}} \qquad (5.10)$$

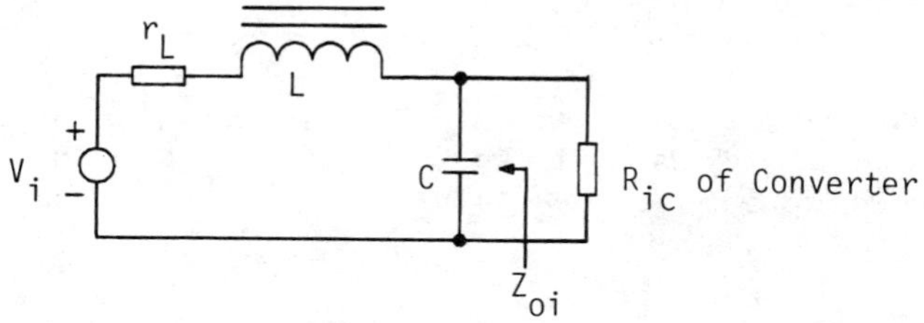

FIG. 5.3 Interaction of R_{io} with Z_{oi} of input filter.

Therefore, if a negative resistance of value $|L/Cr_L|$ is connected across an input filter comprising L and C as shown in Fig. 5.3, oscillation will occur at approximately the resonant (peaking) frequency of the LC filter circuit.

5.3 CONVERTER INPUT IMPEDANCE

For open-loop operation, R_{ic} is always positive; that is, for an increase in input voltage V_i, the system will accept an increased I_i for a constant load R.

In closed-loop operation, the action of the system tends to compensate the line change to maintain a constant power output level (for a constant load R). Under this condition, an increase in input voltage results in a decrease in input current for constant power transfer, assuming no loss in the power conversion process. It is evident, therefore, that with closed-loop operation, the quantity R_{ic} could become negative.

To assure that the output impedance of the input filter would dominate over the combined effect of $R_{ic}//Z_{oi}$, the value of Z_{oi} must be kept well below $|R_{ic}|$.

For a 100% efficient system with a constant load,

$$P_o = P_i = V_i I_i$$

$$R_{ic} = \frac{dV_i}{dI_i} = \frac{d}{dI_i}\left(\frac{P_o}{I_i}\right) = -\frac{P_o}{I_i^2} = -\frac{V_i}{I_i} \tag{5.11}$$

For a converter with a general conversion ratio, $\mu = V_i/V_o = I_o/I_i$, or $V_i = \mu V_o$, and $I_i = I_o/\mu$. Therefore,

$$R_{ic} = -\frac{V_i}{I_i} = -\mu\frac{V_o}{I_i} = -\mu^2\frac{V_o}{I_o} = -\mu^2 R \tag{5.12}$$

This expression relates load R with R_{ic} via the general conversion ratio μ. Table 5.1 tabulates the value of μ for the three different converter configurations [146, 147].

TABLE 5.1 Relationship Between μ and R_{ic}

Conversion ratio	Converter		
	Buck	Boost	Buck-Boost
μ	$\frac{1}{D}$	$1 - D$	$\frac{1 - D}{D}$
$-\mu^2 R$	$-\frac{R}{D^2}$	$-(1 - D)^2 R$	$\frac{-(1 - D)^2 R}{D^2}$

5.4 DESIGN CRITERIA

Equation (5.12) is the low-frequency value of the converter input resistance. Since in most applications the resonant frequency of the input filter is chosen well below that of the output filter to avoid interaction, this expression, though not sufficiently general, is adequate for the establishment of a design criterion; namely, the following relations apply:

$$f_i < f_o \qquad (5.13)$$

where f_i is the input filter resonant frequency and f_o is the output filter resonant frequency;

$$|Z_{oi}|_{max} < |R_{ic}| \qquad (5.14)$$

TABLE 5.2

$H_e(s)$ elements	Buck	Boost	Buck-boost
Ze	$r_L + sL$	$\frac{r_L + sL}{(1 - D)^2}$	$\frac{r_L + sL}{(1 - D)^2}$

A more general exposition of input filter theory can be found in Ref. 146.

In general, if Z_{oi} is made very much smaller than R_{ic}, the stability of the overall system is assured.

If Z_{oi} is made very much smaller than $\mu^2 Z_e$, then the presence of the input filter will not affect the terminal properties of the output filter, where Z_e is tabulated in Table 5.2.

5.5 INPUT FILTER DESIGN CONSIDERATIONS

Without going into the jargons of the orthodox network synthesis, suffice to say that the usual filter requirements for power converters are low-pass filters.

For a given attenuation at a given frequency, it is possible to graphically construct the required filter response first and then calculate or estimate the required element values. For example, a filter is required to produce -18 dB attenuation at a frequency of 100 Hz. The rate of attenuation is not critical.

A single-stage LC filter of this kind has 0 dB of gain/attenuation at zero frequency (dc). The slope of this filter will be -12 dB/octave (or -40 dB/decade).

To construct the asymptotic approximation response of this filter, using logarithmic/linear graph paper.

1. Draw a horizontal line at the 0-dB level.
2. Mark off the point at -18 dB at 100 Hz.
3. Project a line with a slope of -12 dB/octave from the -18 dB point toward the 0-dB level.
4. The line of step 3 cuts the 0-dB level at approximately 340 Hz.
5. The LC product is determined by the formula

$$f_i = \frac{1}{2\pi\sqrt{LC}}$$

6. Refer to appendix B for the optimum decision of the L/C ratio.
7. Calculate L and C.

In general, the resonant frequency of the input filter f_i should be kept at least one decade away from the resonant frequency of the output filter f_o to avoid interaction. In this case, if f_o is 1 kHz, the input filter could be constructed and evaluated as intended. On the other hand, if the output filter f_o is at, say, 200 Hz, then the

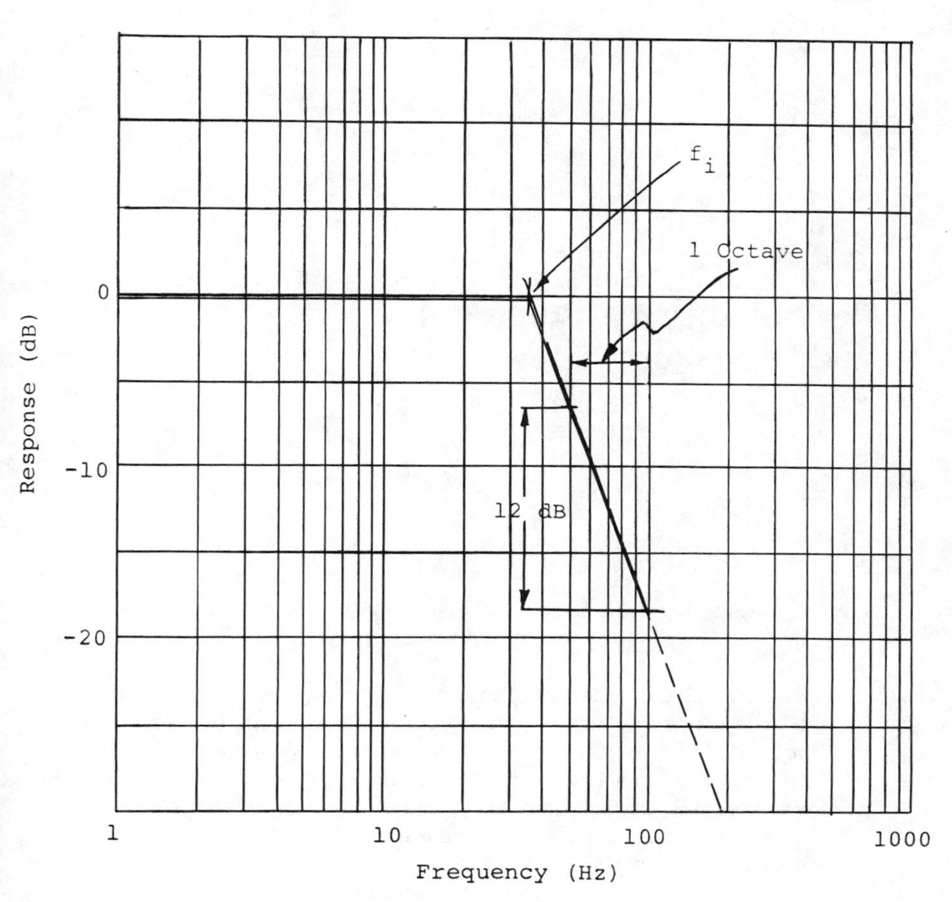

FIG. 5.4 Graphical construction of filter.

input filter frequency f_i should be moved down to around 20 Hz, even though the attenuation characteristic is somewhat overdesigned.

After L and C are determined, damping of the input filter should be considered. If the calculated value of C is already very large, damping may be attempted by using Figs. B.28 to B.30 in Appendix B.

The graphical construction of the sample filter is shown in Fig. 5.4.

More detailed design considerations have been given in Refs. 148-151.

6
Performance Measurement and Evaluation

6.1 INTRODUCTION

For established power supply manufacturers, it is reasonable to assume that they have most of the necessary test equipment to evaluate their products. With the proper equipment, the matter of measuring any particular parameter or component is then reduced to reading the relevant operation manual and following the instruction given for that particular piece of equipment.

This chapter aims to outline the basic approaches required to perform essential measurements by means of a special test set which can be easily made by the reader. With this test set, it is possible to measure the converter input impedance, the output impedance, and the inductance of dc biased inductors.

The closed-loop method of loop gain measurement is illustrated using the basic buck converter as an example.

6.2 THE TEST SET

The primary purpose of the test set is to provide a means of adjusting ac and dc levels and injecting ac signals into the power converter under test at a power level consistent with the maximum output power to be delivered by the converter. The basic components of the test set consist of an adjustable reference voltage for setting the dc test condition, an input terminal to accept and permit the injection of an alternating voltage signal for superimposing onto the

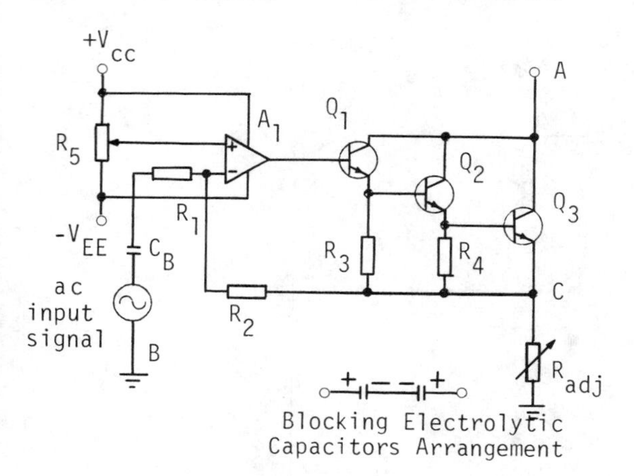

FIG. 6.1 Conceptual schematic of test set.

preset dc level, and a linear power output stage to accommodate the converter or component power requirement. See Fig. 6.1. The voltages $+V_{cc}$ and $-V_{EE}$ are the usual voltage requirement for use with operational amplifiers, typically +18 and -18 V, respectively. Resistor R_5 sets the dc level of the test set. The blocking capacitor C_B should be high in capacitance, typically 100 μF for low-frequency injection. Electrolytic types may be used back to back as shown in Fig. 6.1. Resistor R_{adj} should be of the variable high-wattage type for test current setting. Transistors Q_1 to Q_3 should be able to provide sufficient current gain to drive the required test current, with Q_3 carrying the bulk of the test current. Q_2 and Q_3 could be made up of more than two transistors in parallel with a current-sharing arrangement, if required.

If R_1 is made equal to R_2, the injected ac signal will have an inverted gain of unity. The ac input will appear across points B and C. When an adjustment is made on the ac input level or when first powered up, sufficient time should be allowed for the signal to stabilize before any readings are taken. This is due to the charging time constant of R_1C_B.

6.3 MEASUREMENT OF CONVERTER OUTPUT IMPEDANCE

To ensure that the test set is operating in the linear mode, the dc voltage at point C should be set at approximately half the value of

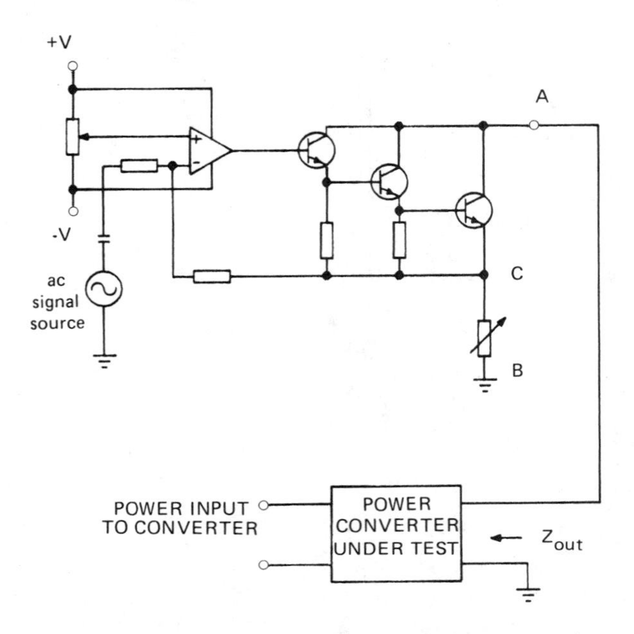

FIG. 6.2 Measurement of converter output impedance.

that at point A. The voltage at point A is supplied by the power converter under test as shown in Fig. 6.2. The test set acts as the load for the power converter.

The output impedance of the power converter under test is calculated by the following relations:

$$\frac{\text{ac voltage at point C}}{R_{adj}} = \frac{\text{ac voltage at point A}}{Z_{out}}$$

Therefore,

$$Z_{out} = \frac{\text{ac voltage at point A}}{\text{ac voltage at point C}} R_{adj} \tag{6.1}$$

With this arrangement, the output impedance of the converter at any frequency is obtainable by varying the frequency of the ac signal source. The setting of the ac signal source should be at a level high enough to provide a few millivolts of ripple at point A at all times. Since voltage ratios are of interest here, an oscilloscope could be used to measure the peak-to-peak voltage values for this exercise.

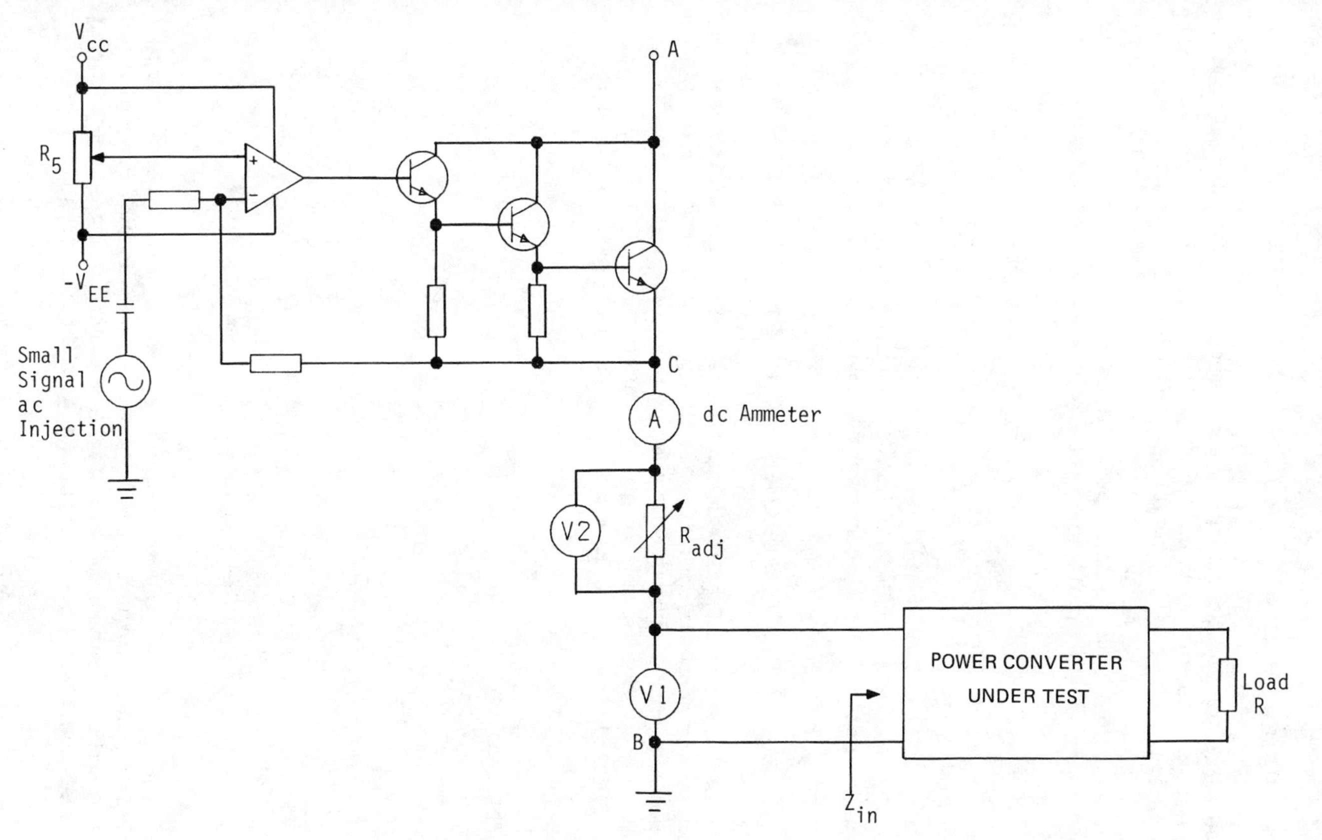

FIG. 6.3 Measurement of converter input impedance.

6.4 MEASUREMENT OF CONVERTER INPUT IMPEDANCE

The setup for input impedance measurement is shown in Fig. 6.3. The input current to the converter under test is monitored by an ammeter. v_1 and v_2 are ac voltage measurements. If an oscilloscope is used, the measurement of v_2 must be made with the oscilloscope isolated from the main power lines.

The converter input impedance is obtained by the following relation:

$$z_{in} = \frac{v_1}{v_2/R_{adj}} = \frac{R_{adj}v_1}{v_2} \tag{6.2}$$

R_{adj} is usually made less than 1 Ω to avoid excessive power consumption and a high component wattage rating.

6.5 INDUCTANCE MEASUREMENT OF DC BIASED INDUCTOR

The arrangement of the inductance measurement for a dc biased inductor is shown in Fig. 6.4. In this procedure the inductor is measured only at one operating frequency and does not require measurement over a wide frequency range.

The inductance is calculated using the measured parameters as follows:

$$\frac{v_2}{v_1} = \frac{R}{\sqrt{R^2 + \omega^2 L^2}}$$

$$\left(\frac{v_2}{v_1}\right)^2 = \frac{R^2}{R^2 + \omega^2 L^2}$$

$$L = \frac{R}{\omega}\sqrt{\left(\frac{v_1}{v_2}\right)^2 - 1} \tag{6.3}$$

In this setup, the inductor takes the place of the converter input terminals as shown in Fig. 6.4. The required dc bias is applied by adjusting R_5 until the ammeter reads the required direct current. The ac signal source is then adjusted to the test frequency at

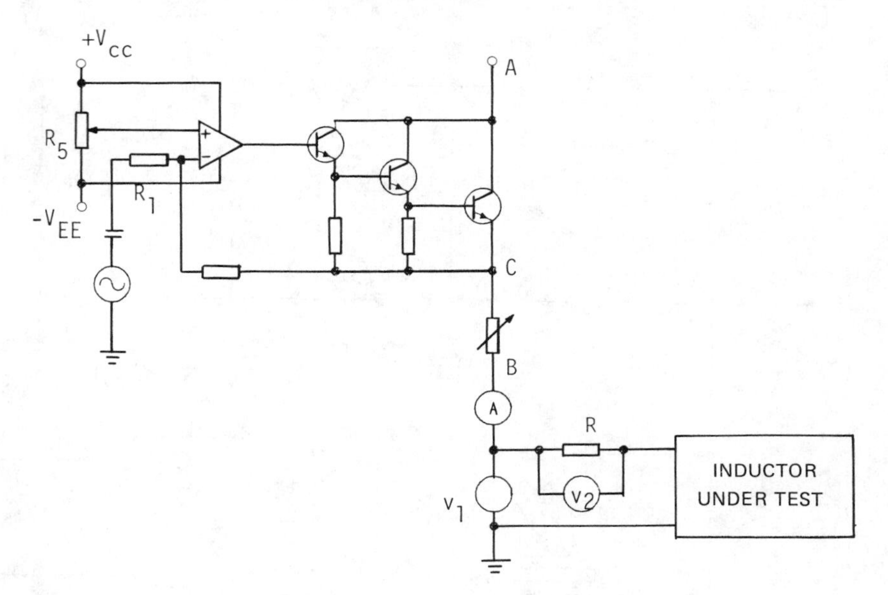

FIG. 6.4 Measurement of dc-biased inductor.

which the inductor is designed to operate. The ac level is then set to obtain a few millivolts of ripple at point A. v_1 and v_2 are then read off an oscilloscope, and the inductance is calculated, using Eq. (6.3).

6.6 MEASUREMENT OF CONVERTER LOOP GAIN

In Chapter 2, the design of the buck converter operating in the continuous conduction mode was outlined in an example in Section 2.6. The theoretical performance of this converter was predicted by means of the calculator program for the buck converter given in Section 2.5. In this section, a method of loop gain measurement of a converter is described.

To ensure stable operation and to check for design discrepancies, it is essential to perform measurements on the gain and phase of the power converter. One method of performing this measurement is to inject a small ac signal into the feedback loop at a suitable location and measure the closed-loop gain of the converter. Figure 6.5 shows an arrangement for loop gain measurement. T_1 is

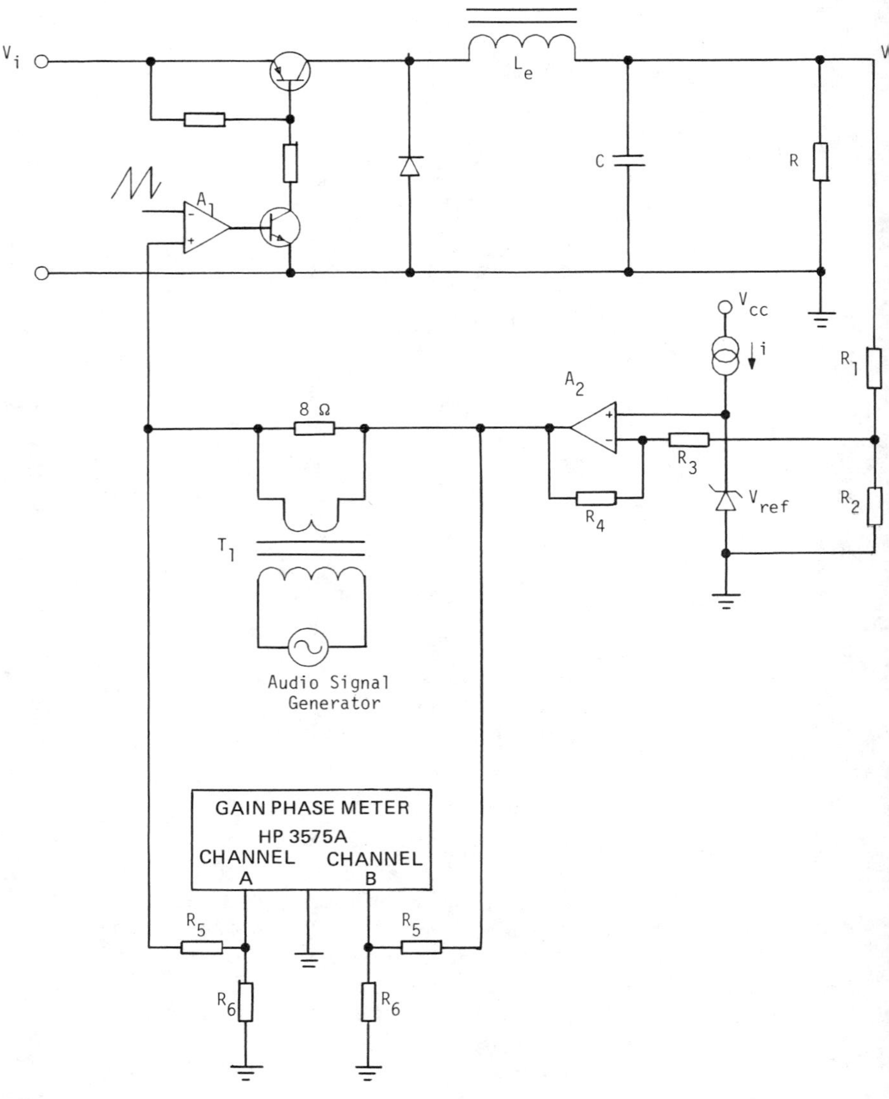

FIG. 6.5 Gain and phase measurement of power converter. (HP3575A is a registered trademark of Hewlett-Packard Company.)

a conventional audio output transformer matched to an 8-Ω loudspeaker load. It should be noted that the bandwidth of this transformer should always be able to accommodate the bandwidth of the converter to be measured.

The output of amplifier A_2 is a convenient and suitable point to break the loop because it is a low-impedance source, and the input to A_1 presents a high input impedance, an ideal condition for voltage injection measurement. The Hewlett-Packard model 3575A* gain/phase meter is capable of making gain and phase measurements under the noisy conditions usually encountered in switching regulators. The voltage divider networks, composed of R_5 and R_6, serve to further reduce unwanted noise to ensure more reliable results.

With the amplitude function knob selected for B/A and the phase reference knob selected for channel A, the loop gain (in decibels) and the phase (in degrees) relationship of the two signals are directly read out from the displays. The audio signal generator is used to set the frequency to provide a number of measurements within the frequency range of interest. The loop gain of a switch mode power converter can also be measured using methods given in Refs. 165 and 166. The closed-loop method of loop gain measurement is particularly useful for high-loop-gain circuits under noisy environments [214, 215].

In the event that the noise within the loop is high enough to render erroneous readout from the gain/phase meter, a selective voltmeter similar to type HP3581C* should be used to obtain meaningful results. The phase information would then be obtainable with methods described in [165, 166].

6.7 TRANSIENT RESPONSE CONSIDERATIONS

The small-signal analysis using state-space averaging techniques provides a convenient method of predicting the stability of regulated power converters. However, under large-signal conditions, such as a step load change, the excitation may be sufficient to start an oscillation for a limited period of time or for an infinite time duration. The nature of this oscillation depends largely on the damping ratio ζ of the system function. A method of obtaining

---

*HP3575A and HP3581C are registered trademarks of Hewlett-Packard Company.

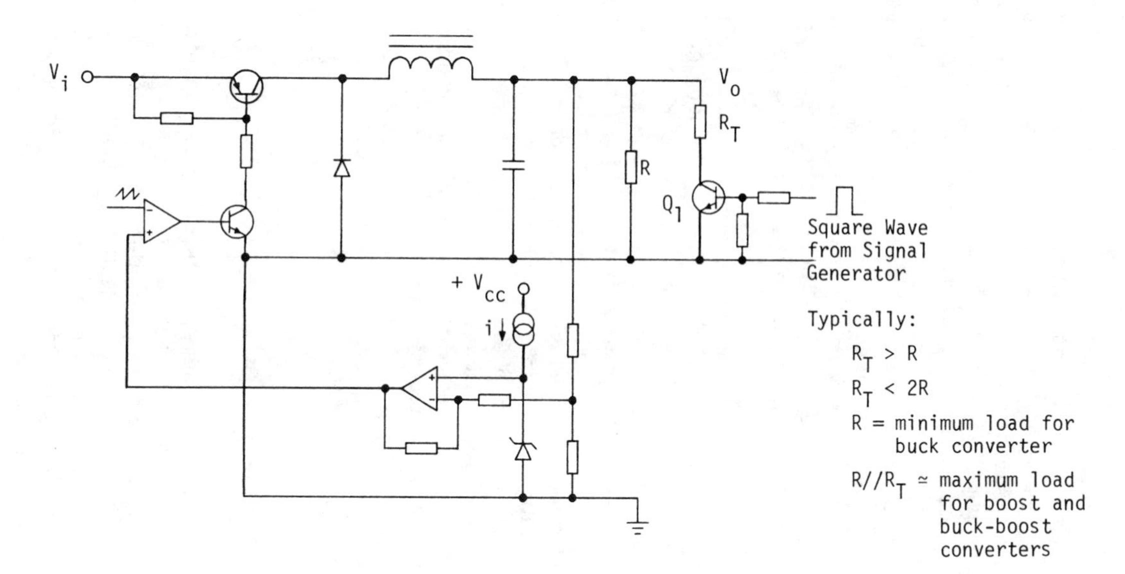

FIG. 6.6 Transient response measurement setup.

the phase margin of a second-order system by means of time response measurement is given in Appendix C. This measurement combined with the state-space small-signal analytical prediction will satisfy most stability design and testing requirements.

In the application of this method to power converters, the arrangement given in Fig. 6.6 is used. The time response is observed using an oscilloscope with probe connected to the output terminals of the converter as shown in Fig. 6.6.

6.8 PREDICTION AND EVALUATION OF CONVERTER EFFICIENCY

In the process of assessing the efficiency of a power converter, it is common practice to measure the input power and the output power to obtain an overall efficiency indication. However, sometimes it is necessary to assess the efficiency of a particular part of the converter to determine if an "optimum" performance for that part of the converter has been achieved. In that event, it becomes necessary to dissect the converter into many parts with an efficiency figure assigned to each part as a design target. This concept is illustrated in Fig. 6.7 using a push-pull converter as an example.

For

$P_o = 500$ W
$V_o = 100$ V
$I_o = 5$ A

Assume a loss of 20 W in the output filter; then

$$P_i = 520 \text{ W}$$

For a 1-V forward voltage drop across the output diodes, the diode loss is 5W; therefore,

$$P_2 = 525 \text{ W}$$

Assume a 96% efficiency for the transformer T_1:

$$P_3 = \frac{P_2}{0.96} = \frac{525}{0.96} = 547 \text{ W}$$

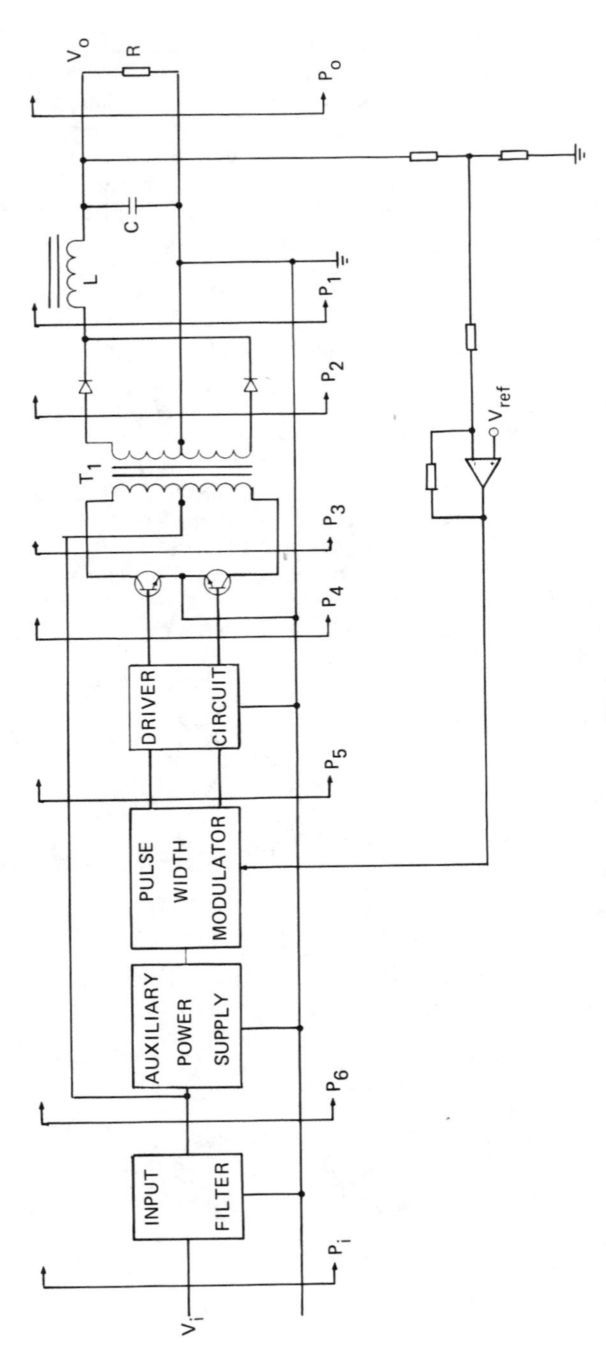

FIG. 6.7 Prediction and evaluation of converter efficiency.

Similarly, if the combined power requirement of the power switches, the driver circuit, the pulse width modulator, and the auxiliary power supply is, say, 40 W, then

$$P_6 = P_3 + 40\ W = 587\ W$$

Assume the input filter dissipates 30 W; then

$$P_i = P_6 + 30 = 587 + 30 = 617\ W$$

The predicted converter efficiency using the preceding assumptions is

$$\frac{P_o}{P_i} \times 100\% = \frac{500}{617} \times 100\% = 81\%$$

Since the input voltage is always known, the input current can be predicted using the preceding power budgeting procedure. Power measurements are then made with an actual breadboard to verify all design targets.

appendix A
State-Space Averaging Analysis

The process of state-space averaging is best explained by first writing the linear differential equations of the power stage for the ON and OFF states, and by then formulating the general matrix equations. The following analysis assumes that the corner frequency of the output low-pass filter is at least a decade or two below the frequency corresponding to half the switching frequency.

For the buck power stage (see Fig. A.1) operating in the continuous conduction mode, assume $R \gg (r + r_L)$; then terms containing R, r, and r_L would assume the value of R only:

For the ON condition (see Fig. A.2a):

$$L \frac{di_L}{dt} = -(r + r_L)i_L - v + v_i \tag{A.1}$$

$$C \frac{dv}{dt} = i_L - \frac{V}{R} \tag{A.2}$$

$$v_o = ri_L + v \tag{A.3}$$

For the OFF condition (see Fig. A.2b):

$$L \frac{di_L}{dt} = -(r + r_L)i_L - v \tag{A.4}$$

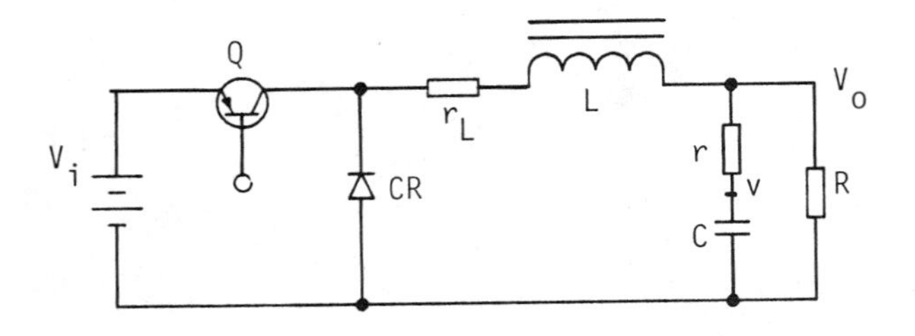

FIG. A.1 Buck power stage.

$$C \frac{dv}{dt} = i_L - \frac{v}{R} \tag{A.5}$$

$$v_o = ri_L + v \tag{A.6}$$

Rearranging the ON state equations, we obtain

$$\frac{di_L}{dt} = -\frac{1}{L}(r + r_L)i_L - \frac{v}{L} + \frac{v_i}{L} \tag{A.7}$$

$$\frac{dv}{dt} = \frac{i_L}{C} - \frac{v}{CR} \tag{A.8}$$

$$v_o = ri_L + v \tag{A.9}$$

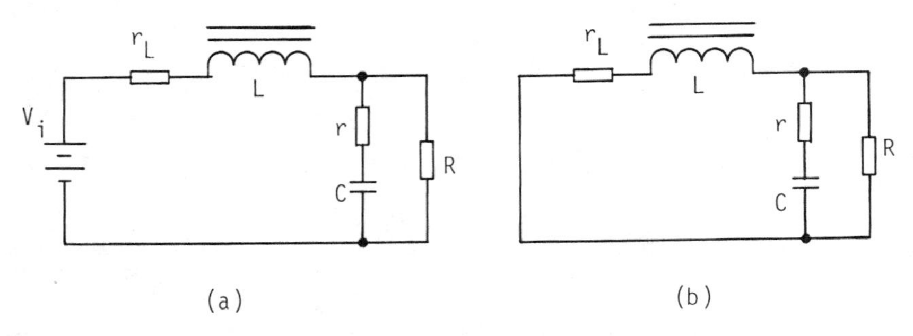

FIG. A.2 Buck power stage topological behavior during ON and OFF states. (a) ON condition. (b) OFF condition.

Designate the state vector as

$$\dot{x} = \begin{bmatrix} \frac{di_L}{dt} \\ \\ \frac{dv}{dt} \end{bmatrix} \tag{A.10}$$

and by similar manipulation of the OFF state equations, two new matrix equations result: For the ON state,

$$\dot{x} = \begin{bmatrix} \frac{di_L}{dt} \\ \\ \frac{dv}{dt} \end{bmatrix} = \begin{bmatrix} -\frac{1}{L}(r + r_L) & -\frac{1}{L} \\ \\ \frac{1}{C} & -\frac{1}{CR} \end{bmatrix} \begin{bmatrix} i_L \\ \\ v \end{bmatrix} + \begin{bmatrix} \frac{1}{L} \\ \\ 0 \end{bmatrix} v_i \tag{A.11}$$

$$v_o = C_1^T x = [r\ 1]x \tag{A.12}$$

and for the OFF state,

$$\dot{x} = \begin{bmatrix} \frac{di_L}{dt} \\ \\ \frac{dv}{dt} \end{bmatrix} = \begin{bmatrix} -\frac{1}{L}(r + r_L) & -\frac{1}{L} \\ \\ \frac{1}{C} & -\frac{1}{CR} \end{bmatrix} \begin{bmatrix} i_L \\ \\ v \end{bmatrix} + \begin{bmatrix} 0 \\ \\ 0 \end{bmatrix} v_i \tag{A.13}$$

$$v_o = C_2^T x = (r,\ 1)x \tag{A.14}$$

Rewrite Eqs. (A.11) to (A.14) in the form:

ON

$$\dot{x} = A_1 x + b_1 v_i \tag{A.15}$$

$$v_o = C_1^T x \tag{A.16}$$

OFF

$$\dot{x} = A_2 x + B_2 v_i \tag{A.17}$$

$$v_o = C_2^T x \tag{A.18}$$

where

$$A_1 = \begin{bmatrix} -\frac{1}{L}(r + r_L) & -\frac{1}{L} \\ \frac{1}{C} & -\frac{1}{CR} \end{bmatrix} \tag{A.19}$$

$$A_2 = \begin{bmatrix} -\frac{1}{L}(r + r_L) & -\frac{1}{L} \\ \frac{1}{C} & -\frac{1}{CR} \end{bmatrix} \tag{A.20}$$

$$b_1 = \begin{bmatrix} \frac{1}{L} \\ 0 \end{bmatrix} \tag{A.21}$$

$$b_2 = \begin{bmatrix} 0 \\ 0 \end{bmatrix} \tag{A.22}$$

$$C_1^T = (r,\ 1) \tag{A.23}$$

$$C_2^T = (r,\ 1) \tag{A.24}$$

Now, by averaging over one period, $(d + d_o)T = T$ or $d + d_o = 1$,

$$\dot{x} = (dA_1 + d_o A_2)x + (db_1 + d_o b_2)v_i \tag{A.25}$$

$$v_o = (dC_1^T + d_o C_2^T)x \tag{A.26}$$

or

$$\dot{x} = Ax + bv_i \tag{A.27}$$

$$v_o = C^T x \tag{A.28}$$

where

$$A = dA_1 + d_o A_2$$

$$= \begin{bmatrix} -\frac{r + r_L}{L} & -\frac{1}{L} \\ \frac{1}{C} & -\frac{1}{CR} \end{bmatrix} \tag{A.29}$$

$$b = db_1 + d_o b_2 = \begin{bmatrix} \frac{d}{L} \\ 0 \end{bmatrix} \tag{A.30}$$

$$C^T = dC_1^T + d_o C_2^T = (r, 1) \tag{A.31}$$

Perturbation is achieved by the introduction of line and duty cycle variations $\hat{v}_i$ and $\hat{d}$, respectively:

$$v_i = v_i + \hat{v}_i \tag{A.32}$$

$$d = D + \hat{d} \tag{A.33}$$

$$d_o = D_o - \hat{d} \tag{A.34}$$

$$x = X + \hat{x} \tag{A.35}$$

$$v_o = V_o + \hat{v}_o \tag{A.36}$$

Therefore, introduction of perturbation to Eq. (A.25) gives

$$\dot{\hat{x}} = [(D + \hat{d})A_1 + (D_o - \hat{d})A_2](X + \hat{x})$$

$$+ [(D + \hat{d})b_1 + (D_o - \hat{d})b_2](V_i + \hat{v}_i)$$

$$= (A_1D + A_1\hat{d} + D_oA_2 - \hat{d}A_2)(X + \hat{x})$$

$$+(Db_1 + \hat{d}b_1 + D_ob_2 - \hat{d}b_2)(V_i + \hat{v}_i)$$

$$= A_1DX + A_1\hat{d}X + D_oA_2X - \hat{d}A_2X + A_1D\hat{x} + A_1\hat{d}\hat{x} + D_oA_2\hat{x}$$

$$- \hat{d}A_2\hat{x} + Db_1V_i + D_ob_2V_i - \hat{d}b_2V_i + Db_1\hat{v}_i + \hat{d}b_1\hat{v}_i$$

$$+ D_ob_2\hat{v}_i - \hat{d}b_2\hat{v}_i + \hat{d}b_1V_i$$

$$\dot{\hat{x}} = (AX + bV_i) + (A\hat{x} + b\hat{v}_i) + [(A_1 - A_2)X + (b_1 - b_2)V_i]\hat{d}$$

$$+ [(A_1 - A_2)\hat{x} + (b_1 - b_2)\hat{v}_i]\hat{d} \qquad \text{(A.37)}$$

Note that for $\hat{d} = 0$, $D = d$, $D_o = d_o$ for the steady-state condition.

Examination of Eq. (A.37) reveals that

$AX + bV_i$ is the steady-state term.
$A\hat{x} + b\hat{v}_i$ is the line variation term.
$[(A_1 - A_2)X + (b_1 - b_2)V_i]\hat{d}$ is the duty cycle variation term.
$[(A_1 - A_2)\hat{x} + (b_1 - b_2)\hat{v}_i]\hat{d}$ is the nonlinear term.

Therefore,

$$\hat{v}_o = \underbrace{C^T\hat{x}}_{\text{line}} + \underbrace{(C_1^T - C_2^T)X\hat{d}}_{\text{duty cycle}} + \underbrace{(C_1^T - C_2^T)\hat{x}\hat{d}}_{\text{nonlinear}} \qquad \text{(A.38)}$$

In this analysis, the perturbation is assumed to be very small such that the perturbed value is very much less than the steady-state value, and only the first-order linear perturbation terms are kept to obtain the linearized small-signal model, i.e.,

$$V_i >> v_i, \quad D >> d, \quad x >> x$$

it follows that the nonlinear terms in Eqs. (A.37) and (A.38) would be extremely small, and dropping these terms would not significantly affect the overall accuracy of the analysis, but would effectively linearize the system.

Now, from the steady-state term of Eq. (A.37),

$$X = -A^{-1}bV_i$$

and

$$V_o = C^TX, \quad \text{steady-state form of Eq. (A.28)}$$

$$= -C^TA^{-1}bV_i$$

or

$$\frac{V_o}{V_i} = -C^TA^{-1}b \quad \textit{steady-state dc transfer function} \tag{A.39}$$

Considering only the line and duty cycle terms, Eq. (A.37) gives

$$\dot{\hat{x}} = A\hat{x} + b\hat{v}_i + [(A_1 - A_2)X + (b_1 - b_2)V_i]\hat{d} \tag{A.40}$$

and Eq. (A.38) gives

$$\hat{v}_o = C^T\hat{x} + (C_1^T - C_2^T)X\hat{d} \tag{A.41}$$

To obtain the line to output transfer function for small signals, put $\hat{d} = 0$ in Eqs. (A.40) and (A.41), then

$$\dot{\hat{x}} = A\hat{x} + b\hat{v}_i \tag{A.42}$$

$$\hat{v}_o = C^T\hat{x} \tag{A.43}$$

Taking the Laplace transform of Eq. (A.42) and setting all initial conditions to zero,

$$s\hat{x}(s) = A\hat{x}(s) + b\hat{v}_i(s) \tag{A.44}$$

$$(sI - A)\hat{x}(s) = b\hat{v}_i(s) \tag{A.45}$$

$$\frac{\hat{x}(s)}{\hat{v}_i(s)} = (sI - A)^{-1}b \tag{A.46}$$

where I is an identity or unit matrix, and from Eq. (A.43),

$$\hat{v}_o(s) = C^T\hat{x}(s) \tag{A.47}$$

Therefore,

$$\frac{\hat{v}_o(s)}{\hat{x}(s)} = C^T \tag{A.48}$$

$$\frac{\hat{v}_o(s)}{\hat{v}_i(s)} = C^T (sI - A)^{-1}b \quad \textit{small-signal line to output transfer function} \tag{1.49}$$

To calculate the duty cycle to output transfer function, put $\hat{v}_i = 0$ in Eqs. (A.40) and (A.41); then

$$\dot{\hat{x}} = A\hat{x} + [(A_1 - A_2)X + (b_1 - b_2)V_i]\hat{d}$$

$$\hat{v}_o = C^T\hat{x} + (C_1^T - C_2^T)X\hat{d}$$

$$\frac{\hat{x}(s)}{\hat{d}(s)} = (sI - A)^{-1}[(A_1 - A_2)X + (b_1 - b_2)V_i] \tag{A.50}$$

$$\frac{\hat{v}_o(s)}{\hat{d}(s)} = C^T(sI - A)^{-1}[(A_1 - A_2)X + (b_1 - b_2)V_i] + (C_1^T - C_2^T)X \tag{A.51}$$

Equations (A.1) to (A.14) show a procedure of setting up the state matrices for the case of the buck power stage operating in the continuous conduction mode.

Equations (A.15) to (A.18), (A.25) to (A.28), and (A.32) to (A.51) are general state-space equations applicable to all basic power stage configurations.

Equation (A.51) is the duty cycle to output transfer function.

To obtain the steady-state dc transfer function for the buck power stage operating in the continuous conduction mode, Eq. (A.39) is used:

$$\frac{V_o}{V_i} = -c^T A^{-1} b$$

$$= -(r, 1) \begin{bmatrix} \frac{r_L + r}{L} & -\frac{1}{L} \\ \frac{1}{C} & -\frac{1}{CR} \end{bmatrix}^{-1} \begin{bmatrix} \frac{d}{L} \\ 0 \end{bmatrix} \qquad (A.52)$$

It will be observed that in Eq. (A.52) the first term is essentially that of Eq. (A.31), the second term is the inverse of Eq. (A.29), and the third term is Eq. (A.30).

It is also recalled [206] that for a matrix in the form of

$$A = \begin{bmatrix} a_{11} & a_{12} \\ a_{21} & a_{22} \end{bmatrix}$$

its inverse is equal to

$$A^{-1} = \frac{1}{a_{11}a_{22} - a_{12}a_{21}} \begin{bmatrix} a_{22} & -a_{12} \\ -a_{21} & a_{11} \end{bmatrix}$$

Therefore,

$$\frac{V_o}{V_i} = \frac{-(r, 1)}{\frac{r_L + r}{L}\frac{1}{CR} + \frac{1}{LC}} \begin{bmatrix} -\frac{1}{CR} & \frac{1}{L} \\ -\frac{1}{C} & -\frac{r_L + r}{L} \end{bmatrix} \begin{bmatrix} \frac{d}{L} \\ 0 \end{bmatrix} \qquad (A.53)$$

$$= \frac{-(r, 1)}{\frac{r_L + r}{L}\frac{1}{CR} + \frac{1}{LC}} \begin{bmatrix} \frac{-d}{LCR} \\ \frac{-d}{LC} \end{bmatrix} = \frac{dr + dR}{r_L + r + R}$$

$$\frac{V_o}{V_i} = d\frac{r + R}{r_L + r + R} \cong D \tag{A.54}$$

Similarly, the small-signal line to output transfer function for the buck converter is [from Eq. (A.49)]

$$\frac{\hat{v}_o(s)}{\hat{v}_i(s)} = C^T(sI - A)^{-1}b$$

$$= (r,\ 1)\begin{bmatrix} s + \frac{r_L + r}{L} & \frac{1}{L} \\ -\frac{1}{C} & s + \frac{1}{CR} \end{bmatrix}^{-1}\begin{bmatrix} \frac{D}{L} \\ 0 \end{bmatrix}$$

$$= \frac{(r,\ 1)}{\left(s + \frac{r_L + r}{L}\right)\left(s + \frac{1}{CR}\right) + \frac{1}{LC}}\begin{bmatrix} s + \frac{1}{CR} & -\frac{1}{L} \\ \frac{1}{C} & s + \frac{r_L + r}{L} \end{bmatrix}\begin{bmatrix} \frac{D}{L} \\ 0 \end{bmatrix}$$

$$= \frac{(r,\ 1)}{\left(s + \frac{r_L + r}{L}\right)\left(s + \frac{1}{CR}\right) + \frac{1}{LC}}\begin{bmatrix} \left(s + \frac{1}{CR}\right)\frac{D}{L} \\ \frac{D}{LC} \end{bmatrix}$$

$$= \frac{\frac{Dr}{L}\left(s + \frac{1}{CR}\right) + \frac{D}{LC}}{s^2 + \left(\frac{r_L + r}{L} + \frac{1}{CR}\right)s + \frac{r_L + r}{LCR} + \frac{1}{LC}}$$

$$= \frac{D\left[\frac{r}{L}\left(s + \frac{1}{CR}\right) + \frac{1}{LC}\right]}{s^2 + \left(\frac{r_L + r}{L} + \frac{1}{CR}\right)s + \frac{r_L + r}{LCR} + \frac{1}{LC}}$$

$$= \frac{D}{LC}\,\frac{rC\left(s + \frac{1}{CR}\right) + 1}{s^2 + s\left(\frac{r_L + r}{L} + \frac{1}{CR}\right) + \frac{r_L + r}{LCR} + \frac{1}{LC}}$$

$$\frac{\hat{v}_o(s)}{\hat{v}_i(s)} = D\,\frac{1}{LC}\,\frac{rCs + \frac{r}{R} + 1}{s^2 + s\left(\frac{r_L + r}{L} + \frac{1}{CR}\right) + \frac{r_L + r}{LCR} + \frac{1}{LC}} \tag{A.55}$$

from which the first term is identified as the steady-state dc transfer function and the rest of the expression constitutes the effective output filter transfer function $H_e(s)$.

From Eq. (A.51), the duty cycle to output small-signal transfer function is

$$\frac{\hat{v}_o(s)}{\hat{d}(s)} = C^T(sI - A)^{-1}[(A_1 - A_2)X + (b_1 - b_2)V_i] + (C_1^T - C_2^T)X$$

$$= (r,\ 1)\begin{bmatrix} s + \frac{1}{L}(r_L + r) & \frac{1}{L} \\ -\frac{1}{C} & s + \frac{1}{CR} \end{bmatrix}^{-1}\begin{bmatrix} \frac{1}{L} \\ 0 \end{bmatrix} V_i + 0$$

$$= \frac{V_i(r,\ 1)}{K_o}\begin{bmatrix} s + \frac{1}{CR} & -\frac{1}{L} \\ \frac{1}{C} & s + \frac{r_L + r}{L} \end{bmatrix}\begin{bmatrix} \frac{1}{L} \\ 0 \end{bmatrix}$$

where

$$K_o = s^2 + s\left(\frac{r_L + r}{L} + \frac{1}{CR}\right) + \frac{r_L + r}{LCR} + \frac{1}{LC}$$

$$\frac{\hat{v}_o(s)}{\hat{d}(s)} = \frac{V_i[r\ 1]}{K_o}\begin{bmatrix} \left(s + \frac{1}{CR}\right)\frac{1}{L} \\ \frac{1}{LC} \end{bmatrix}$$

$$= \frac{V_i\left[\frac{r}{L}\left(s + \frac{1}{CR}\right) + \frac{1}{LC}\right]}{K_o} = \frac{V_i\,\frac{rC(s + 1/CR) + 1}{LC}}{K_o}$$

$$\frac{\hat{v}_o(s)}{\hat{d}(s)} = \frac{V_o}{D}\frac{1}{LC}\frac{rCs + \frac{r}{R} + 1}{s^2 + s\left(\frac{r_L + r}{L} + \frac{1}{CR}\right) + \frac{r_L + r}{LCR} + \frac{1}{LC}} \tag{A.56}$$

$$\frac{\hat{v}_o(s)}{\hat{d}(s)} = \frac{V_o}{D} H_e(s) \tag{A.57}$$

By adopting the previous analysis procedures, the steady-state dc transfer function for the boost power stage (see Fig. A.3) can be derived as follows:

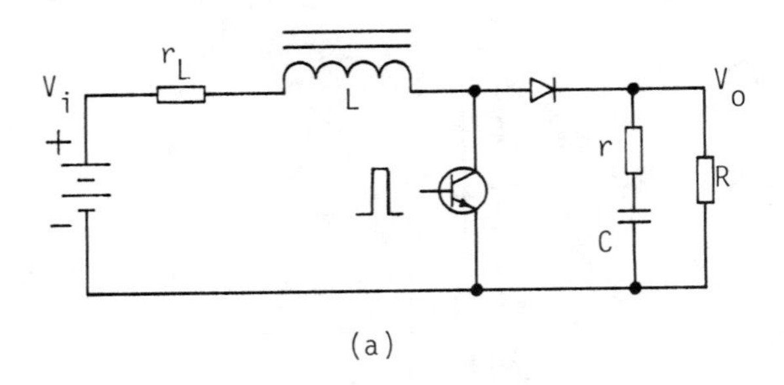

(a)

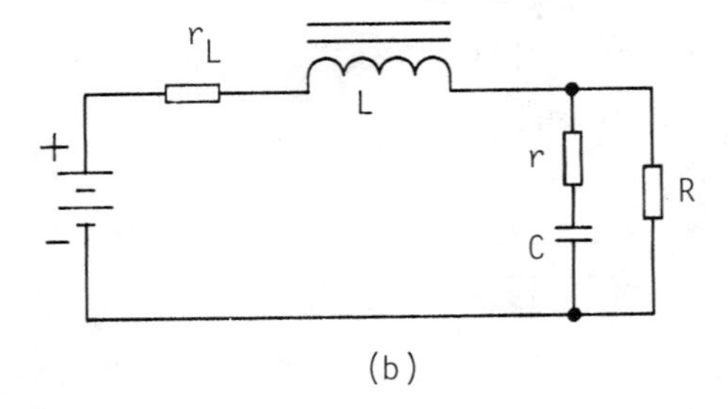

(b)

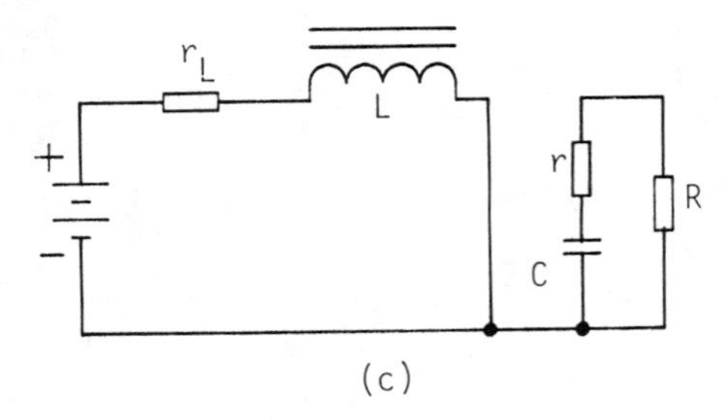

(c)

FIG. A.3 Boost converter. (a) Basic circuit. (b) OFF condition. (c) ON condition.

$$\frac{V_o}{V_i} = -c^T A^{-1} b \tag{A.39}$$

$$= -\begin{bmatrix} D_o \dfrac{rR}{r+R} & \dfrac{R}{r+R} \end{bmatrix} \begin{bmatrix} \dfrac{r_L + D_o \dfrac{rR}{r+R}}{-L} & \dfrac{-D_o R}{L(r+R)} \\[2ex] \dfrac{D_o R}{(r+R)C} & \dfrac{-1}{(r+R)C} \end{bmatrix}^{-1} \begin{bmatrix} \dfrac{1}{L} \\[2ex] 0 \end{bmatrix}$$

$$\frac{V_o}{V_i} = \frac{-\begin{bmatrix} D_o \dfrac{rR}{r+R} & \dfrac{R}{r+R} \end{bmatrix}}{\left\{\dfrac{r_L + D_o[rR/(r+R)]}{L}\right\}\left[\dfrac{1}{(r+R)C}\right] + \dfrac{D_o^2 R^2}{LC(r+R)^2}}$$

$$\times \begin{bmatrix} \dfrac{-1}{(r+R)C} & \dfrac{D_o R}{L(r+R)} \\[2ex] \dfrac{-D_o R}{(r+R)C} & \dfrac{r_L + \dfrac{D_o rR}{r+R}}{-L} \end{bmatrix} \begin{bmatrix} \dfrac{1}{L} \\[2ex] 0 \end{bmatrix}$$

$$= \frac{-\begin{bmatrix} \dfrac{D_o rR}{r+R} & \dfrac{R}{r+R} \end{bmatrix}}{\dfrac{r_L + D_o[rR/(r+R)]}{LC(r+R)} + \dfrac{D_o^2 R^2}{LC(r+R)^2}} \begin{bmatrix} \dfrac{-1}{LC(r+R)} \\[2ex] \dfrac{-D_o R}{LC(r+R)} \end{bmatrix}$$

$$= \frac{D_o rR + D_o R^2}{r_L(r+R) + D_o rR + D_o^2 R^2}$$

$$\frac{V_o}{V_i} = \frac{1}{D_o} \frac{R}{\dfrac{r_L}{D_o^2} + \dfrac{r//R}{D_o} + \dfrac{R^2}{r+R}} \tag{A.58}$$

$$\frac{V_o}{V_i} \simeq \frac{1}{D_o} \tag{A.59}$$

The small-signal line to output transfer function of the boost converter, from Eq. (A.49), is

$$\frac{\hat{v}_o(s)}{\hat{v}_i(s)} = c^T(sI - A)^{-1}b \tag{A.49}$$

$$= (D_o r, 1)\begin{bmatrix} s + \dfrac{r_L + D_o r}{L} & \dfrac{D_o}{L} \\ -\dfrac{D_o}{C} & s + \dfrac{1}{CR} \end{bmatrix}^{-1}\begin{bmatrix} \dfrac{1}{L} \\ 0 \end{bmatrix}$$

$$= \frac{(D_o r, 1)}{\left(s + \dfrac{r_L + D_o r}{L}\right)\left(s + \dfrac{1}{CR}\right) + \dfrac{D_o^2}{LC}}$$

$$\times \begin{bmatrix} s + \dfrac{1}{CR} & -\dfrac{D_o}{L} \\ \dfrac{D_o}{C} & s + \dfrac{r_L + D_o r}{L} \end{bmatrix}\begin{bmatrix} \dfrac{1}{L} \\ 0 \end{bmatrix}$$

$$= \frac{1}{K_1}(D_o r, 1)\begin{bmatrix} \dfrac{1}{L}\left(s + \dfrac{1}{CR}\right) \\ \dfrac{D_o}{LC} \end{bmatrix}$$

where

$$K_1 = \left(s + \frac{r_L + D_o r}{L}\right)\left(s + \frac{1}{CR}\right) + \frac{D_o^2}{LC}$$

$$= \frac{(D_o r/L)[s + (1/CR)] + D_o/LC}{K_1}$$

$$= \frac{D_o}{LC}\,\frac{srC + (r/R) + 1}{K_1}$$

and substituting L_e for $L/(1 - D)^2$

$$\frac{\hat{v}_o(s)}{\hat{v}_i(s)} = \frac{1}{D_o L_e C} \tag{A.60}$$

$$\times \frac{srC + (r/R) + 1}{s^2 + s\left(\dfrac{(r_L/D_o^2) + (r/D_o)}{L_e} + \dfrac{1}{CR}\right) + \dfrac{(r_L/D_o^2) + (r/D_o)}{L_e CR} + \dfrac{1}{L_e C}}$$

$$= \frac{1}{D_o} H_e(s) \tag{A.61}$$

For the derivation of the duty cycle to output small-signal transfer function of the boost converter, the following matrices will be used:

$$(A_1 - A_2) = \begin{bmatrix} \dfrac{r}{L} & \dfrac{1}{L} \\ \dfrac{-1}{C} & 0 \end{bmatrix} \tag{A.62}$$

$$(b_1 - b_2) = \begin{bmatrix} 0 \\ 0 \end{bmatrix} \tag{A.63}$$

$$c_1^T - c_2^T = (-r,\ 0) \tag{A.64}$$

$$c^T(sI - A)^{-1} = \frac{1}{K_1}\left[D_o r\left(s + \frac{1}{CR}\right) + \frac{D_o}{C} \quad - \frac{D_o^2 r}{L} \right.$$
$$\left. + s + \frac{r_L + D_o r}{L}\right] \tag{A.65}$$

$$X = -A^{-1}bV_i = \frac{V_i}{[(r_L + D_o r)/R] + D_o^2} \begin{bmatrix} \dfrac{1}{R} \\ D_o \end{bmatrix} \tag{A.66}$$

For $R \gg (r_L + D_o r)$,

$$X \cong \frac{V_i}{D_o^2}\begin{bmatrix}\frac{1}{R}\\ D_o\end{bmatrix} = \frac{V_o}{D_o}\begin{bmatrix}\frac{1}{R}\\ D_o\end{bmatrix} = V_o\begin{bmatrix}\frac{1}{D_o R}\\ 1\end{bmatrix} \quad (A.67)$$

$$\frac{\hat{v}_o(s)}{\hat{d}(s)} = \frac{[D_o r\ 1]}{K_1}\begin{bmatrix} s + \frac{1}{CR} & -\frac{D_o}{L} \\ \frac{D_o}{C} & s + \frac{r_L + D_o r}{L}\end{bmatrix}\begin{bmatrix}\frac{r}{L} & \frac{1}{L}\\ -\frac{1}{C} & 0\end{bmatrix}$$

$$\times V_o\begin{bmatrix}\frac{1}{D_o R}\\ 1\end{bmatrix} + [-r\ 0]V_o\begin{bmatrix}\frac{1}{D_o R}\\ 1\end{bmatrix}$$

$$= \left[\frac{V_o}{K_1}\quad D_o r\left(s + \frac{1}{CR}\right) + \frac{D_o^2}{C} \quad -\frac{D_o r}{L} + s + \frac{r_L + D_o r}{L}\right]$$

$$\times \begin{bmatrix}\frac{r}{D_o RL} + \frac{1}{L}\\ -\frac{1}{D_o RC}\end{bmatrix} - \frac{V_o r}{D_o R}$$

After some manipulation and substituting $L/(1 - D)^2$ for L_e, we obtain

$$\frac{\hat{v}_o(s)}{\hat{d}(s)} = \frac{V_o}{D_o}\,\frac{1}{L_e C}$$

$$\times \frac{\left(rsC + \frac{r}{R} + 1\right)\left(\frac{r}{D_o R} + 1\right) + \frac{r}{R} - \frac{sL_e}{R} - \frac{r_L}{RD_o^2} - \frac{r}{D_o R}}{s^2 + s\left[\frac{(r_L/D_o^2) + (r/D_o)}{L_e} + \frac{1}{CR}\right] + \frac{(r_L/D_o^2) + (r/D_o)}{L_e CR} + \frac{1}{L_e C}}$$

$$- \frac{V_o r}{D_o R_o}$$

After making suitable approximations for practical cases,

$$\frac{\hat{v}_o(s)}{\hat{d}(s)} \simeq \frac{V_o}{D_o}\left(1 - \frac{sL_e}{R}\right) H_e(s) \tag{A.68}$$

A somewhat similar analysis [62] of the buck-boost power stage yields the following relations:

$$\frac{V_o}{V_i} \simeq \frac{D}{1 - D} \tag{A.69}$$

$$\frac{\hat{v}_o(s)}{\hat{v}_i(s)} = \frac{D}{1 - D} H_e(s) \tag{A.70}$$

where $H_e(s)$ is identical to that of the boost converter, as given in Eq. (A.61):

$$\frac{\hat{v}_o(s)}{\hat{d}(s)} = \frac{V_o}{D(1 - D)}\left[1 - \frac{sDL}{R(1 - D)^2}\right] H_e(s) \tag{A.71}$$

Comparison of Eqs. (A.57), (A.68), and (A.71) shows that a distinct effective output filter transfer function $H_e(s)$ is present in each of the three basic power converter configurations. With this knowledge in mind, it is possible to identify individual converter characteristics from functions other than $H_e(s)$. One of the first noticeable characteristics is that of the buck converter, which has no frequency-dependent functions other than $H_e(s)$. This means that, for all known buck and buck-deprived converters, the converter loop gain would be represented by a product $H_e(s)$ and a constant term, neglecting the effect of storage time modulation [76, 90]. The loop gain could then be tailored by error amplifier gain compensation and/or output filter damping.

For the boost and buck-boost converters, there is a zero in the output to duty ratio small-signal transfer function in each case. By virtue of the negative sign within the expression, this zero is located in the right half of the complex frequency plane.

Also worthy of attention is the effective output filter transfer function $H_e(s)$ for the boost and buck-boost converters; it is identical and is depicted in Fig. A.4. From this figure, it is evident that the tuned frequency of this filter is *duty ratio dependent* (unlike the buck and buck-derived converters, which has only the nominal element values to contend with). This means that the actual loop gain responses at high and low duty ratios would exhibit two distinct resonant frequencies for the "same" filter elements, due to constant periodic changes between two different topologies during switching.

Therefore, the graphs provided in Appendix B are directly applicable to loop gain analysis of buck and buck-derived-type converters, whereas, for the boost and the buck-boost converters, due to the presence of the right half-plane zero, these graphs would only serve as guidelines for tailoring loop responses up to some point just beyond the dominant poles (pole locations due to D_H and D_L) of the output filter $H_e(s)$.

The equations obtained for $\hat{v}_o/\hat{d}$ are easily related to $\hat{v}_o/\hat{v}_c$ by substituting $\hat{d} = \hat{v}_c/V_m$ (Middlebrook [71] and Pressman [48, p. 319]) so that

$$\frac{\hat{v}_o(s)}{\hat{v}_c(s)} = \frac{1}{V_m}\,\frac{\hat{v}_o(s)}{\hat{d}(s)} \qquad (A.72)$$

This result is reflected in Chapter 2 in Eqs. (2.1) to (2.4).

It is also evident that, with the knowledge of V_m (which is usually supplied by the manufacturers of the modulator circuits), it

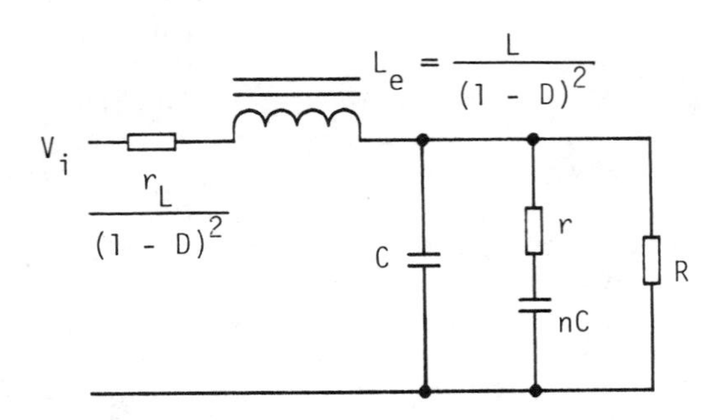

FIG. A.4 Effective output filter of boost and buck-boost converters.

is possible to evaluate circuits built with a whole variety of modulator control circuits, as long as they are of the ramp and comparator types.

Because of the omission of r in some elements of the A matrix, the results obtained for Eqs. (A.54), (A.55), (A.57), (A.61), and (A.68) to (A.71) are correct only to first-order accuracy. Note that the A matrix for Eq. (A.58) includes all appropriate parasitic components. Equation (A.59) shows the result with the second-order effect neglected and is correct only to first-order accuracy.

appendix B
Graphical Design Aids

The following graphs are provided as design aids so that the designer can do most performance prediction without requiring detailed calculations. Should the designer wish to check further, calculator programs are provided for this purpose in Chapter 2.

Figures B.1 to B.3 show the undamped LC filter whose response varies with load. For ease of reference and comparison, all the filters in this appendix were plotted with a corner frequency of 1 kHz. While normalized curves could have been used, these curves are nevertheless accurate for the intended purposes.

Figures B.1 to B.16 have the characteristics shown for L/C = 1. Figures B.17 to B.27 show responses with ratios of L/C = 0.1. Figure B.28 shows an output filter with a "trap" circuit. In this figure all inductors and capacitors are made equal for the graphical plot. In practice, however, the series tuned inductor could be made larger in inductance (since it does not carry any direct current) and the series tuned capacitor made smaller in capacitance, but of the same type or make as that used for C. The responses of the circuit are shown in Figs. B.29 and B.30. Figures B.31 to B.33 have the characteristics shown for L/C = 10.

All graphics in this appendix were produced with the Hewlett-Packard desk-top computer HP-9845C.

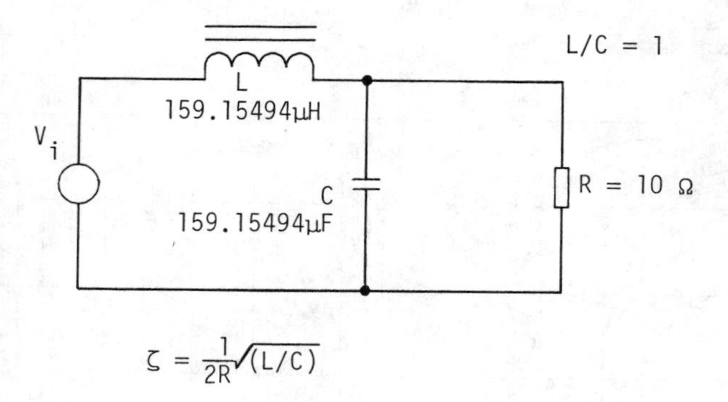

$$\zeta = \frac{1}{2R}\sqrt{(L/C)}$$

FIG. B.1 Output filter without damping.

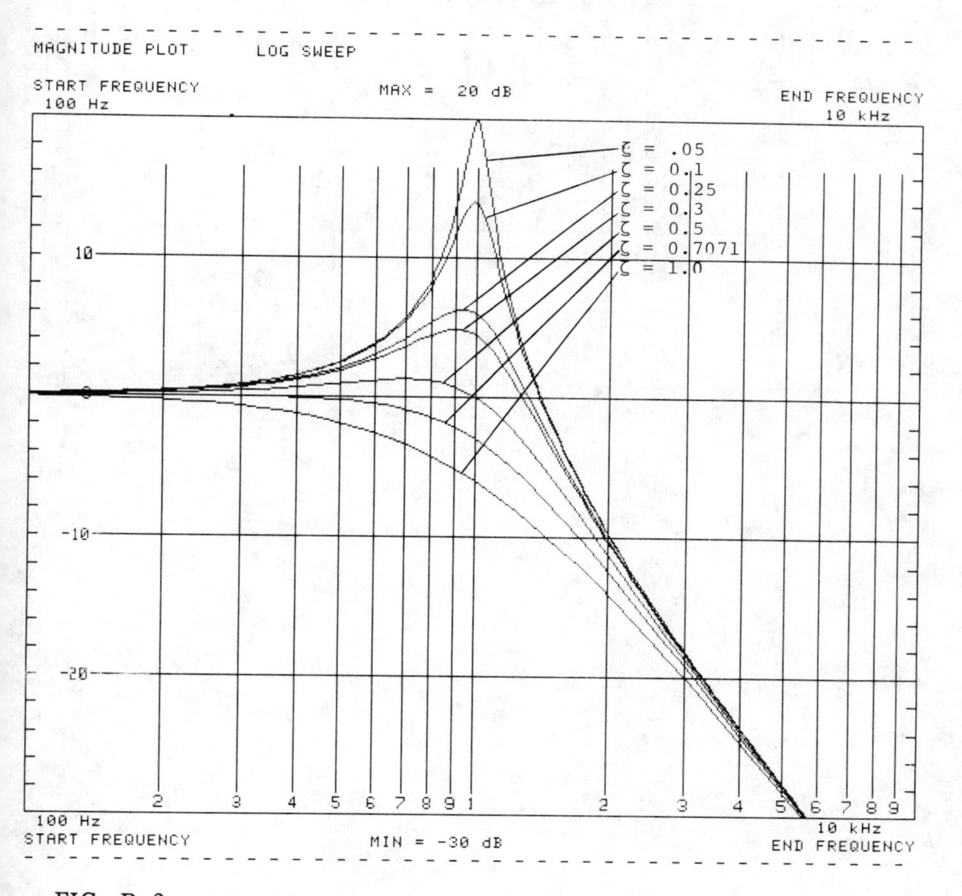

FIG. B.2

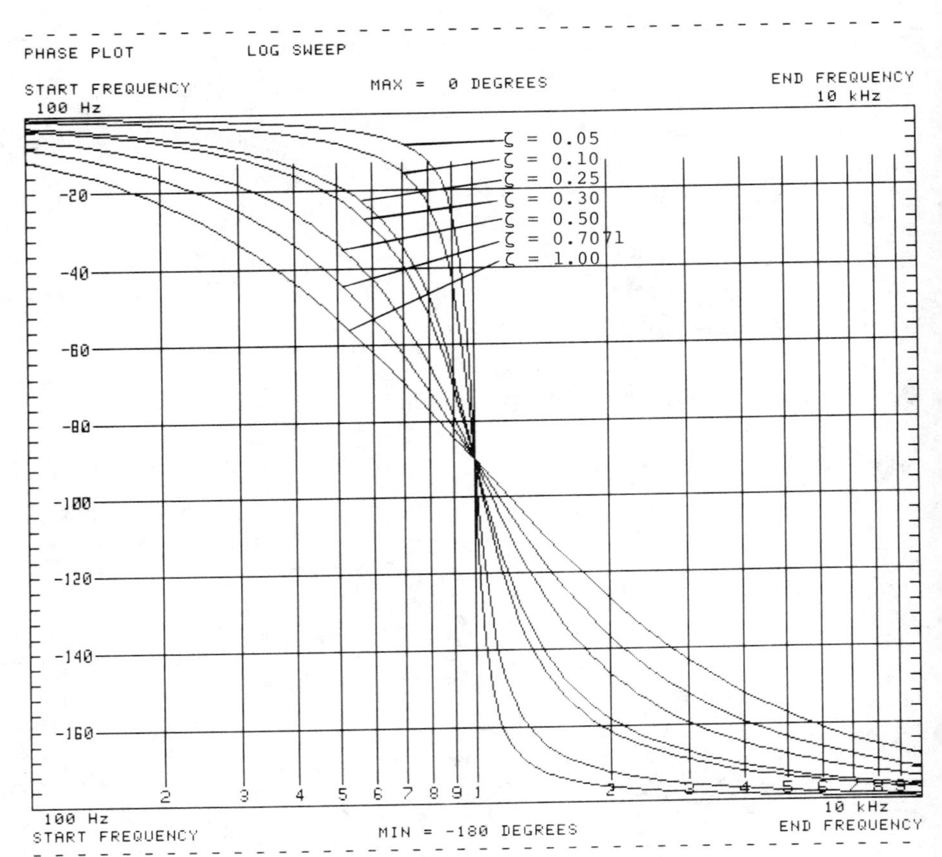
PHASE PLOT
LOG SWEEP
START FREQUENCY
100 Hz
MAX = 0 DEGREES
END FREQUENCY
10 kHz
ζ = 0.05
ζ = 0.10
ζ = 0.25
ζ = 0.30
ζ = 0.50
ζ = 0.7071
ζ = 1.00
-20
-40
-60
-80
-100
-120
-140
-160
2
3
4
5
6
7
8
9
1
100 Hz
START FREQUENCY
MIN = -180 DEGREES
10 kHz
END FREQUENCY

FIG. B.3

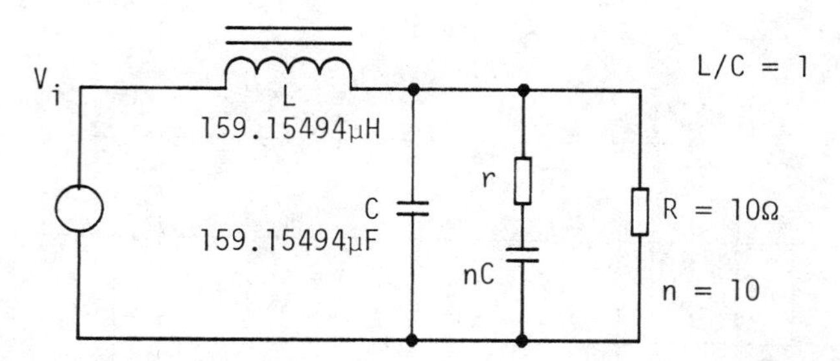

FIG. B.4 Output filter with damping.

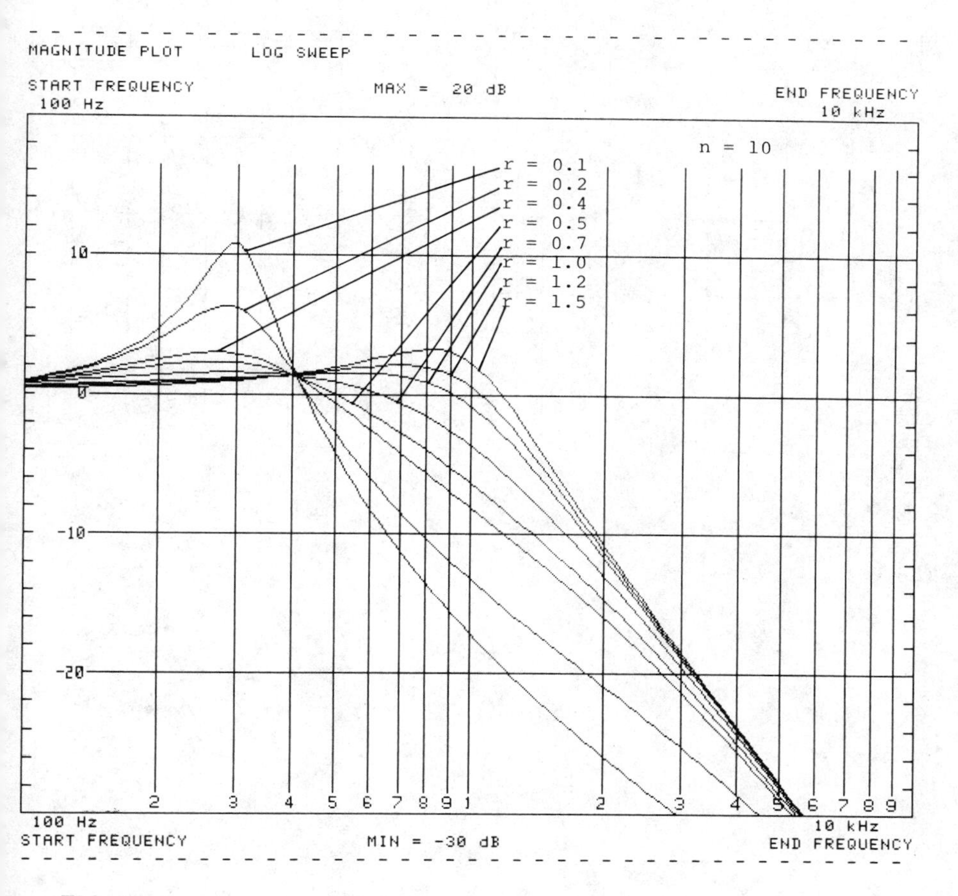

FIG. B.5

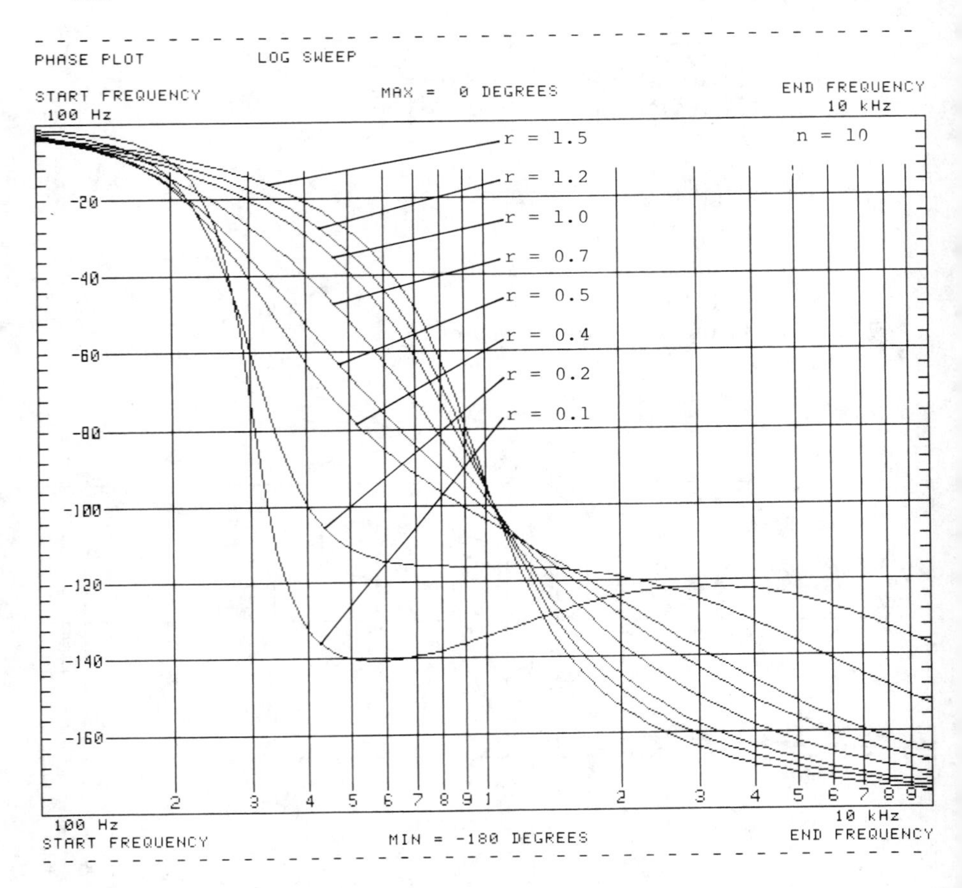
PHASE PLOT
LOG SWEEP
START FREQUENCY
100 Hz
MAX = 0 DEGREES
END FREQUENCY
10 kHz
n = 10
r = 1.5
r = 1.2
r = 1.0
r = 0.7
r = 0.5
r = 0.4
r = 0.2
r = 0.1
-20
-40
-60
-80
-100
-120
-140
-160
2
3
4
5
6
7
8
9
1
2
3
4
5
6
7
8
9
100 Hz
START FREQUENCY
MIN = -180 DEGREES
10 kHz
END FREQUENCY

FIG. B.6

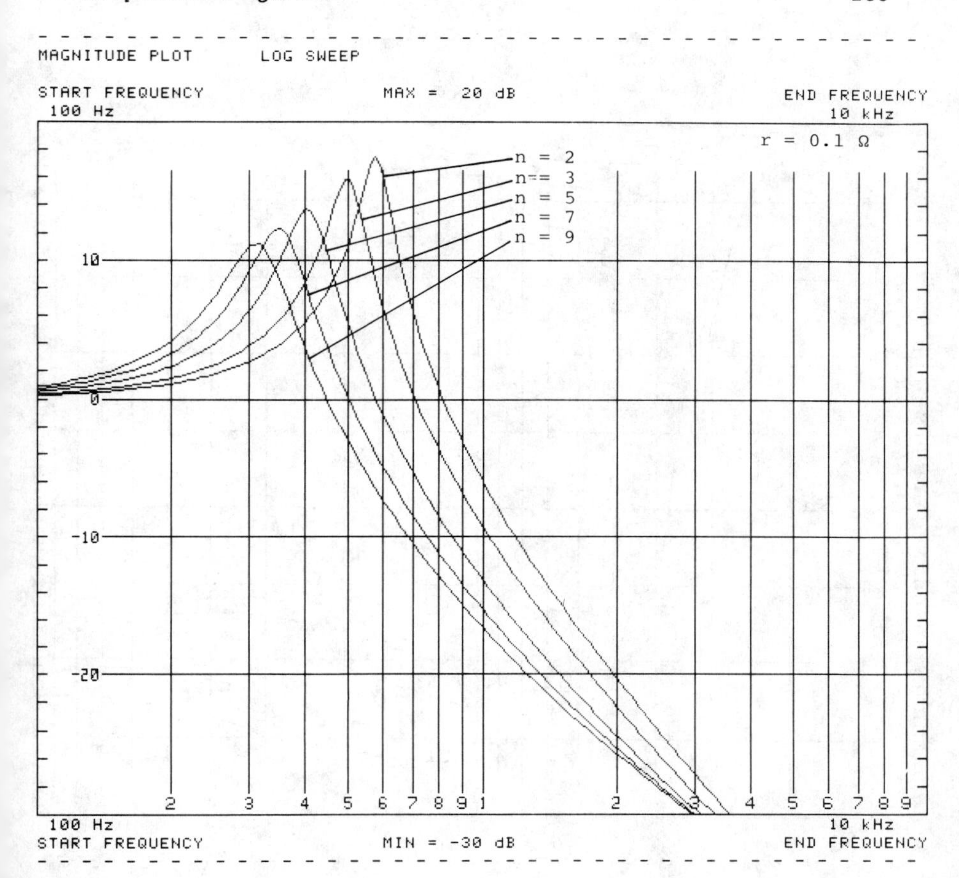
MAGNITUDE PLOT
LOG SWEEP
START FREQUENCY
100 Hz
MAX = 20 dB
END FREQUENCY
10 kHz
r = 0.1 Ω
n = 2
n = 3
n = 5
n = 7
n = 9
10
0
-10
-20
2
3
4
5
6
7
8
9
1
2
3
4
5
6
7
8
9
100 Hz
START FREQUENCY
MIN = -30 dB
10 kHz
END FREQUENCY

FIG. B.7

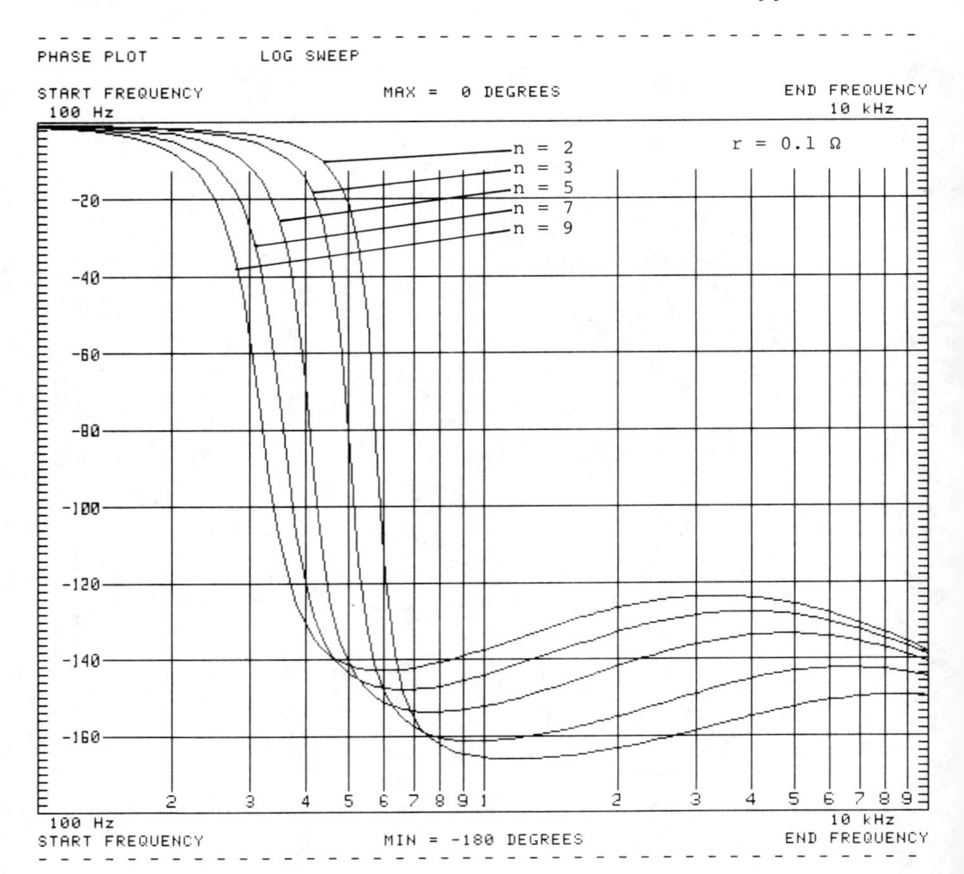
PHASE PLOT
LOG SWEEP
START FREQUENCY
100 Hz
MAX = 0 DEGREES
END FREQUENCY
10 kHz
r = 0.1 Ω
n = 2
n = 3
n = 5
n = 7
n = 9
-20
-40
-60
-80
-100
-120
-140
-160
2 3 4 5 6 7 8 9 1 2 3 4 5 6 7 8 9
100 Hz
START FREQUENCY
MIN = -180 DEGREES
10 kHz
END FREQUENCY

FIG. B.8

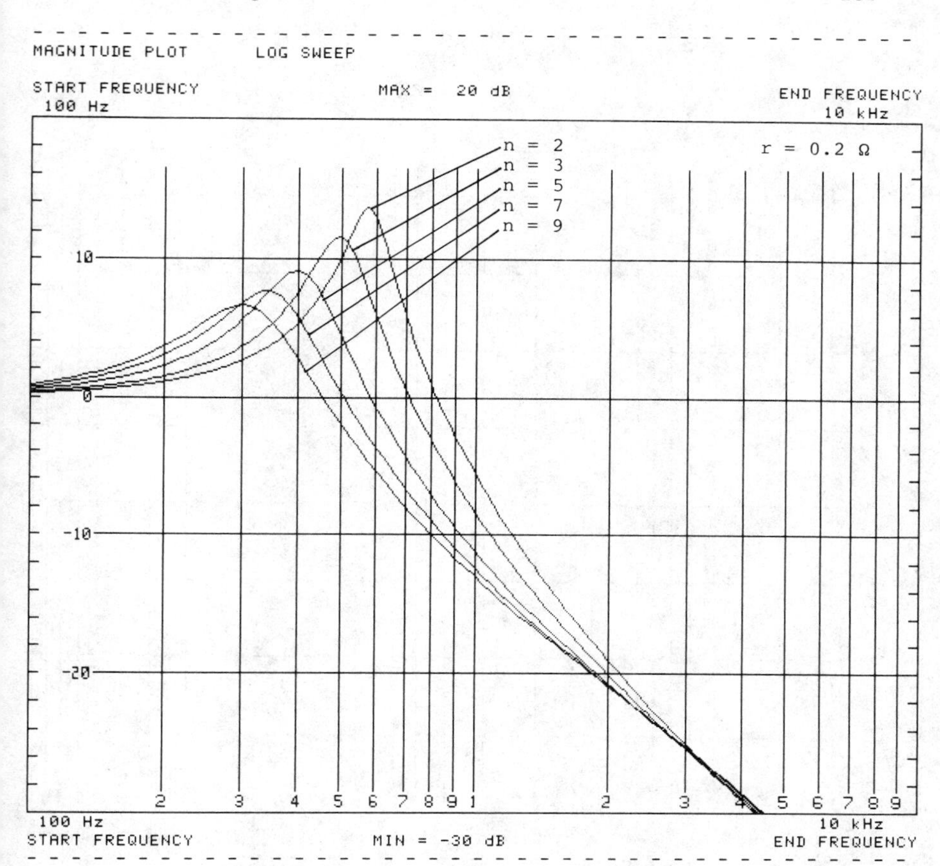
MAGNITUDE PLOT LOG SWEEP
START FREQUENCY
100 Hz
MAX = 20 dB
END FREQUENCY
10 kHz
n = 2
n = 3
n = 5
n = 7
n = 9
r = 0.2 Ω
10
0
-10
-20
2 3 4 5 6 7 8 9 1 2 3 4 5 6 7 8 9
100 Hz
START FREQUENCY
MIN = -30 dB
10 kHz
END FREQUENCY

FIG. B.9

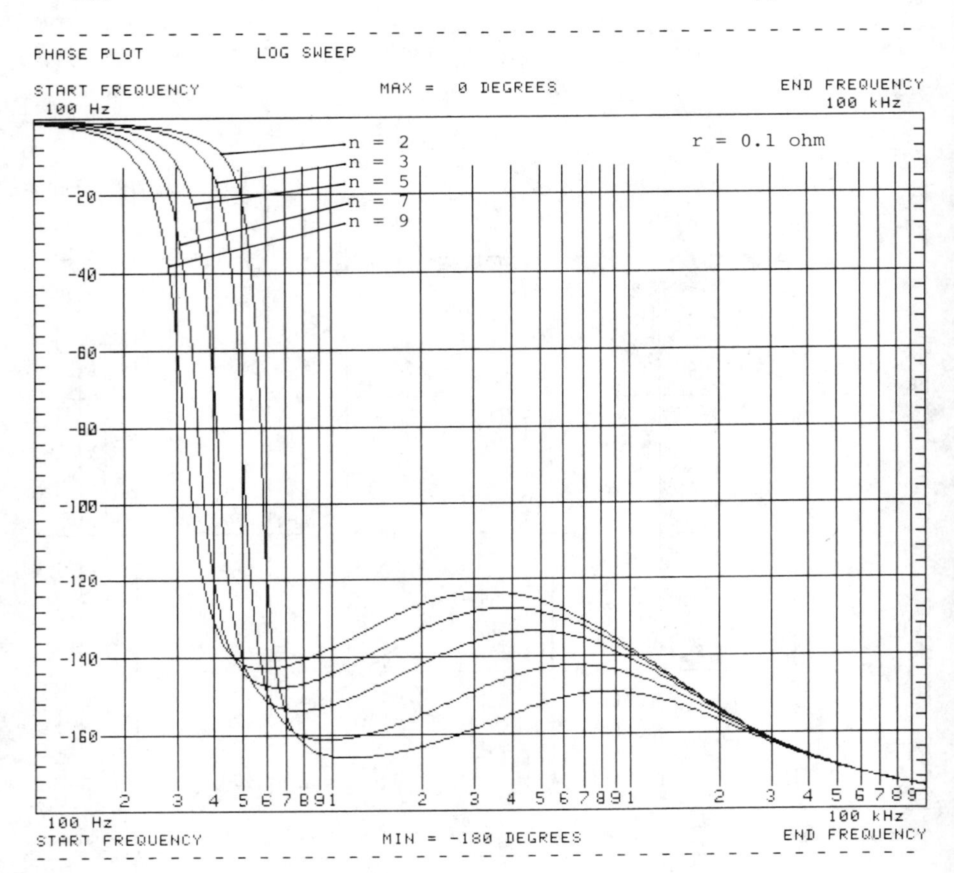
PHASE PLOT
LOG SWEEP
START FREQUENCY
100 Hz
MAX = 0 DEGREES
END FREQUENCY
100 kHz
r = 0.1 ohm
n = 2
n = 3
n = 5
n = 7
n = 9
-20
-40
-60
-80
-100
-120
-140
-160
2 3 4 5 6 7 8 9 1 2 3 4 5 6 7 8 9 1 2 3 4 5 6 7 8 9
100 Hz
START FREQUENCY
MIN = -180 DEGREES
100 kHz
END FREQUENCY

FIG. B.10

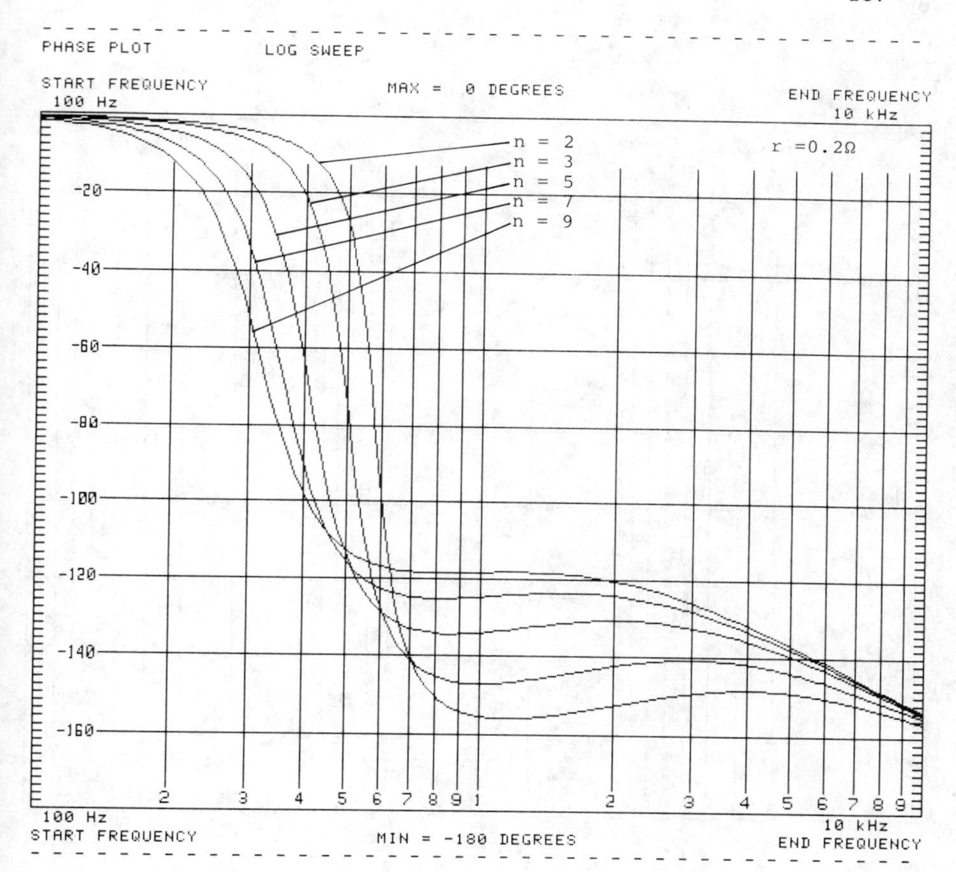
PHASE PLOT
LOG SWEEP
START FREQUENCY
100 Hz
MAX = 0 DEGREES
END FREQUENCY
10 kHz
n = 2
n = 3
n = 5
n = 7
n = 9
r =0.2Ω
-20
-40
-60
-80
-100
-120
-140
-160
2 3 4 5 6 7 8 9 1 2 3 4 5 6 7 8 9
100 Hz
START FREQUENCY
MIN = -180 DEGREES
10 kHz
END FREQUENCY

FIG. B.11

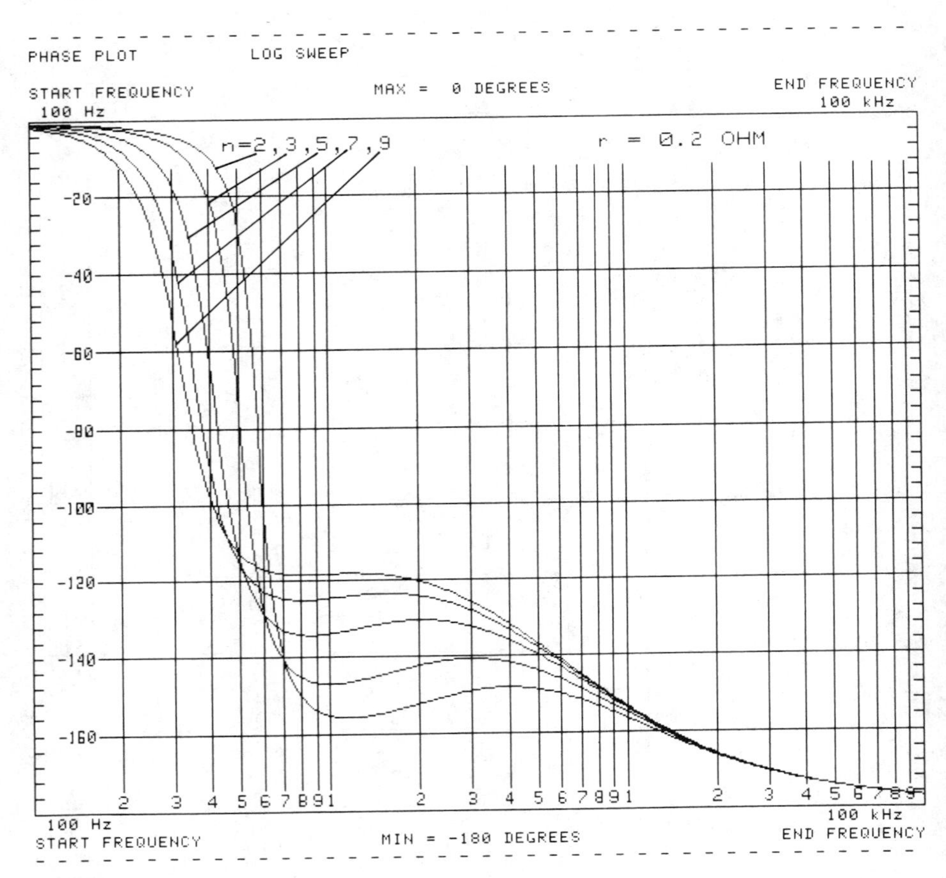
PHASE PLOT
LOG SWEEP
START FREQUENCY
100 Hz
MAX = 0 DEGREES
END FREQUENCY
100 kHz
n=2,3,5,7,9
r = 0.2 OHM
-20
-40
-60
-80
-100
-120
-140
-160
2 3 4 5 6 7 8 9 1 2 3 4 5 6 7 8 9 1 2 3 4 5 6 7 8 9
100 Hz
START FREQUENCY
MIN = -180 DEGREES
100 kHz
END FREQUENCY

FIG. B.12

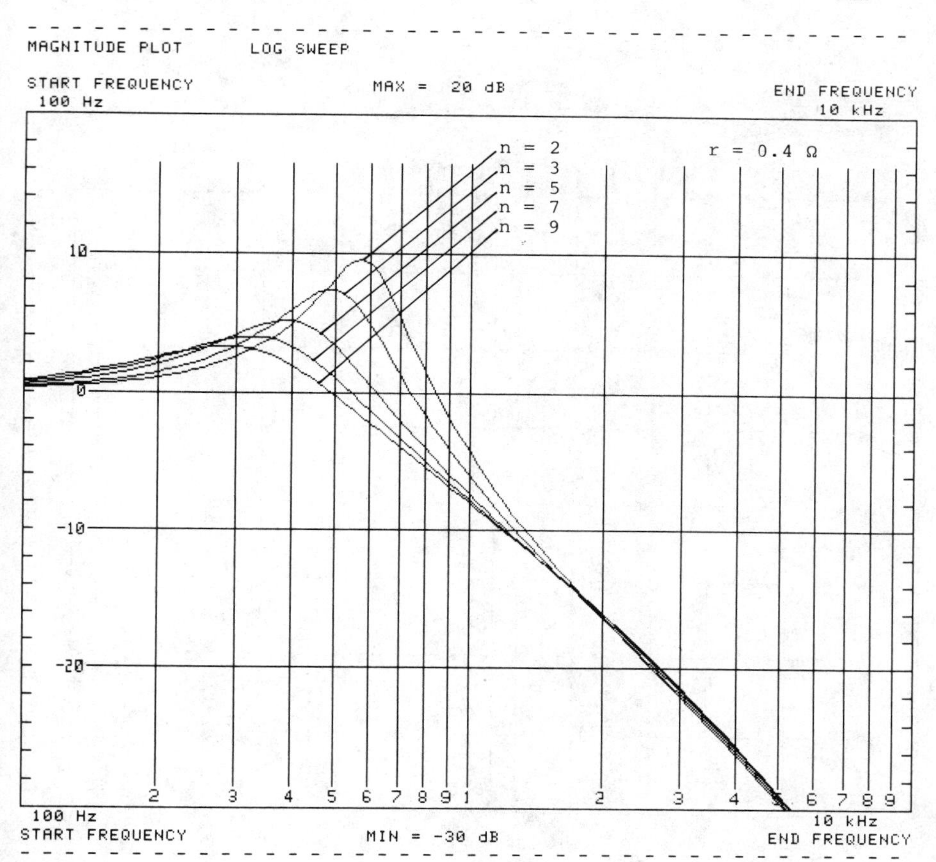
MAGNITUDE PLOT
LOG SWEEP
START FREQUENCY
100 Hz
MAX = 20 dB
END FREQUENCY
10 kHz
n = 2
n = 3
n = 5
n = 7
n = 9
r = 0.4 Ω
10
0
-10
-20
2
3
4
5
6
7
8
9
1
100 Hz
START FREQUENCY
MIN = -30 dB
10 kHz
END FREQUENCY

FIG. B.13

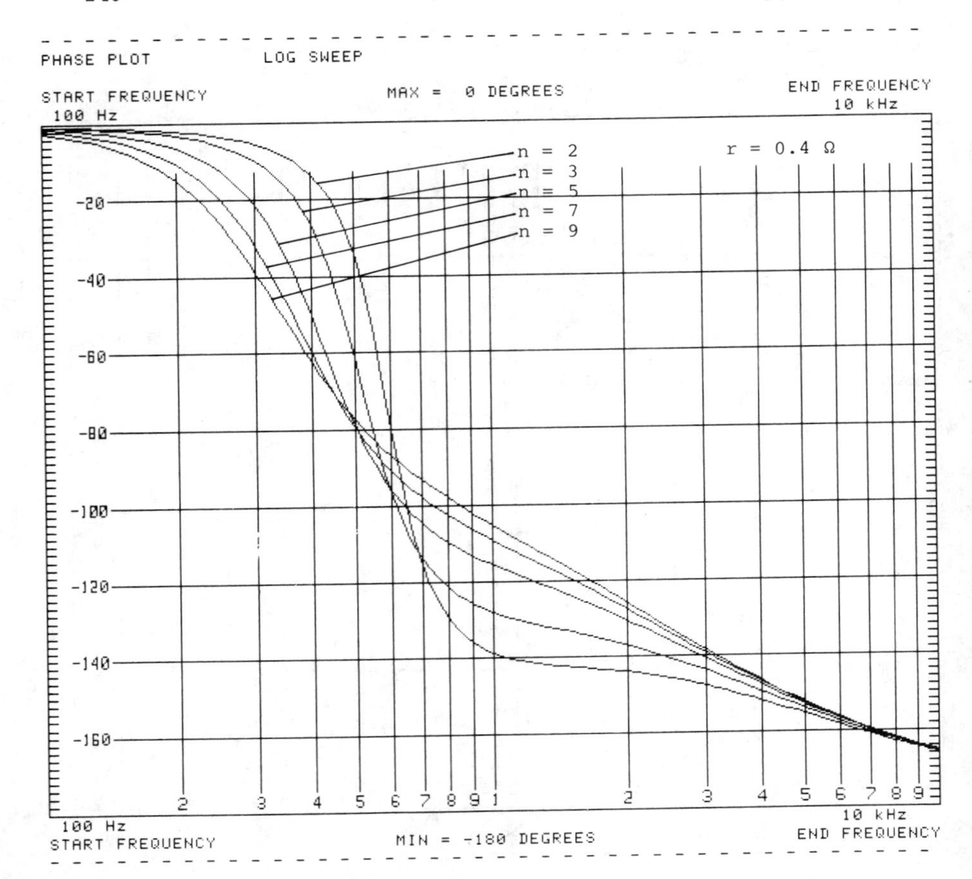
PHASE PLOT
LOG SWEEP
START FREQUENCY
100 Hz
MAX = 0 DEGREES
END FREQUENCY
10 kHz
n = 2
n = 3
n = 5
n = 7
n = 9
r = 0.4 Ω
-20
-40
-60
-80
-100
-120
-140
-160
2 3 4 5 6 7 8 9 1 2 3 4 5 6 7 8 9
100 Hz
START FREQUENCY
MIN = -180 DEGREES
10 kHz
END FREQUENCY

FIG. B.14

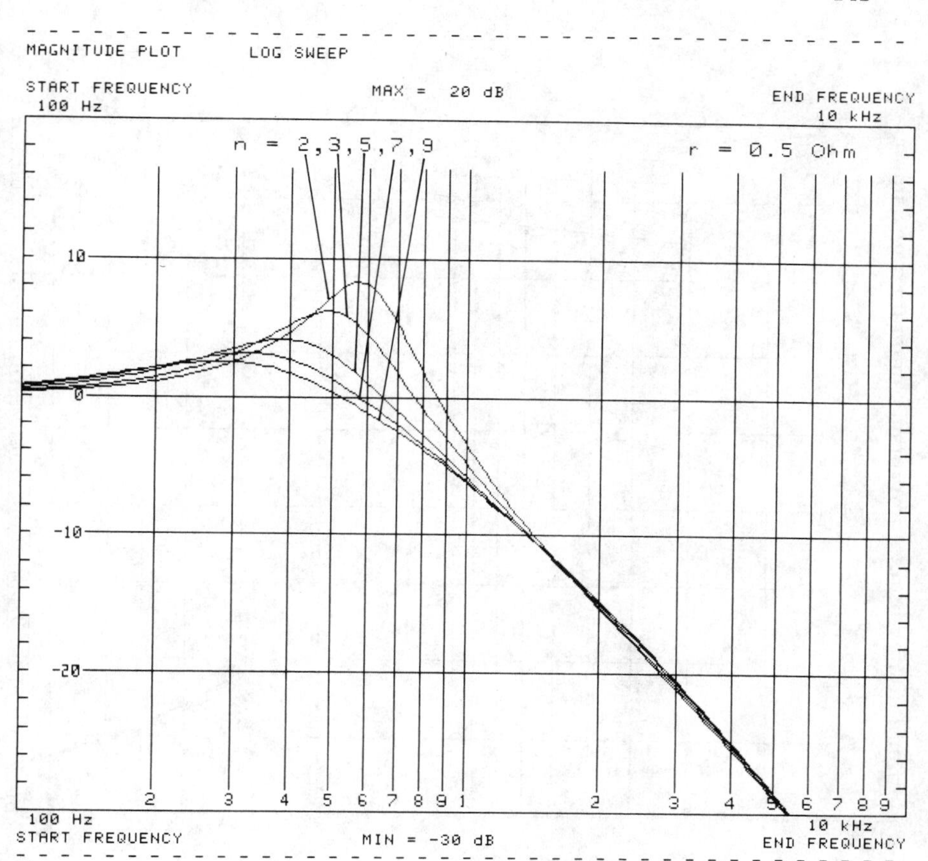
MAGNITUDE PLOT LOG SWEEP
START FREQUENCY
100 Hz
MAX = 20 dB
END FREQUENCY
10 kHz
n = 2,3,5,7,9
r = 0.5 Ohm
10
0
-10
-20
2 3 4 5 6 7 8 9 1 2 3 4 5 6 7 8 9
100 Hz
START FREQUENCY
MIN = -30 dB
10 kHz
END FREQUENCY

FIG. B.15

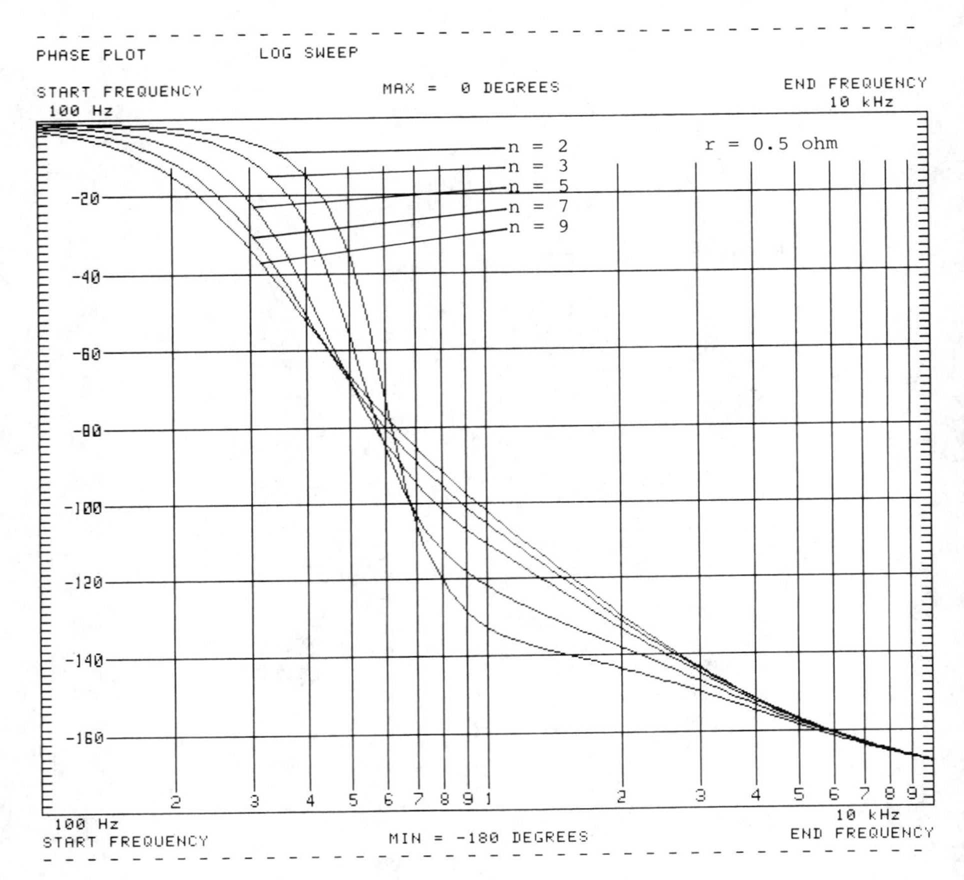
PHASE PLOT
LOG SWEEP
START FREQUENCY
100 Hz
MAX = 0 DEGREES
END FREQUENCY
10 kHz
r = 0.5 ohm
n = 2
n = 3
n = 5
n = 7
n = 9
-20
-40
-60
-80
-100
-120
-140
-160
2 3 4 5 6 7 8 9 1 2 3 4 5 6 7 8 9
100 Hz
START FREQUENCY
MIN = -180 DEGREES
10 kHz
END FREQUENCY

FIG. B.16

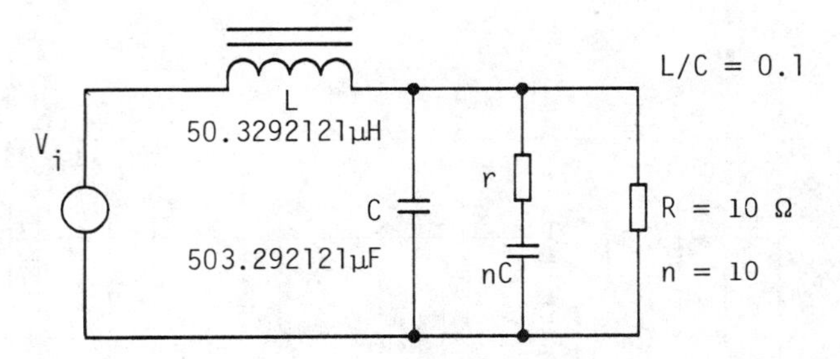

FIG. B.17 Output filter with damping.

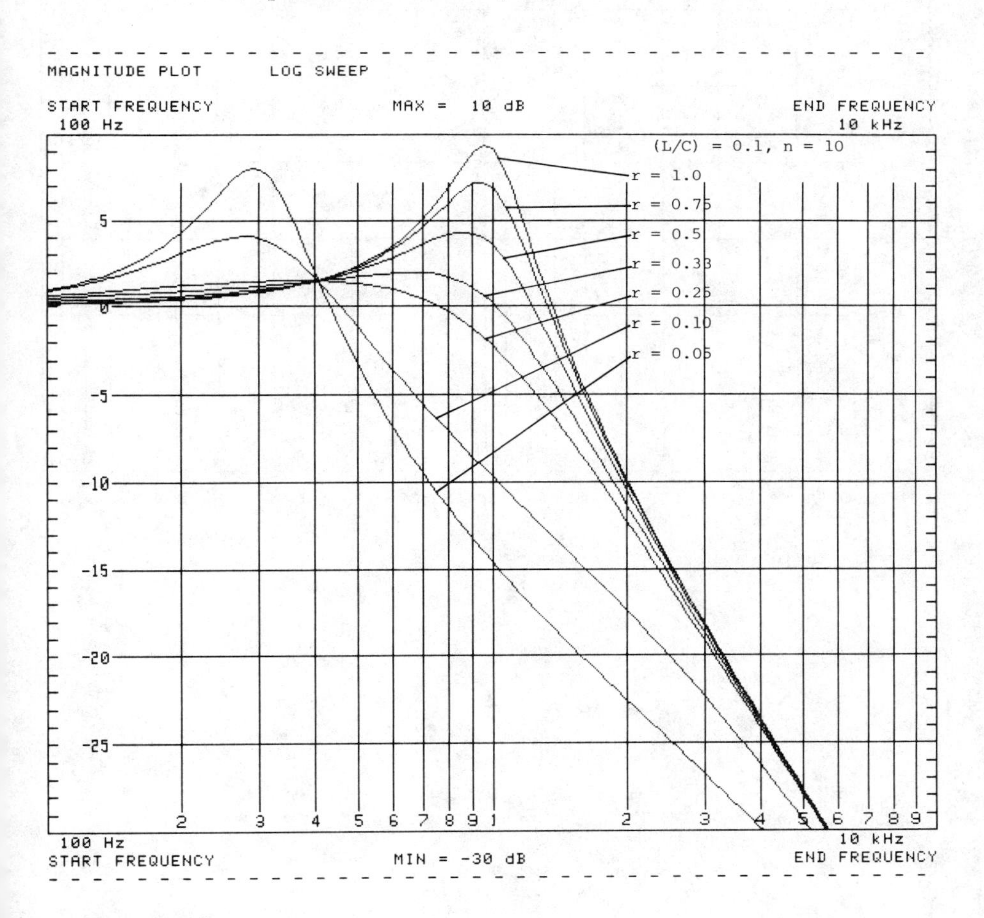

FIG. B.18

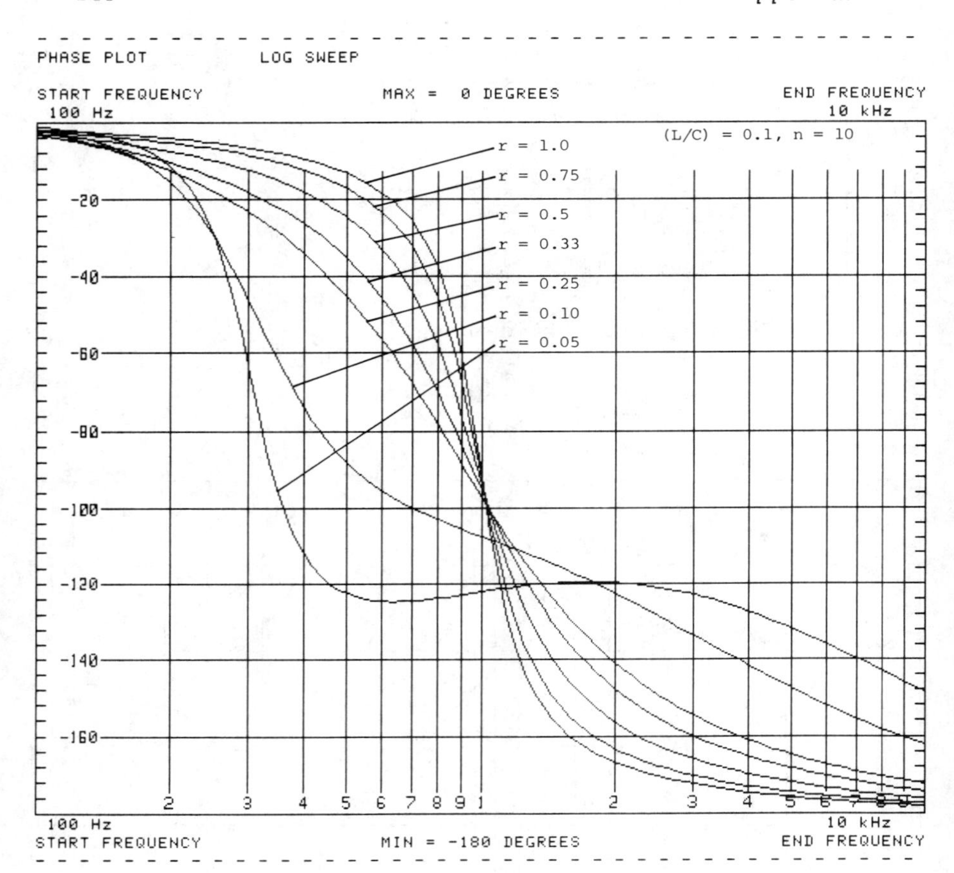
PHASE PLOT
LOG SWEEP
START FREQUENCY
100 Hz
MAX = 0 DEGREES
END FREQUENCY
10 kHz
(L/C) = 0.1, n = 10
r = 1.0
r = 0.75
r = 0.5
r = 0.33
r = 0.25
r = 0.10
r = 0.05
-20
-40
-60
-80
-100
-120
-140
-160
2 3 4 5 6 7 8 9 1 2 3 4 5 6 7 8 9
100 Hz
START FREQUENCY
MIN = -180 DEGREES
10 kHz
END FREQUENCY

FIG. B.19

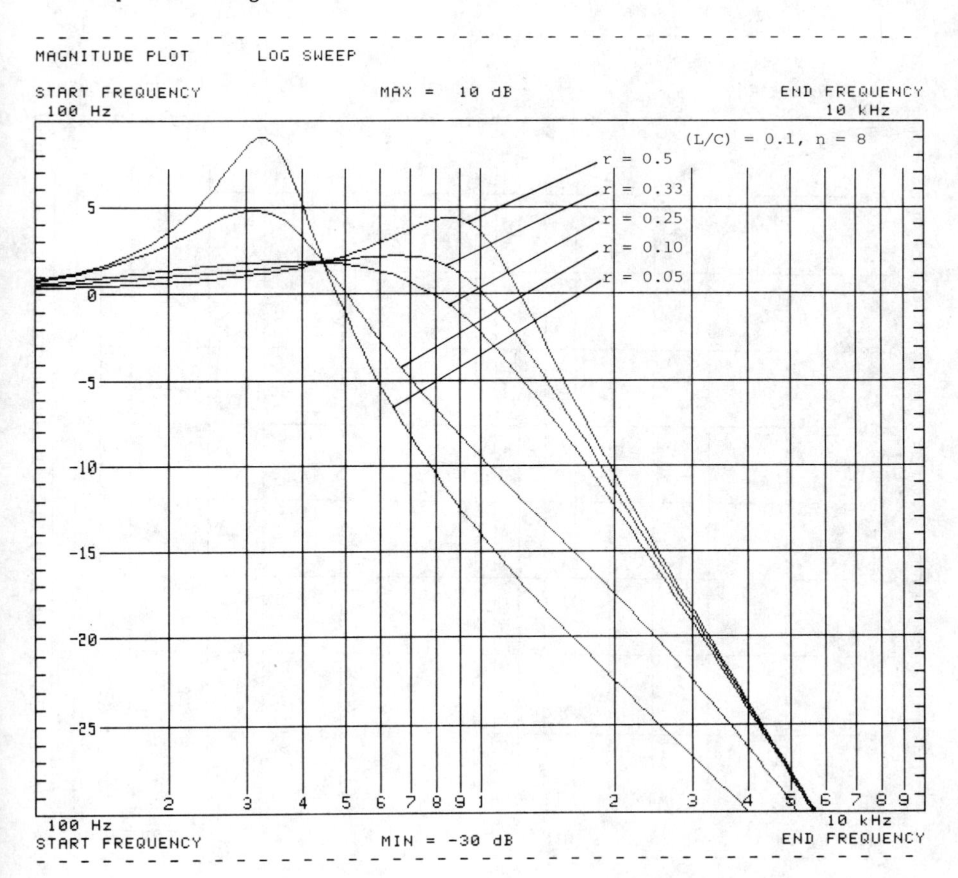

MAGNITUDE PLOT LOG SWEEP
START FREQUENCY
100 Hz
MAX = 10 dB
END FREQUENCY
10 kHz
(L/C) = 0.1, n = 8
r = 0.5
r = 0.33
r = 0.25
r = 0.10
r = 0.05
5
0
-5
-10
-15
-20
-25
2 3 4 5 6 7 8 9 1 2 3 4 5 6 7 8 9
100 Hz
START FREQUENCY
MIN = -30 dB
10 kHz
END FREQUENCY

FIG. B.20

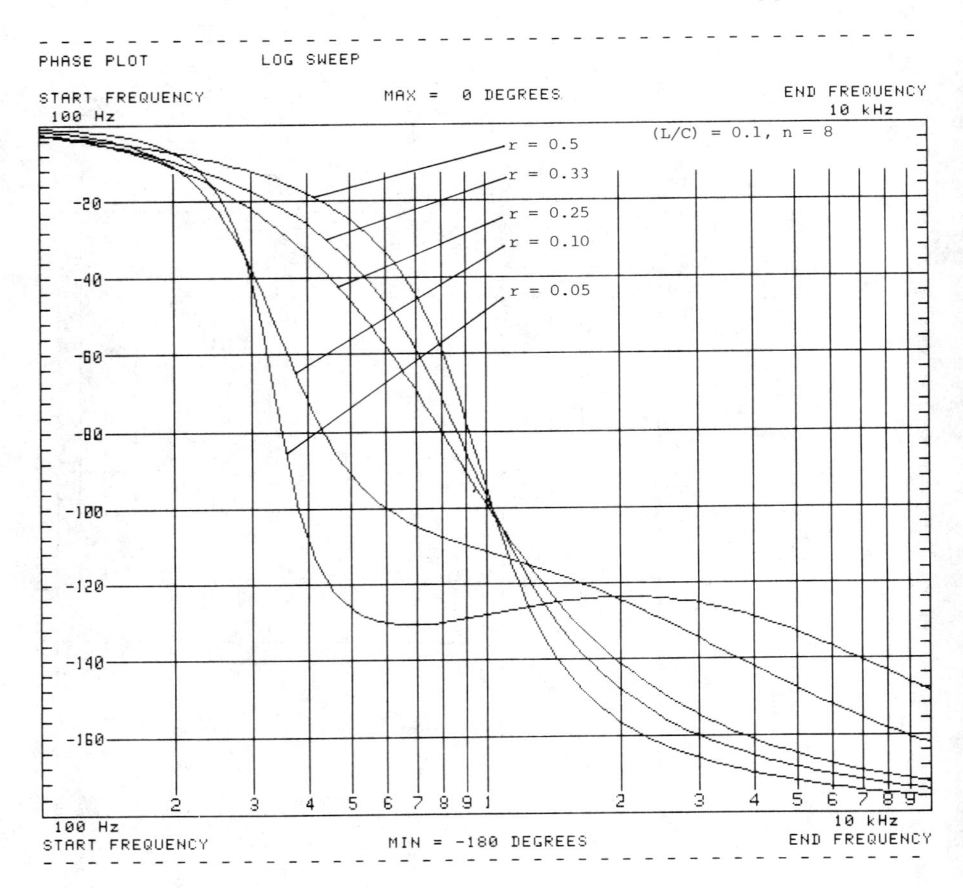
PHASE PLOT
LOG SWEEP
START FREQUENCY
100 Hz
MAX = 0 DEGREES
END FREQUENCY
10 kHz
(L/C) = 0.1, n = 8
r = 0.5
r = 0.33
r = 0.25
r = 0.10
r = 0.05
-20
-40
-60
-80
-100
-120
-140
-160
2
3
4
5
6
7
8
9
1
2
3
4
5
6
7
8
9
100 Hz
START FREQUENCY
MIN = -180 DEGREES
10 kHz
END FREQUENCY

FIG. B.21

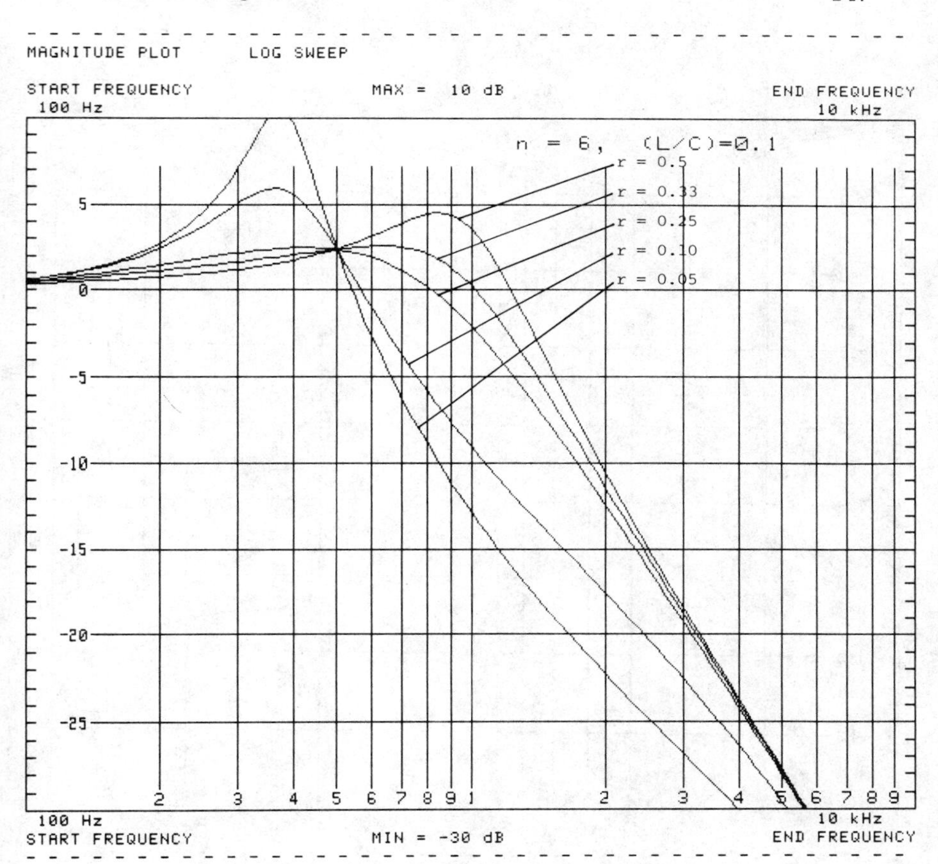
MAGNITUDE PLOT LOG SWEEP
START FREQUENCY
100 Hz
MAX = 10 dB
END FREQUENCY
10 kHz
n = 6, (L/C)=0.1
r = 0.5
r = 0.33
r = 0.25
r = 0.10
r = 0.05
5
0
-5
-10
-15
-20
-25
2 3 4 5 6 7 8 9 1 2 3 4 5 6 7 8 9
100 Hz
START FREQUENCY
MIN = -30 dB
10 kHz
END FREQUENCY

FIG. B.22

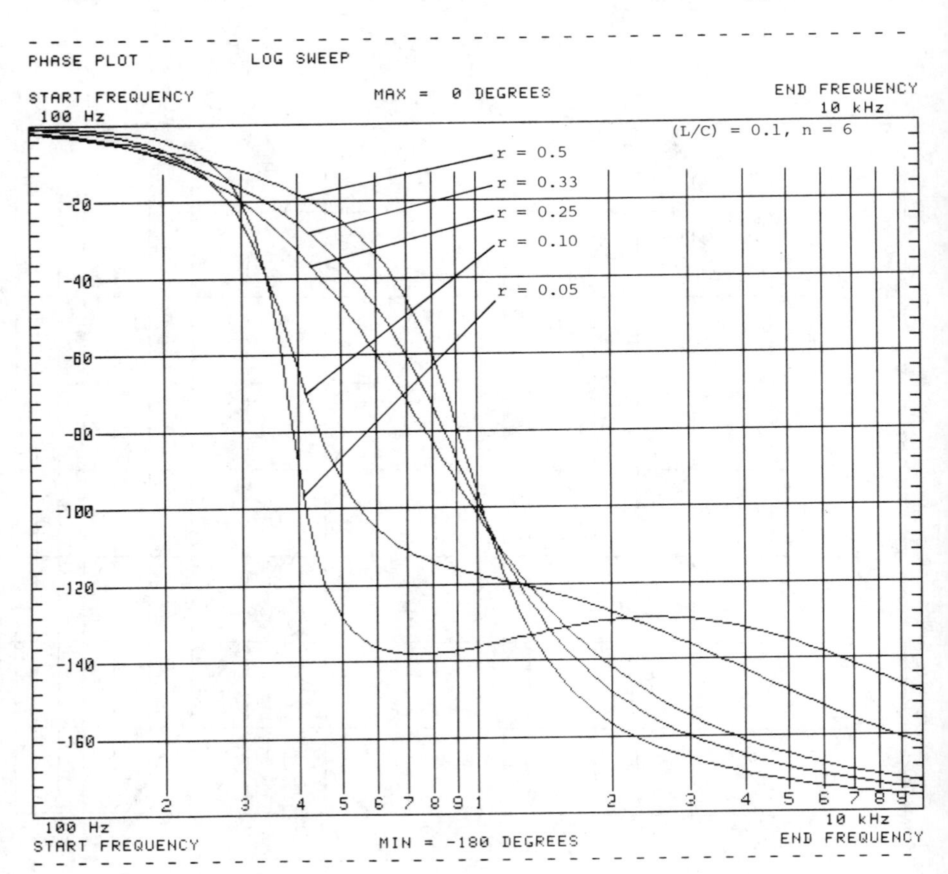
PHASE PLOT
LOG SWEEP
START FREQUENCY
100 Hz
MAX = 0 DEGREES
END FREQUENCY
10 kHz
(L/C) = 0.1, n = 6
r = 0.5
r = 0.33
r = 0.25
r = 0.10
r = 0.05
-20
-40
-60
-80
-100
-120
-140
-160
2
3
4
5
6
7
8
9
1
2
3
4
5
6
7
8
9
100 Hz
START FREQUENCY
MIN = -180 DEGREES
10 kHz
END FREQUENCY

FIG. B.23

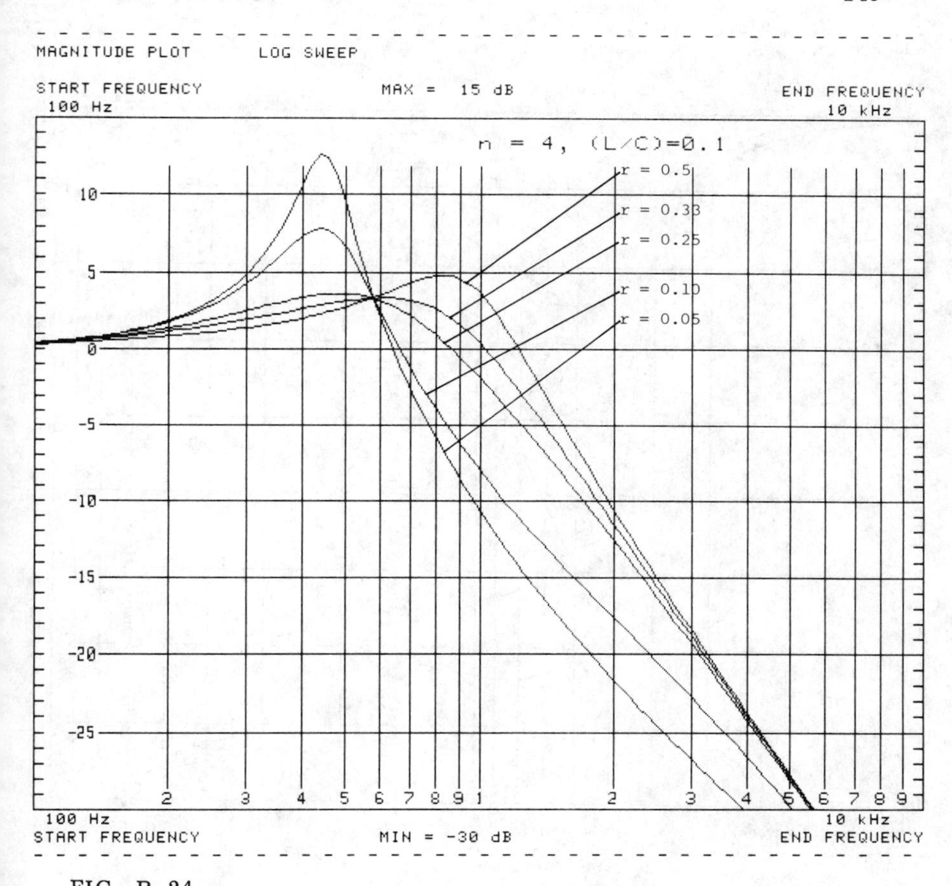
MAGNITUDE PLOT
LOG SWEEP
START FREQUENCY
100 Hz
MAX = 15 dB
END FREQUENCY
10 kHz
n = 4, (L/C)=0.1
r = 0.5
r = 0.33
r = 0.25
r = 0.10
r = 0.05
10
5
0
-5
-10
-15
-20
-25
2 3 4 5 6 7 8 9 1 2 3 4 5 6 7 8 9
100 Hz
START FREQUENCY
MIN = -30 dB
10 kHz
END FREQUENCY

FIG. B.24

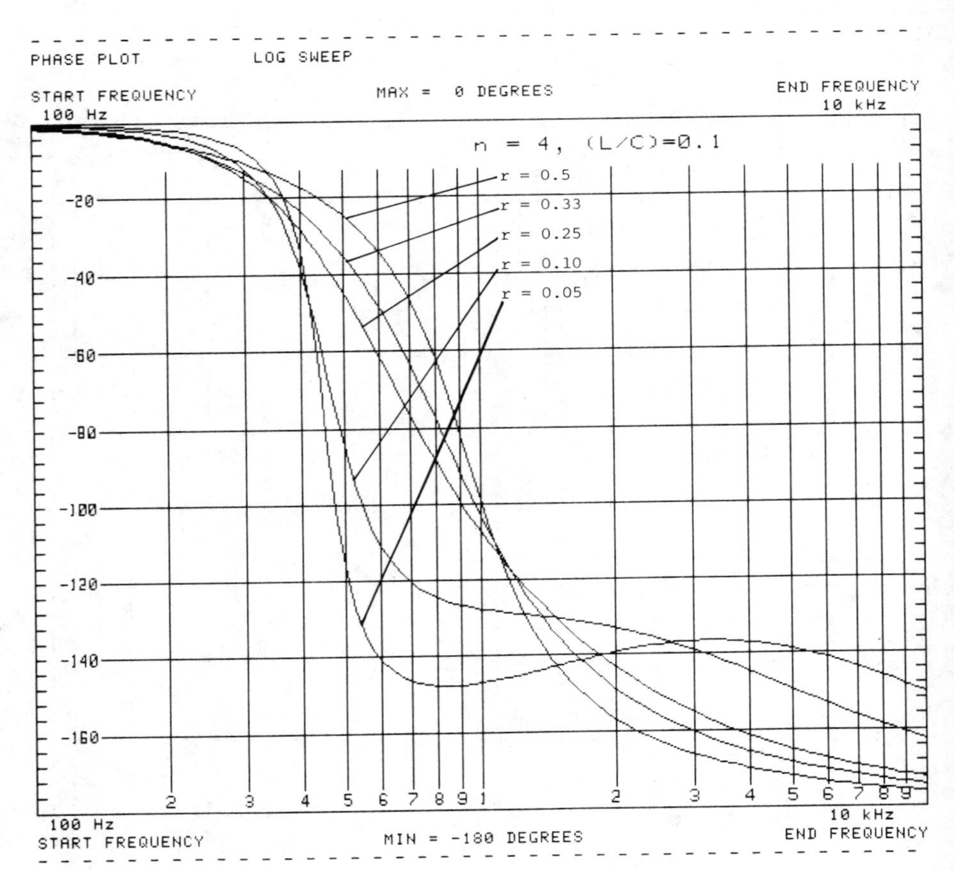
PHASE PLOT
LOG SWEEP
START FREQUENCY
100 Hz
MAX = 0 DEGREES
END FREQUENCY
10 kHz
n = 4, (L/C)=0.1
r = 0.5
r = 0.33
r = 0.25
r = 0.10
r = 0.05
-20
-40
-60
-80
-100
-120
-140
-160
2 3 4 5 6 7 8 9 1 2 3 4 5 6 7 8 9
100 Hz
START FREQUENCY
MIN = -180 DEGREES
10 kHz
END FREQUENCY

FIG. B.25

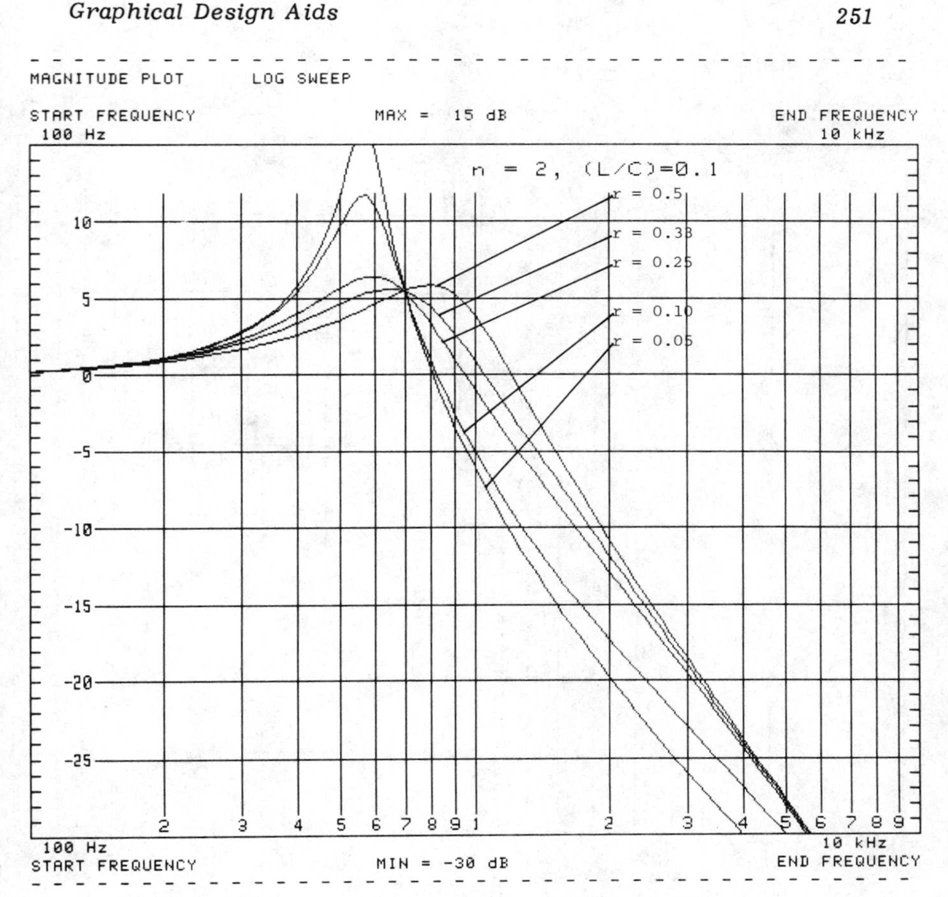
MAGNITUDE PLOT
LOG SWEEP
START FREQUENCY
100 Hz
MAX = 15 dB
END FREQUENCY
10 kHz
n = 2, (L/C)=0.1
r = 0.5
r = 0.33
r = 0.25
r = 0.10
r = 0.05
10
5
0
-5
-10
-15
-20
-25
2 3 4 5 6 7 8 9 1 2 3 4 5 6 7 8 9
100 Hz
START FREQUENCY
MIN = -30 dB
10 kHz
END FREQUENCY

FIG. B.26

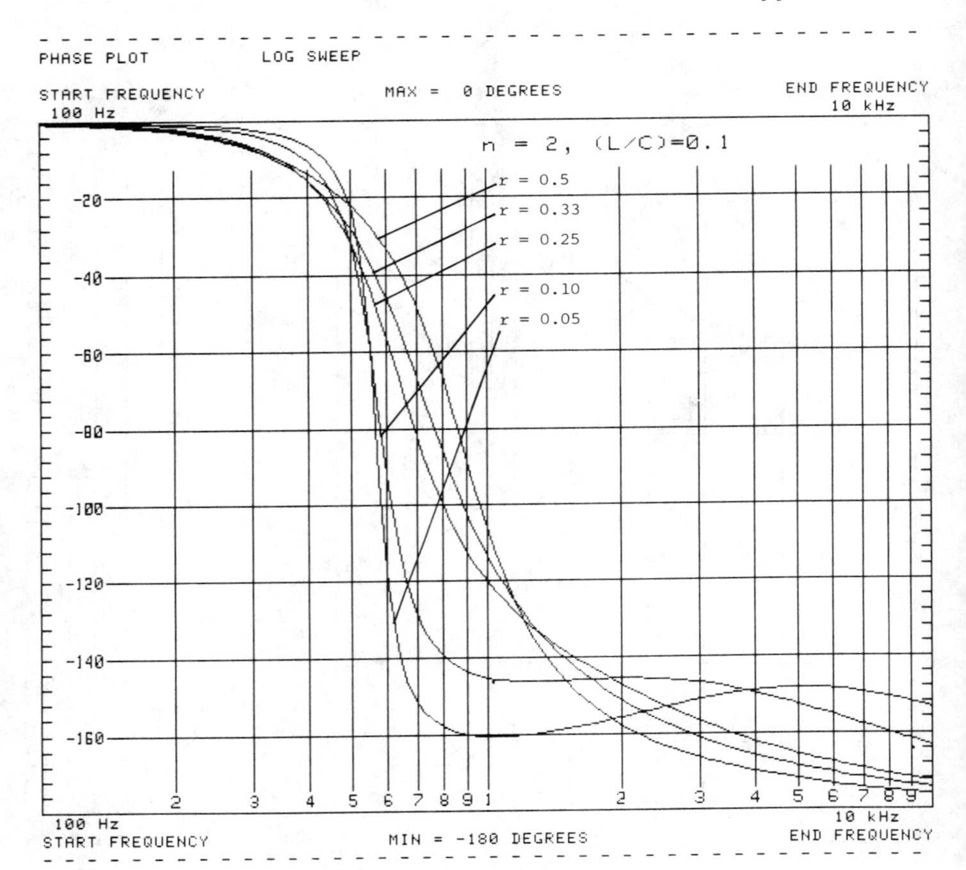
PHASE PLOT
LOG SWEEP
START FREQUENCY
100 Hz
MAX = 0 DEGREES
END FREQUENCY
10 kHz
n = 2, (L/C)=0.1
r = 0.5
r = 0.33
r = 0.25
r = 0.10
r = 0.05
-20
-40
-60
-80
-100
-120
-140
-160
2 3 4 5 6 7 8 9 1 2 3 4 5 6 7 8 9
100 Hz
START FREQUENCY
MIN = -180 DEGREES
10 kHz
END FREQUENCY

FIG. B.27

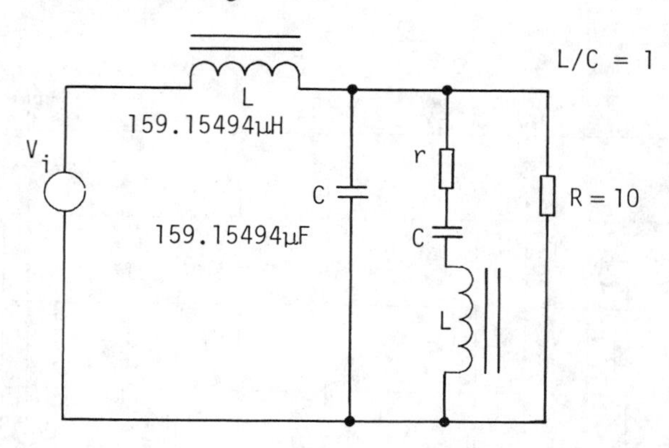

FIG. B.28 Output filter with "trap" circuit.

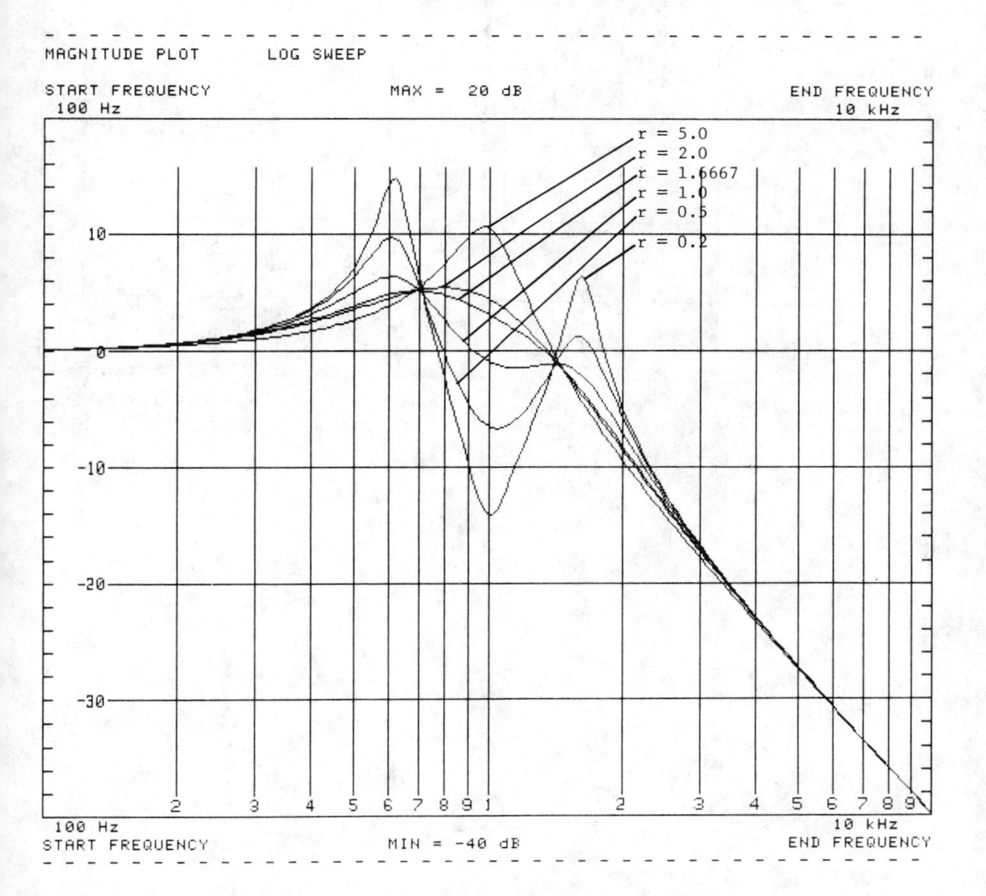

FIG. B.29

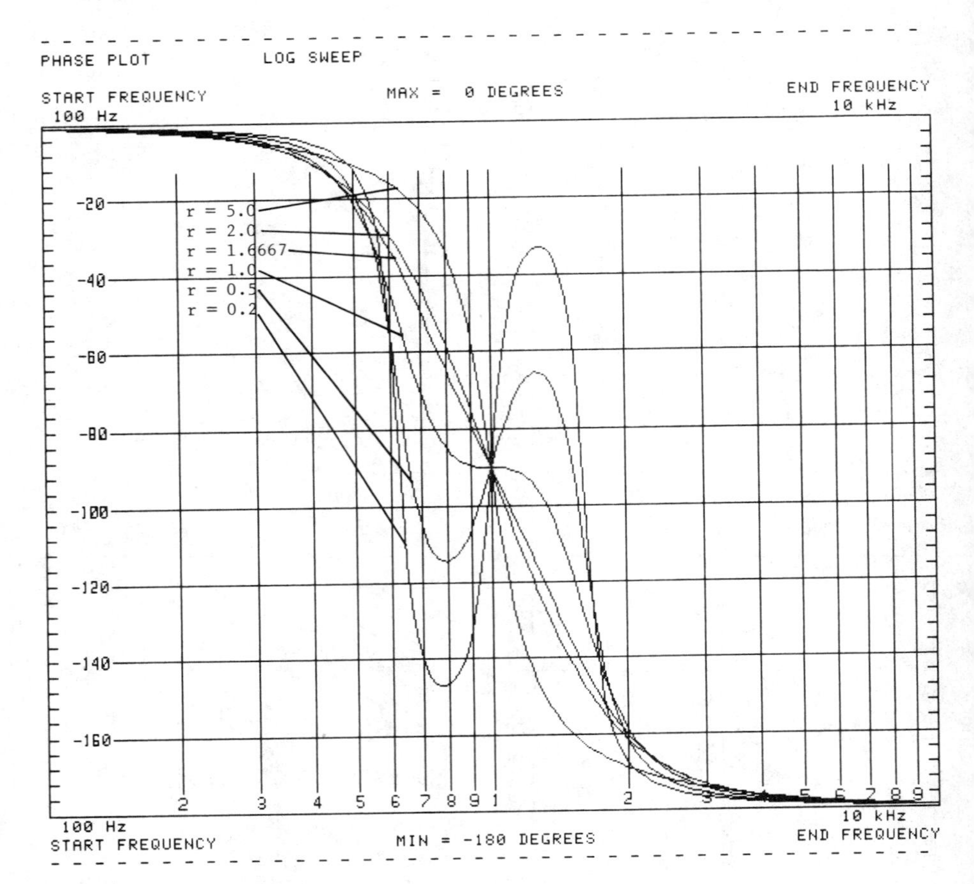
PHASE PLOT
LOG SWEEP
START FREQUENCY
100 Hz
MAX = 0 DEGREES
END FREQUENCY
10 kHz
-20
-40
-60
-80
-100
-120
-140
-160
r = 5.0
r = 2.0
r = 1.6667
r = 1.0
r = 0.5
r = 0.2
2 3 4 5 6 7 8 9 1 2 3 4 5 6 7 8 9
100 Hz
START FREQUENCY
MIN = -180 DEGREES
10 kHz
END FREQUENCY

FIG. B.30

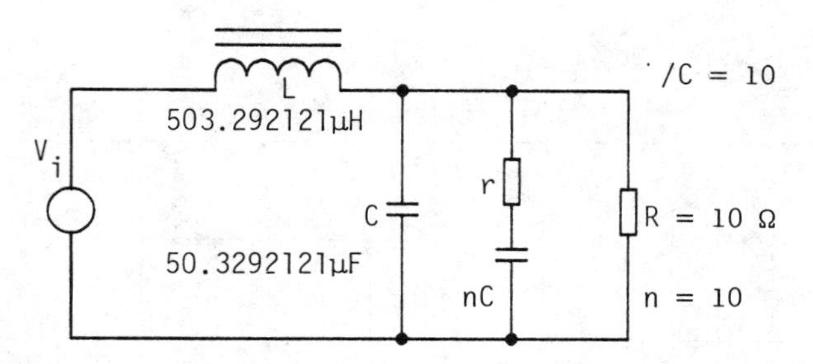

FIG. B.31

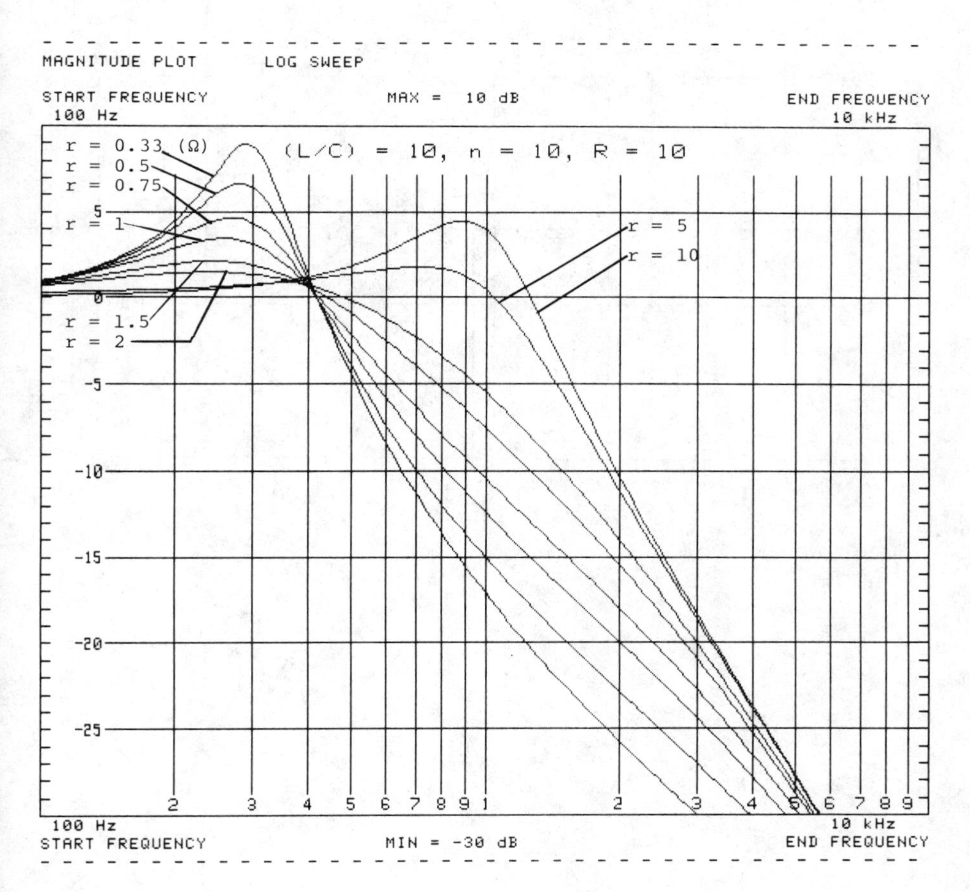

FIG. B.32

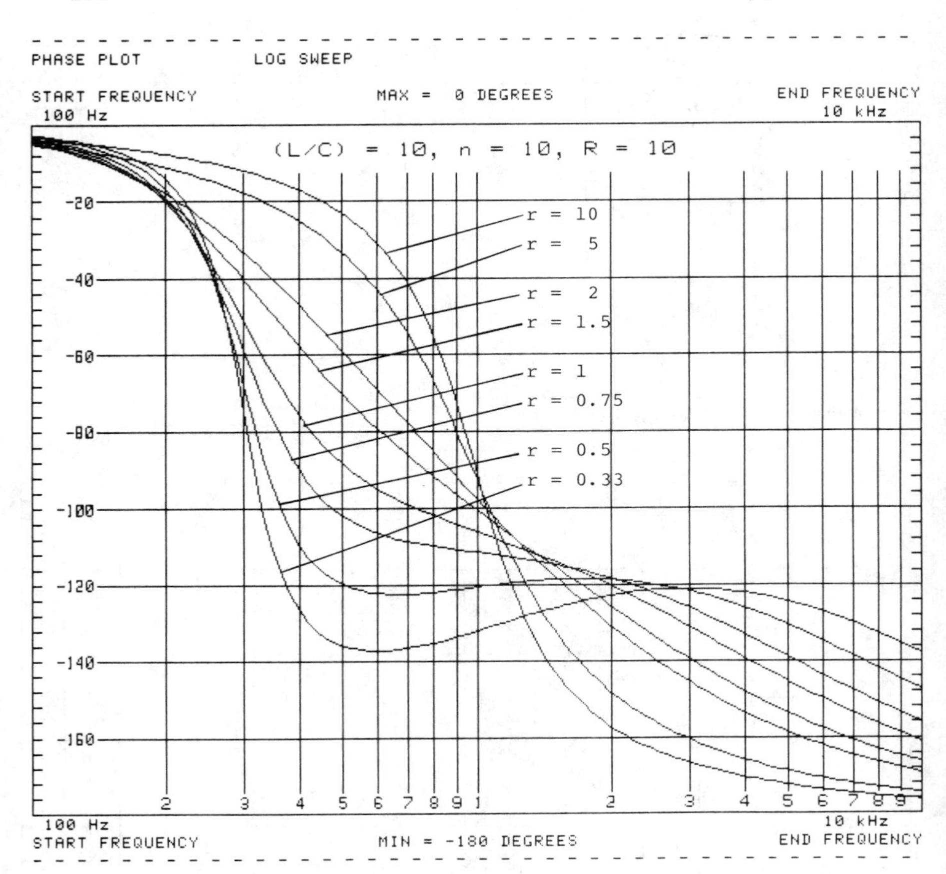
PHASE PLOT
LOG SWEEP
START FREQUENCY
100 Hz
MAX = 0 DEGREES
END FREQUENCY
10 kHz
(L/C) = 10, n = 10, R = 10
r = 10
r = 5
r = 2
r = 1.5
r = 1
r = 0.75
r = 0.5
r = 0.33
-20
-40
-60
-80
-100
-120
-140
-160
2 3 4 5 6 7 8 9 1 2 3 4 5 6 7 8 9
100 Hz
START FREQUENCY
MIN = -180 DEGREES
10 kHz
END FREQUENCY

FIG. B.33

appendix C
Using Transient Response to Determine System Stability

By definition, the transfer function of a second-order control system can be written as

$$\frac{C(s)}{R(s)} = \frac{\omega_n^2}{s^2 + 2\zeta\omega_n + \omega_n^2} \tag{C.1}$$

where ζ is the damping ratio and ω_n is the natural frequency of oscillation of the system.

For unit step input,

$$R(s) = \frac{1}{s}$$

which reduces Eq. (C.1) to

$$C(s) = \frac{\omega_n^2}{s(s^2 + 2\zeta\omega_n + \omega_n^2)} \tag{C.2}$$

The inverse Laplace transform of Eq. (C.2) is

$$C(t) = \mathcal{L}^{-1}\left[\frac{\omega_n^2}{s(s^2 + 2\zeta\omega_n + \omega_n^2)}\right] \tag{C.3}$$

$$= 1 - \frac{e^{-\zeta\omega_n t}}{\sqrt{1 - \zeta^2}} \sin\left(\omega_n\sqrt{1 - \zeta^2}\, t + \tan^{-1}\frac{\sqrt{1 - \zeta^2}}{\zeta}\right) \tag{C.4}$$

The step response for a second-order control system is plotted in Fig. C.1.

The transfer function of Eq. (C.1) has poles at

$$s = -\zeta\omega_n \pm \sqrt{(\zeta^2 - 1)\omega_n^2} \tag{C.5}$$

For $\zeta < 1$,

$$s = -\zeta\omega_n \pm j\omega_n\sqrt{1 - \zeta^2} \tag{C.6}$$

This case is of greatest interest to control system designers.

To find the maxima and minima along the time axis, differentiate Eq. (C.4) with respect to t:

$$\frac{dC(t)}{dt} = \frac{-e^{-\zeta\omega_n t}}{\sqrt{1 - \zeta^2}} \omega_n\sqrt{1 - \zeta^2} \cos(\omega t + \vartheta)$$

$$+ \frac{\zeta\omega_n e^{-\zeta\omega_n t}}{\sqrt{1 - \zeta^2}} \sin(\omega t + \vartheta) \tag{C.7}$$

where

$$\vartheta = \tan^{-1}\frac{\sqrt{1 - \zeta^2}}{\zeta} \tag{C.8}$$

Equating Eq. (C.7) to zero gives

$$-\sqrt{1 - \zeta^2} \cos(\omega t + \vartheta) + \zeta \sin(\omega t + \vartheta) = 0 \tag{C.9}$$

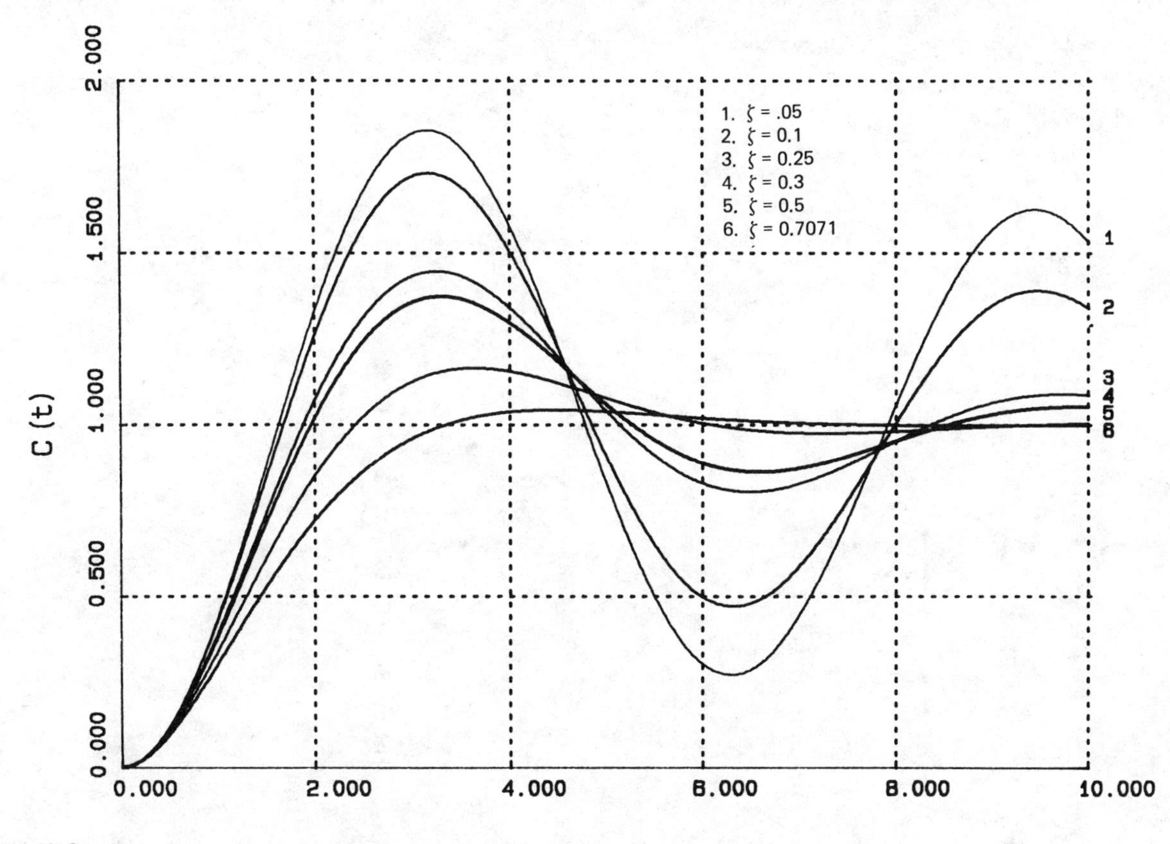

FIG. C.1 Step responses of a second-order control system.

Therefore,

$$\tan(\omega t + \vartheta) = \frac{\sqrt{1 - \zeta^2}}{\zeta} \tag{C.10}$$

From Eqs. (C.8) and (C.10),

$$\tan\left(\omega t + \tan^{-1}\frac{\sqrt{1 - \zeta^2}}{\zeta}\right) = \frac{\sqrt{1 - \zeta^2}}{\zeta} \tag{C.11}$$

But

$$\omega t = n\pi, \quad n = 0, 1, 2, \ldots \tag{C.12}$$

Therefore,

$$t = \frac{n\pi}{\omega_n\sqrt{1 - \zeta^2}} \tag{C.13}$$

Therefore, for n = 1, the first maximum of the system output is

$$t_1 = \frac{\pi}{\omega_n\sqrt{1 - \zeta^2}} \tag{C.14}$$

Substituting Eq. (C.13) into Eq. (C.4) for t gives

$$\text{for } n = 1\text{:} \quad C(t)_1 = 1 + e^{-\zeta\pi/\sqrt{1 - \zeta^2}} \quad \text{maxima}$$

$$\text{for } n = 2\text{:} \quad C(t)_2 = 1 - e^{-2\zeta\pi/\sqrt{1 - \zeta^2}} \quad \text{minima}$$

$$\text{for } n = 3\text{:} \quad C(t)_3 = 1 + e^{-3\zeta\pi/\sqrt{1 - \zeta^2}} \quad \text{maxima}$$

Therefore [212],

$$\log_e\left[\frac{C(t)_1 - C(t)_3}{C(t)_2}\right] = \frac{-\zeta\pi}{\sqrt{1 - \zeta^2}} \tag{C.15}$$

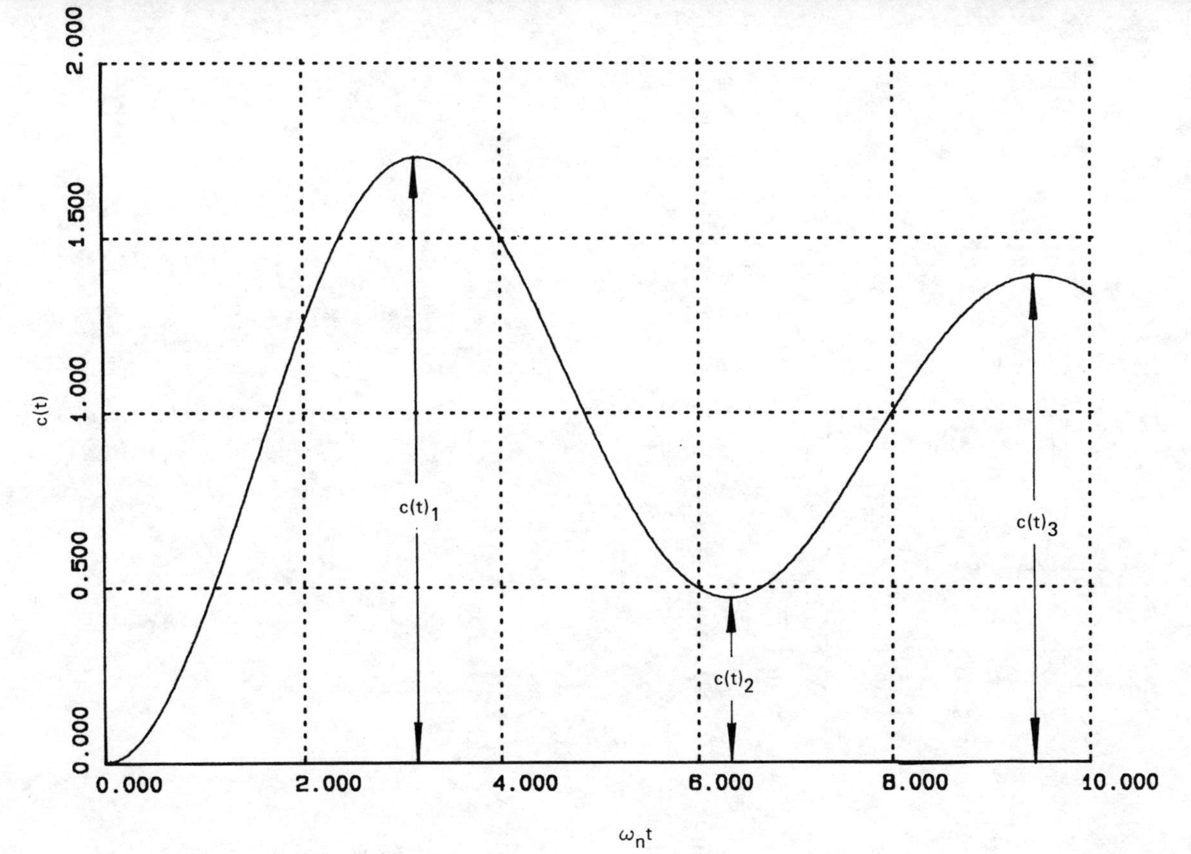

FIG. C.2 Maximum and minimum values for calculating ζ.

or

$$\log_{10}\left[\frac{C(t)_1 - C(t)_3}{C(t)_2}\right] = \frac{-\zeta\pi}{2.3\sqrt{1 - \zeta^2}} \tag{C.16}$$

Solving for ζ gives

$$\zeta = \frac{\left|\log_{10}\left[\frac{C(t)_1 - C(t)_3}{C(t)_2}\right]\right|}{\sqrt{\log_{10}^2\left[\frac{C(t)_1 - C(t)_3}{C(t)_2}\right] + 1.86571}} \tag{C.17}$$

Equation (C.13) is used to find ω_n using the calculated values of t_1, t_3, and ζ [from Eq. (C.17)]:

$$\omega_n = \frac{2\pi}{(t_3 - t_1)\sqrt{1 - \zeta^2}} \tag{C.18}$$

See Fig. C.2.

appendix D

Derivation of Current Density and Area Product Relationships

The relationship of current density J to the area product A_p of a magnetic component for a given temperature rise is derived as follows [120].

The total surface area of the component A_t is (see Figs. D.1 to D.4)

$$A_t = K_4 l^2 \qquad (D.1)$$

The area product is

$$A_p = K_2 l^4 \qquad (D.2)$$

$$l^4 = \frac{A_p}{K_2} \qquad (D.3)$$

$$l^2 = \sqrt{\frac{A_p}{K_2}} \qquad (D.4)$$

Substituting Eq. (D.4) into Eq. (D.1) for l^2 gives

$$A_t = K_4 \sqrt{\frac{A_p}{K_2}} \qquad (D.5)$$

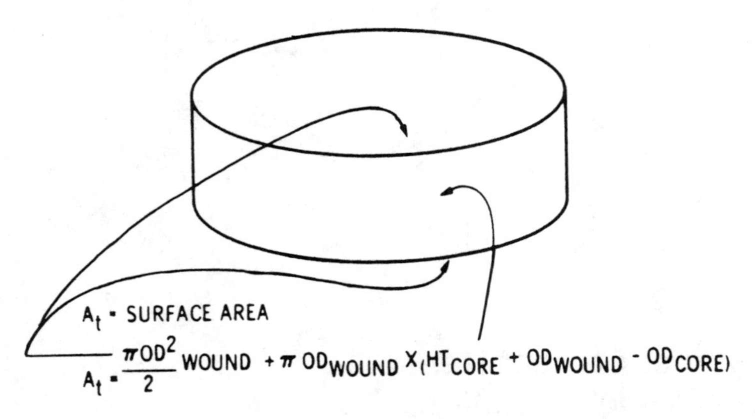

FIG. D.1 Tape-wound core, powder core, and pot core surface area A_t.

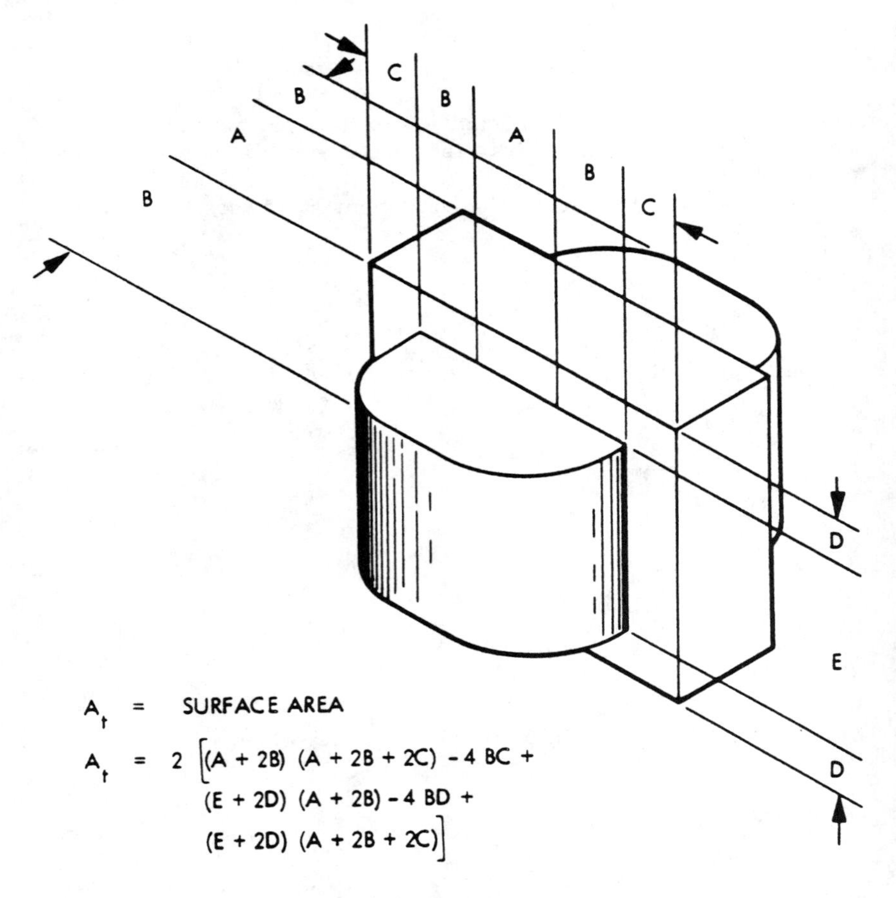

FIG. D.2 EI lamination surface area A_t.

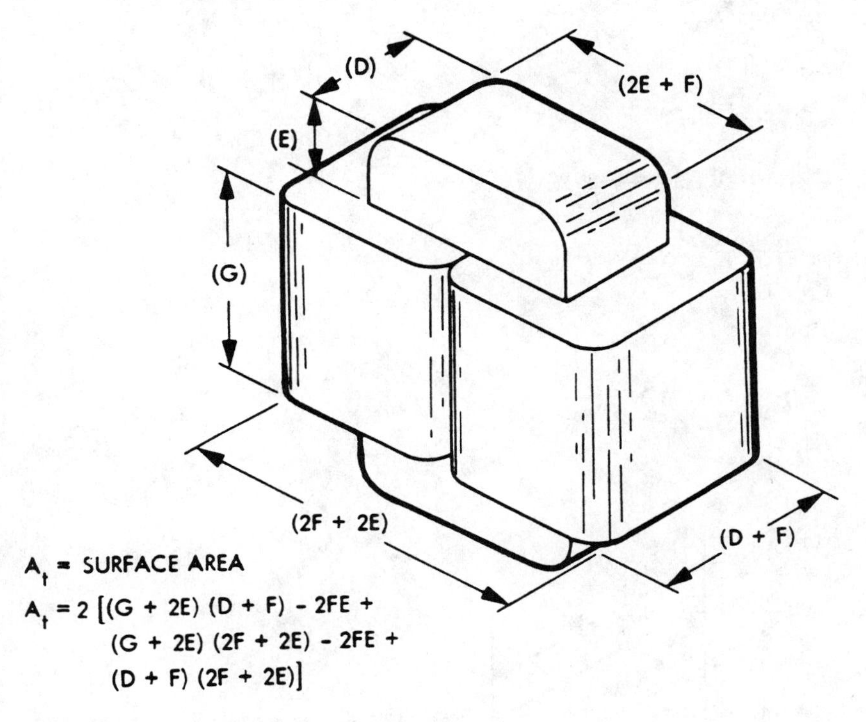

FIG. D.3 C-core surface area A_t.

Introducting the core configuration constant K_s, we obtain

$$K_s = \frac{K_4}{\sqrt{K_2}} \tag{D.6}$$

$$A_t = K_s \sqrt{A_p} \tag{D.7}$$

The copper loss is

$$P_{cu} = I^2 R \tag{D.8}$$

$$I = A_w J \tag{D.9}$$

Therefore,

$$P_{cu} = A_w^2 J^2 R \tag{D.10}$$

The winding resistance is

$$R_w = \frac{MLT}{A_w} N_p \tag{D.11}$$

Therefore,

$$P_{cu} = A_w^2 J^2 N_p \frac{MLT}{A_w} \tag{D.12}$$

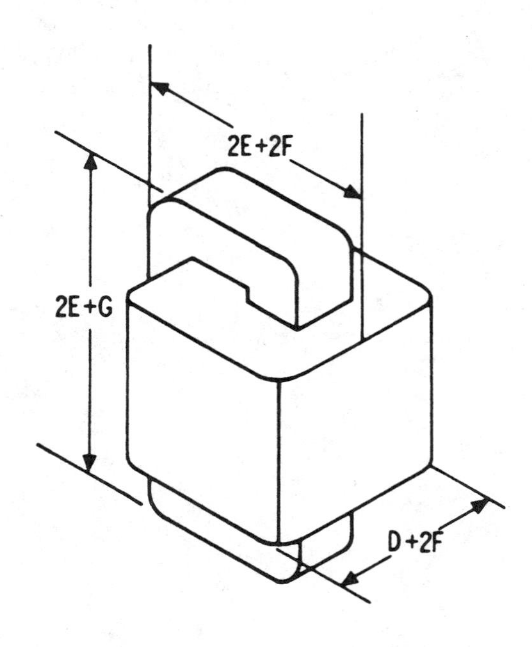

$$A_t = 2\left\{ 2(E+F)\left[(D+2F)+(G+2E)\right]+(G+2E)(D+2F)-8EF\right\}$$

FIG. D.4 Single-coil C-core surface area A_t.

$$P_{cu} = A_w J^2 N_p MLT \tag{D.13}$$

But MLT has a dimension of length

$$MLT = K_5 A_p^{0.25} \tag{D.14}$$

Therefore,

$$P_{cu} = A_w J^2 K_5 A_p^{0.25} N_p \tag{D.15}$$

$$A_w N = K_6 W_a = K_3 \sqrt{A_p} \tag{D.16}$$

$$P_{cu} = K_6 A_p^{0.5} K_5 A_p^{0.25} J^2 \rho \tag{D.17}$$

$$K_7 = K_6 K_5 \rho \tag{D.18}$$

Assume that $P_{cu} = P_{fe}$:

$$P_{cu} = K_7 A_p^{0.75} J^2 = P_{fe} \tag{D.19}$$

The sum of copper and iron losses is

$$P_\Sigma = P_{cu} + P_{fe} \tag{D.20}$$

The change in temperature is

$$\Delta T = K_8 \frac{P_\Sigma}{A_t} \tag{D.21}$$

$$\Delta T = \frac{2K_8 K_7 J^2 A_p^{0.75}}{K_s A_p^{0.5}} \tag{D.22}$$

$$K_9 = \frac{2K_8 K_7}{K_s} \tag{D.23}$$

TABLE D.1 Core Configuration Constants

Core	Losses	K_j (25°C)	K_j (50°C)	(x)	K_s	K_w	K_v
Pot core	$P_{cu} = P_{fe}$	433	632	-0.17	33.8	48.0	14.5
Powder core	$P_{cu} \gg P_{fe}$	403	590	-0.12	32.5	58.8	13.1
Lamination	$P_{cu} = P_{fe}$	366	534	-0.12	41.3	68.2	19.7
C-core	$P_{cu} = P_{fe}$	323	468	-0.14	39.2	66.6	17.9
Single-coil	$P_{cu} \gg P_{fe}$	395	569	-0.14	44.5	76.6	25.6
Tape-wound core	$P_{cu} = P_{fe}$	250	365	-0.13	50.9	82.3	25.0

$J = K_j A_p^{(x)}$ $\quad$ $A_t = K_s A_p^{0.50}$

$W_t = K_w A_p^{0.75}$ $\quad$ $Vol = K_v A_p^{0.75}$

$$\Delta T = K_g J^2 A_p^{0.25} \tag{D.24}$$

$$J^2 = \frac{\Delta T}{K_g A_p^{0.25}} \tag{D.25}$$

$$K_{10} = \frac{\Delta T}{K_9} \tag{D.26}$$

$$J^2 = K_{10} A_p^{-0.25} \tag{D.27}$$

$$J = K_j A_p^{-0.125} \tag{D.28}$$

or

$$J = K_j A_p^{x} \tag{D.29}$$

where x is given in Table D.1.

appendix E
SPICE Models for Computer Simulation

```
*___________________________________________________________
*                    MACRO: 2
*            Buck Step_Down Transformer
*               Mode : Continuous
*___________________________________________________________
*
.SUBCKT BUCK 1 2 3 4 5
*   NODE DEFINITION
*   Input : 1,2
*   Output : 3,4
*   Duty Cycle : 5
*   N=1 is transformer turn ratio
*   1:D*N is the effective turn ratio
*   Gain of GIN is N/ROUT
*
Rduty 5 0 1MEG
Rout  6 3 .01
Gin   1 2 POLY(2) 6 3 5 0 0 0 0 0 100.0
Eout  6 4 POLY(2) 1 2 5 0 0 0 0 0 1.0
.ENDS BUCK
*___________________________________________________________
*                    MACRO: 3
*            Boost Step_Up Transformer
*               Mode : Continuous
*___________________________________________________________
*
.SUBCKT BOOST 1 2 3 4 5
*   Node Definition
*   Input : 1,2
*   Output : 3,4
*   Duty Cycle : 5
```

```
*   N=1 is transformer turn ratio
*   (1-D):N is the effective turn ratio
*
Rduty 5 0 1MEG
Rin   1 6 .01
Rout  3 4 100MEG
Gout  4 3 POLY(2) 1 6 8 0 0 0 0 0 100.0
Ein   6 2 POLY(2) 3 4 8 0 0 0 0 0 1.0
Rduty1 8 0 1MEG
Vduty 8 7 DC 1
Eduty 7 0 5 0 -1.0
.ENDS BOOST
*________________________________________________________________
*                         MACRO: 4
*                  SG1524 voltage regulator
*________________________________________________________________
*
*    NODES 1 & 2 : ERROR AMP INPUTS
*    NODE 9      : ERROR AMP OUTPUT
*    NODE 11     : DUTY CYCLE OUTPUT
*    NODE 16     : 5V REFERENCE OUTPUT
*
.SUBCKT SG1524 1 2 9 11 16
RI 2 1 100K
R1 9 0 5MEG
RO 20 0 1K
RC 9 25 1G
VO 25 0 DC .7
VA 23 0 DC 4
VB 24 0 DC .6
CI 2 1 1P
C1 9 0 159P
G1 0 9 2 1 1.6M
ED 11 0 20 0 .01
VR 16 0 DC 5
GD 0 20 9 25 17.05M
IX 0 20 DC 1.613M
VP 21 0 DC 49.4
VM 22 0 DC 1.6
D1 20 21 DSWIT1
D2 22 20 DSWIT1
D3 9 23 DSWIT1
D4 24 9 DSWIT1
.MODEL DSWIT1 D(RS=.1)
.ENDS SG1524
```

```
*________________________________________________________________
*                          MACRO: 5
*                    Dc-to-Dc Transformer
 *_______________________________________________________________
*
.SUBCKT XFMR 1 2 3 4
*   Node Definition
*   Input : 1,2
*   Output : 3,4
*   Gain of Primary current source = N/Rsecond
*   Gain of Secondary current source = N/Rsecpar
*   N = .62
*
Rprimary 5 6 .01
Rcorloss 6 2 10K
Leakind  1 5 200U
Rsecond  7 3 .1
Rsecpar  7 4 .1
Gcurpri  6 2 7 3 6.2
Gcursec  4 7 6 2 6.2
.ENDS   XFMR
*________________________________________________________________
```

appendix F
Pulse Handling Capability of Wire-Wound Resistors

Reprinted with permission of Dale Electronics, Columbus, Nebraska.

INTRODUCTION:

Power wirewound resistors have steady state power and voltage ratings which limit the temperature of the unit to less than + 275°C or + 350°C. For short durations of 5 seconds or less these ratings are satisfactory, however, the resistors are capable of handling much higher levels of power and voltage. For instance, the RS-5, 10 Ohm has a continuous rating of 5 watts but for a duration of 1 millisecond the unit can handle 24,600 watts and for 1 microsecond the unit can handle 24,600,000 watts. The reason for this seemingly high power capability is the fact that energy, which is the product of power and time, is what creates heat; not just power alone.

SHORT PULSES (LESS THAN 100 MILLISECOND)

For short pulses, it is necessary to determine the energy the customer is planning to apply along with the resistance value. The energy of the pulse is then compared with the energy capability shown in the energy - resistance chart. The energy per Ohm shown in the left hand column is the amount required to raise the wire temperature to + 350°C with no heat loss in the core, coating or leads. This assumption will be fairly accurate for microsecond pulses and gets more conservative as the pulses get longer. For this reason, the energy - resistance chart is limited to pulses up to 100 millisecond. For a single square wave pulse, energy is calculated as follows:

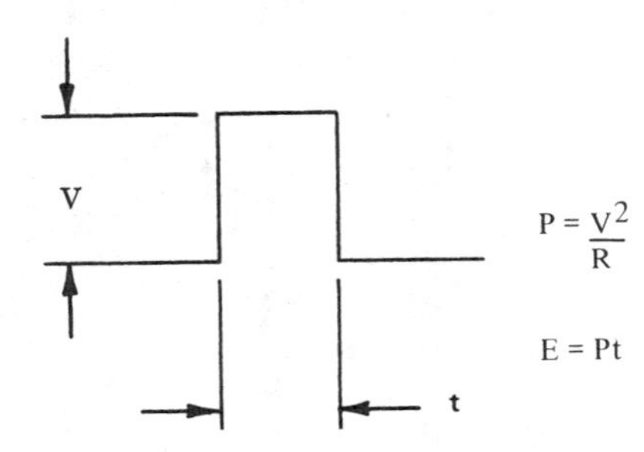

$$P = \frac{V^2}{R}$$

$$E = Pt$$

Where: P = Pulse Power (Watts)
V = Pulse Voltage (Volts)
R = Resistance (Ohms)
t = Pulse Duration (Seconds)
E = Energy (Watt-seconds or Joules)

(1) After the energy has been calculated, divide by the resistance to obtain watt-seconds per ohm.

(2) Go to the energy-resistance chart and choose the energy per ohm value which is equal to or greater than that which was calculated.

(3) Follow across the chart to the right until the resistance value or one higher than called for in the application is reached.

(4) The resistor styles shown at the top of this column will then be the smallest size to handle the pulse.

Note: Any style shown to the right of this column could be used and would provide an additional safety factor.

Example: A single square wave pulse having an amplitude of 100 volts for 1 millisecond is applied to a 10 Ohm resistor. What would be the smallest resistor style in the RS line that would handle this pulse?

$$P = \frac{V^2}{R} = \frac{100^2}{10} = \frac{10{,}000}{10} = 1000 \text{ watts}$$

$$E = Pt = (1000)\ (.001) = 1 \text{ watt-seconds}$$

$$\frac{E}{R} = \frac{1}{10} = \frac{.1 \text{ watt - second}}{\text{ohm}}$$

(1) From the energy-resistance chart, page 7, the next higher energy is .153 watt-seconds per ohm.

(2) Following this row to the right, the next highest value above 10 Ohm is 21.1 Ohms.

(3) At the top of the column, the RS-2B is indicated as the style which would handle this pulse.

Another frequently used short pulse application is the capacitor discharge circuit. Here a capacitor is charged to a given voltage and then discharged through a wirewound resistor. The energy for this pulse is calculated as follows:

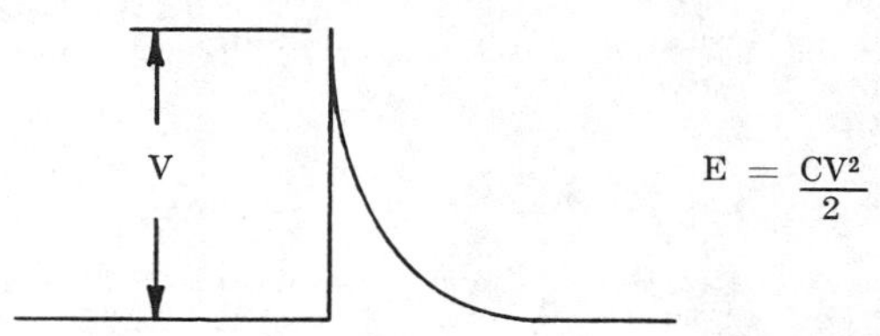

$$E = \frac{CV^2}{2}$$

Where: C = Capacitance (Farads)

V = Peak Voltage (Volts)

E = Energy (Watt-seconds or Joules)

(1) After the energy is computed the value is divided by the resistance to obtain watt-seconds per ohm.

(2) The energy-resistance chart is then used the same way as in the case of the square-wave pulse to find the resistor style.

Example: A 2 microfarad capacitor charged to 400 volts is being discharged into a 1K resistor. What is the smallest RS that will handle this pulse?

(1) The energy of the pulse is:

$$E = \frac{CV^2}{2} = \frac{(2x10^{-6})\ (400)^2}{2} = (1x10^{-6})\ (16x10^4) = .16 \text{ watt-second}$$

(2) Divide the energy by the resistance:

$$\frac{E}{R} = \frac{.16}{1x10^3} = .16 \times 10^{-3} = 160 \times 10^{-6} \frac{\text{watt - second}}{\text{ohm}}$$

(3) The next highest energy per ohm found in the energy - resistance chart is 221 x 10^{-6}.

(4) Going across that row to the right, the next highest value above 1K is 1420 Ohms.

(5) The RS style shown at the top of that column is the RS-1A.

EQUALLY SPACED REPETITIVE PULSES

When calculating pulse handling capability for repetitive pulses, the average power as well as the individual pulse energy must be considered. Calculations for repetitive pulses are as follows:

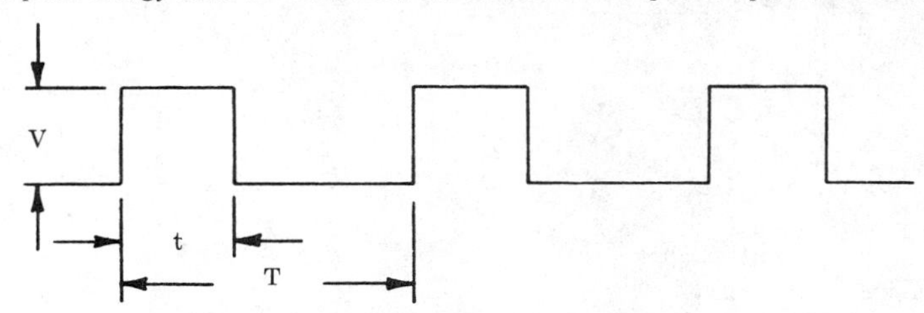

(1) The pulse power $P = \frac{V^2}{R}$ is calculated for a single pulse.

(2) The single pulse energy E = Pt is then computed.

(3) The average power is calculated as follows:

$$P_A = \frac{Pt}{T}$$

P_A = Average Power (watts)

P = Pulse Power (watts)

t = Pulse Width (seconds)

T = Cycle Time (seconds)

(4) A new energy based on the summation of pulse energy and the contribution due to average power is then computed as follows:

$$E_{AP} = E\left(1 + \frac{P_A}{P_R}\right)$$

Where: E_{AP} = Pulse energy + Average Power energy (watt-seconds)
E = Pulse energy (watt-seconds)
P_A = Average power (watts)
P_R = Rated power (watts)

Note: P_R is the continuous power rating of a resistor style that must be chosen as a starting point.

(5) Divide E_{AP} by R to find the energy per ohm.

(6) Go to the energy-resistance chart and find the resistor style chosen for P_R.

(7) Follow down that column until the resistance value equal to or greater than the one being used is reached.

(8) Follow across that row to the left and note the energy per ohm.

(9) If the energy per ohm in the chart is greater than that calculated, the resistor style chosen is satisfactory.

(10) If the energy per ohm in the chart is less than that calculated, a larger style must be chosen and the above calculations repeated.

Example: A series of equally spaced square wave pulses having an amplitude of 200 volts, a pulse width of 20 milliseconds and cycle time of 20 seconds is applied to a 100 Ohm resistor. Will the RS-5 handle this pulse?

(1) The pulse power is:

$$P = \frac{V^2}{R} = \frac{(200)^2}{100} = \frac{4 \times 10^4}{1 \times 10^2} = 400 \text{ watts}$$

(2) The pulse energy is:

$$E = Pt = (400)(.02) = 8 \text{ watt-seconds}$$

(3) The average power is:

$$P_A = \frac{Pt}{T} = \frac{(400)(20 \times 10^{-3})}{20} = .4 \text{ watts}$$

(4) The energy due to the pulse and average power is:

$$E_{AP} = E\left(1 + \frac{P_A}{P_R}\right)$$

$$= 8\left(1 + \frac{.4}{5}\right)$$

$$= 8.64 \text{ watt seconds}$$

(5) The energy per ohm is:

$$\frac{E_{AP}}{R} = \frac{8.64}{100} = \frac{.0864 \text{ watt-second}}{\text{ohm}}$$

(6) In the energy resistance chart, the next higher value above 100 Ohm is 116 Ohm.

(7) Following this row to the left, the energy per ohm is .090 which is sufficient to handle the application.

NON-INDUCTIVE RESISTORS

Non-inductive power resistors consist of two windings each of which is twice the finished resistance value. For this reason, the energy capability will nearly always be greater than a standard wound unit. The procedure used in calculating energy capability for non-inductive styles is as follows:

(1) Compute the energy per ohm by dividing the energy by 4 times the resistance value.

(2) Go to the energy-resistance chart and choose the energy per ohm value which is equal to or greater than that which was calculated.

(3) Follow the chart to the right until twice the resistance value or one higher than called for is reached.

(4) The non-inductive equivalent shown at the top of the column will then be the smallest size to handle the pulse.

Example: What is the smallest NS style resistor required to handle a .2 watt-second pulse applied to a 500 Ohm value?

(1) The energy per ohm is $\frac{E}{4R} = \frac{.2}{2000} = 100 \times 10^{-6} \frac{\text{Joules}}{\text{ohm}}$

(2) In the energy-resistance chart the next higher energy per ohm is 145×10^{-6}.

(3) Following this row to the right the next higher value than twice the resistance value of 1000 Ohms is 1960 Ohms.

(4) At the top of this column the RS-1A is indicated. The non-inductive equivalent is the NS-1A.

LONG PULSES (100 MILLISECONDS TO 5 SECONDS)

For long pulses much of the heat is dissipated in the core, leads, and coating. As a result, the calculations used for short pulses are far too conservative. For long pulse applications the short time overload ratings in the catalog are used. For RS, G, NS, and GN styles this overload is either 10 or 5 times rated power for 5 seconds, depending on the style.

(1) To find the power overload for a 5 second pulse, multiply the power rating by either 5 or 10 depending on size.

(2) To find the overload capability for 1 to 5 seconds, convert the overload power to energy by multiplying by 5 seconds then convert back to power by dividing by the pulse width in seconds.

(3) For pulse durations between 100 millisecond and 1 second use the overload power computed for 1 second.

Example: How much power can an RS-5 handle for 2 seconds?

(1) Rated power for an RS-5 is 5 watts.

(2) From the catalog an RS-5 will take 10 times rated power for 5 seconds.
$5 \times 10 = 50$ watts

(3) For 5 seconds the energy capability is $50 \times 5 = 250$ watt-seconds.

(4) For 2 seconds the power capability is $\frac{250}{2} = 125$ watts.

VOLTAGE LIMITATIONS

Short pulses — No overload voltage rating has ever been established for wirewound resistors when pulsed for short durations. Sandia Corporation has, however, performed a study on our NS and RS resistors using 20 microsecond pulses. This study indicates that this type unit will take about 20,000 volts per inch as long as the energy shown in the energy-resistance chart is not exceeded.

Long pulses — For pulses 100 millisecond to 5 seconds the recommended maximum overload is $\sqrt{10}$ times the maximum working voltage for 4 watt size and larger and $\sqrt{5}$ times the maximum working voltage for sizes smaller than 4 watt.

Energy-Resistance Chart

ENERGY PER OHM JOULES OR WATT-SECOND	EGS-1 RS-¼ G-1	EGS-2 RS-½ G-2	EGS-3 RS-1A G-3	ESS-2B RS-2B G-5	RS-2C G-5C	RS-2 G-6	RS-5 EGS-10 RS-5-69 G-10	RS-7 G-12	RS-10 ESS-10 RS-10-38 G-15
13.9×10^{-6}	3480	4920	10.4K	24.5K	32.3K	47.1K	90.9K	154K	265K
20.3×10^{-6}	2589	3659	7580	18.69K	24.19K	31.79K	69.4K	114.9K	197K
28.7×10^{-6}	1999	2829	5840	14.19K	18.29K	26.99K	51.7K	88K	152K
39.5×10^{-6}	1549	2189	4630	10.89K	13.69K	20.69K	40.4K	68.59K	111K
53.1×10^{-6}	1239	1749	3630	8600	11.39K	16.69K	31.4K	54.39K	93.5K
70.0×10^{-6}		1414	2920	6980	9250	13.59K	25.9K	44.19K	75.5K
90.6×10^{-6}	1000	1149	2740	6550	7560	11.09K	24.5K	36.79K	71.5K
145×10^{-6}	670	947	1960	4650	6260	8910	17.3K	29.5K	50.6K
221×10^{-6}	492	684	1420	3370	4560	6570	12.7K	20.59K	37.4K
324×10^{-6}	355	502	1040	2460	3270	4820	9220	15.69K	26.9K
460×10^{-6}	272	384	792	1860	2480	3640	7000	11.89K	20.4K
632×10^{-6}	206	291	615	1340	1920	2840	5460	9240	15.7K
850×10^{-6}	167	236	487	1150	1530	2260	4310	7320	12.4K
1.12×10^{-3}	131	186	393	935	1201	1800	3850	5900	10K
2.07×10^{-3}	96.3	136	283	671	910	1250	2840	4260	7540
3.54×10^{-3}	65.1	92	192	454	601	875	1690	2870	4920
5.67×10^{-3}	45.7	64.5	134	313	424	617	1160	2030	3460

Energy Resistance Chart (Continued)

ENERGY PER OHM JOULES OR WATT-SECOND	EGS-1 RS-¼ G-1	EGS-2 RS-½ G-2	EGS-3 RS-1A G-3	ESS-2B RS-2B G-5	RS-2C G-5C	RS-2 G-6	RS-5 EGS-10 RS-5-69 G-10	RS-7 G-12	RS-10 ESS-10 RS-10-38 G-15
8.65×10^{-3}	33.2	47	97.7	227	307	444	843	1470	2510
12.7×10^{-3}	23.8	33.6	71.1	168	222	310	622	1073	1840
20.4×10^{-3}	17.9	25.3	51.8	122	163	237	447	777	1340
33.2×10^{-3}	12.2	17.2	36.1	85.5	113	165	320	544	932
56.7×10^{-3}	8.22	11.6	24.2	57.8	76.3	111	215	364	618
.055	6.06	8.56	17.6	42.1	55.5	70.3	156	263	451
.090	4.47	6.32	13.3	31.6	40.5	51.0	116	201	343
.153	2.98	4.07	8.52	21.1	27.9	40.8	78.5	133	229
.245	2.18	3.09	6.28	14.8	19.6	28.6	55.4	95.0	160
.374	1.50	2.13	4.57	10.8	14.2	21.0	40.2	68.2	117
.589	1.12	1.59	3.27	7.86	10.3	14.9	29.0	49.0	84.1
.943	.780	1.10	2.31	5.46	7.22	10.6	20.3	34.4	59.3
1.52	.542	.773	1.61	3.80	5.13	7.40	14.1	24.2	41.6
2.46	383	.538	1.13	2.69	3.56	5.47	10.11	17.2	29.4
3.76	271	.394	.829	1.99	2.61	3.81	7.36	12.4	21.4
5.98			.591	1.41	1.84	2.15	5.24	8.87	15.1
2.00	.178	.244	.423	1.02			3.86	6.44	11.2

Energy-Resistance Chart (Continued)

ENERGY PER OHM JOULES OR WATT-SECOND	EGS-1 RS-¼ G-1	EGS-2 RS-½ G-2	EGS-3 RS-1A G-3	ESS-2B RS-2B G-5	RS-2C G-5C	RS-2 G-6	RS-5 EGS-10 RS-5-69 G-10	RS-7 G-12	RS-10 ESS-10 RS-10-38 G-15
2.61	.147	.201	.307	.773	1.36	1.59	2.86	4.79	8.1
4.23	.116	.159	.268	.681	1.00	1.35	2.52	4.09	7.0
5.23		.114	.210	.529	.784	1.04	2.00	3.19	5.4
8.04			.189	.475	.709	.949	1.84	2.99	4.9
13.4			.152	.383	.569	.764	1.43	2.29	3.9
20.9			.121	.297	.439	.591	1.12	1.79	3.1
33.2				.237	.354	.468	.875	1.48	2.52
42.1				.188	.278	.369	.716	1.16	2.00
25.1				.168	.248	.332	.630	1.03	1.74
41.8				.121	.179	.209	.487	.77	1.26
67.7					.139	.196	.380	.60	.992
168.5					.102	.147	.276	.47	.699
100.4							.171	.29	.485
166.8							.114		.310
271								.179	.252
674								.110	.184

appendix G
Capacitor Life Prediction Guidelines

1. The attached equations and tables can be used to predict the life of aluminum electrolytic capacitors at derated voltages and temperatures. Failures are defined as parameter drift beyond the limits outlined in the life test section of the appropriate product bulletin.

2. Based on dc aluminum electrolytic capacitor tests, the inherent relationships between temperature, voltage and life were established. A failure rate for each product was established from testing at maximum rated conditions. From this failure rate a base life time was established.

3. The multipliers found in the tables for each product type were derived from acceleration factors for voltage and temperature deration [equations (3), (4), and (5)].

4. The expected life for each product type is determined by computing the capacitor hot spot temperature [equation (1) or equation (2)], and the ratio of use voltage to rated voltage. From this, the base life multiplier can be found in the appropriate table. The multiplier times the base life (found in Table 1) yields the expected life.

5. The computation of expected life assumes a constant or decreasing failure rate and that the wearout portion of the product life has not been reached. The expected life is the statistical time required to generate one failure in 25 units based on a 60% confidence level.

6. Multipliers resulting in expected life times in excess of 10 years may not be valid due to secondary failure modes not considered in the construction of these tables.

Reprinted with permission of Mepco/Electra, Inc., A North-American Phillips Company, Morristown, New Jersey.

CALCULATION OF CORE TEMP

(1) $$\text{Core Temp } (^{\circ}\text{C})^{*} = (\text{CRF})(103)\left(\frac{I^{2}\,\text{ESR}}{\text{AREA}}\right)^{.833} + \text{AMB.}$$

or

(2) $$\text{Core Temp } (^{\circ}\text{C})^{*} = (\text{CRF})(\text{Case Temp.} - \text{AMB.}) + \text{AMB.}$$

D = Dia. (in.)
L = Case Length (in.)
CRF = Core Rise Factor = 1.068 + .31154 x Can Dia.
AREA = Surface Area of Can = $\frac{\pi D^2}{4} + \pi DL$
I = Ripple Current (Amps)
AMB = Ambient Temperature (°C)
ESR = Equivalent Series Resistance (ohms)

ACCELERATION FACTORS

(3) $A_1 = 2^{(T\ \text{Max-Core})/10}$ (Due to Chemical Kinetics)

(4) $$A_2 = \frac{I_L \text{ at Rated Voltage \& Temperature}}{I_L \text{ at Derated Voltage \& Temperature}}$$

(5) $A = A_1 \times A_2$

BASE LIFE

TYPE	LIFE HOURS	AMBIENT TEMP. DEGREES C	DESIGN CORE TEMP. DEGREES C	LIFE MULTIPLIER TABLE
3050	1000	85	95	2
3120	2000	85	105	4
3186	500	85	95	2
3188	1500	85	105	4
3191	1000	85	100	3
3192	1000	85	115	5
3428	2000	85	110	6

*Based on free convection in still air.

% RATED VOLTAGE

CORE TEMPERATURE (°C)	50.	55.	60.	65.	70.	75.	80.	85.	90.	95.	100.
95.	5.3	4.7	4.1	3.5	3.0	2.6	2.2	1.8	1.5	1.2	1.0
94.	5.7	5.1	4.4	3.8	3.3	2.8	2.3	1.9	1.6	1.3	1.1
93.	6.2	5.5	4.8	4.2	3.6	3.0	2.5	2.1	1.7	1.4	1.2
92.	6.8	6.0	5.2	4.5	3.8	3.2	2.7	2.3	1.9	1.5	1.2
91.	7.4	6.5	5.7	4.9	4.2	3.5	2.9	2.4	2.0	1.6	1.3
90.	8.1	7.1	6.2	5.3	4.5	3.8	3.2	2.6	2.2	1.8	1.4
89.	8.8	7.7	6.7	5.7	4.9	4.1	3.4	2.8	2.3	1.9	1.5
88.	9.6	8.4	7.3	6.2	5.3	4.4	3.7	3.0	2.5	2.0	1.7
87.	10.5	9.1	7.9	6.7	5.7	4.8	4.0	3.3	2.7	2.2	1.8
86.	11.4	9.9	8.6	7.3	6.2	5.1	4.3	3.5	2.9	2.4	1.9
85.	12.4	10.8	9.3	7.9	6.6	5.6	4.6	3.8	3.1	2.5	2.1
84.	13.5	11.7	10.1	8.5	7.2	6.0	5.0	4.1	3.3	2.7	2.2
83.	14.7	12.7	10.9	9.2	7.8	6.5	5.4	4.4	3.6	2.9	2.4
82.	16.0	13.8	11.8	10.0	8.4	7.0	5.8	4.7	3.9	3.2	2.6
81.	17.4	15.0	12.8	10.8	9.1	7.5	6.2	5.1	4.2	3.4	2.7
80.	19.0	16.3	13.9	11.7	9.8	8.1	6.7	5.5	4.5	3.6	2.9
79.	20.6	17.7	15.1	12.7	10.6	8.8	7.2	5.9	4.8	3.9	3.2
78.	22.5	19.2	16.3	13.7	11.4	9.5	7.8	6.4	5.2	4.2	3.4
77.	24.4	20.9	17.7	14.8	12.3	10.2	8.4	6.8	5.6	4.5	3.7
76.	26.6	22.7	19.2	16.0	13.3	11.0	9.0	7.4	6.0	4.9	3.9
75.	28.9	24.6	20.8	17.3	14.4	11.9	9.7	7.9	6.4	5.2	4.2
74.	31.4	26.7	22.5	18.8	15.5	12.8	10.5	8.5	6.9	5.6	4.5
73.	34.2	29.0	24.3	20.3	16.8	13.8	11.3	9.2	7.5	6.0	4.9
72.	37.2	31.4	26.4	21.9	18.1	14.9	12.1	9.9	8.0	6.5	5.2
71.	40.4	34.1	28.5	23.7	19.5	16.0	13.1	10.6	8.6	7.0	5.6
70.	43.9	37.0	30.9	25.6	21.1	17.3	14.1	11.4	9.3	7.5	6.0
69.	47.7	40.1	33.4	27.7	22.7	18.6	15.2	12.3	10.0	8.0	6.5
68.	51.9	43.5	36.2	29.9	24.5	20.0	16.3	13.2	10.7	8.6	6.9
67.	56.4	47.2	39.1	32.3	26.5	21.6	17.6	14.2	11.5	9.3	7.5
66.	61.2	51.1	42.3	34.9	28.5	23.3	18.9	15.3	12.4	10.0	8.0
65.	66.5	55.4	45.8	37.7	30.8	25.1	20.3	16.5	13.3	10.7	8.6
64.	72.3	60.0	49.5	40.7	33.2	27.0	21.9	17.7	14.3	11.5	9.2
63.	78.5	65.1	53.6	43.9	35.8	29.1	23.6	19.0	15.3	12.3	9.9
62.	85.2	70.5	57.9	47.4	38.6	31.3	25.3	20.5	16.5	13.3	10.6
61.	92.5	76.3	62.6	51.2	41.6	33.7	27.3	22.0	17.7	14.2	11.4
60.	100.4	82.7	67.7	55.2	44.9	36.3	29.3	23.6	19.0	15.3	12.3
59.	109.0	89.5	73.2	59.6	48.3	39.1	31.6	25.4	20.4	16.4	13.2
58.	118.3	97.0	79.1	64.3	52.1	42.1	33.9	27.3	22.0	17.6	14.1
57.	128.3	105.0	85.5	69.4	56.1	45.3	36.5	29.4	23.6	18.9	15.2
56.	139.2	113.6	92.4	74.9	60.5	48.8	39.3	31.6	25.3	20.3	16.3
55.	151.0	123.0	99.8	80.7	65.2	52.5	42.2	33.9	27.2	21.8	17.5
54.	163.8	133.1	107.8	87.1	70.2	56.5	45.4	36.4	29.2	23.4	18.8
53.	177.7	144.0	116.5	93.9	75.6	60.8	48.8	39.2	31.4	25.1	20.1
52.	192.7	155.9	125.8	101.3	81.5	65.4	52.5	42.1	33.7	27.0	21.6
51.	208.9	168.6	135.8	109.2	87.7	70.4	56.4	45.2	36.2	29.0	23.2
50.	226.4	182.4	146.6	117.8	94.5	75.7	60.7	48.6	38.9	31.1	24.9
49.	245.4	197.2	158.3	127.0	101.7	81.5	65.2	52.2	41.7	33.4	26.7
48.	266.0	213.3	170.9	136.8	109.5	87.6	70.1	56.1	44.8	35.8	28.6
47.	288.2	230.6	184.5	147.5	117.9	94.3	75.3	60.2	48.1	38.5	30.7
46.	312.3	249.3	199.1	159.0	127.0	101.4	81.0	64.7	51.7	41.3	33.0
45.	338.3	269.5	214.8	171.3	136.7	109.0	87.0	69.5	55.5	44.3	35.4
44.	366.4	291.3	231.8	184.6	147.1	117.3	93.5	74.6	59.6	47.5	38.0
43.	396.9	314.8	250.1	198.9	158.3	126.1	100.5	80.2	63.9	51.0	40.7
42.	429.8	340.1	269.7	214.2	170.4	135.6	108.0	86.1	68.7	54.8	43.7
41.	465.3	367.5	291.0	230.8	183.3	145.8	116.1	92.5	73.7	58.8	46.9
40.	503.7	397.0	313.8	248.6	197.3	156.8	124.7	99.3	79.1	63.1	50.3
39.	545.3	428.8	338.4	267.7	212.2	168.5	134.0	106.6	84.9	67.7	54.0
38.	590.1	463.1	364.9	288.3	228.3	181.2	143.9	114.5	91.2	72.6	57.9
37.	638.6	500.2	393.4	310.5	245.6	194.7	154.6	122.9	97.8	77.9	62.1
36.	691.0	540.0	424.1	334.3	264.2	209.3	166.1	132.0	105.0	83.6	66.6
35.	747.5	583.1	457.2	359.9	284.2	225.0	178.4	141.7	112.7	89.7	71.5
34.	808.6	629.4	492.7	387.4	305.7	241.8	191.7	152.2	121.0	96.3	76.7
33.	874.6	679.4	531.0	417.1	328.7	259.9	205.9	163.4	129.8	103.3	82.2
32.	945.8	733.2	572.3	448.9	535.5	279.2	221.1	175.4	139.3	110.8	88.2
31.	1022.7	791.3	616.6	483.1	380.1	300.1	237.5	188.3	149.5	118.9	94.6
30.	1105.7	853.8	664.4	519.9	408.7	322.4	255.0	202.1	160.5	127.6	101.5
29.	1195.3	921.2	715.7	559.5	439.5	346.4	273.9	217.0	172.2	136.8	108.9
28.	1291.9	993.7	771.0	602.0	472.5	372.2	294.1	232.9	184.8	146.8	116.8
27.	1396.3	1071.9	830.4	647.8	507.9	399.9	315.8	250.0	198.3	157.5	125.3
26.	1508.8	1156.1	894.4	696.9	546.0	429.6	339.1	268.3	212.8	169.0	134.3
25.	1630.2	1246.8	963.2	749.7	586.9	461.5	364.1	288.0	228.3	181.2	144.1

	% RATED VOLTAGE										
CORE TEMPERATURE (°C)	50.	55.	60.	65.	70.	75.	80.	85.	90.	95.	100.
100.	4.9	4.4	3.9	3.4	2.9	2.5	2.1	1.8	1.5	1.2	1.0
99.	5.4	4.8	4.2	3.7	3.2	2.7	2.3	1.9	1.6	1.3	1.1
98.	5.8	5.2	4.6	4.0	3.4	2.9	2.5	2.1	1.7	1.4	1.2
97.	6.4	5.7	5.0	4.3	3.7	3.2	2.7	2.2	1.8	1.5	1.2
96.	6.9	6.2	5.4	4.7	4.0	3.4	2.9	2.4	2.0	1.6	1.3
95.	7.6	6.7	5.9	5.1	4.4	3.7	3.1	2.6	2.1	1.8	1.4
94.	8.2	7.3	6.4	5.5	4.7	4.0	3.4	2.8	2.3	1.9	1.5
93.	9.0	7.9	6.9	6.0	5.1	4.3	3.6	3.0	2.5	2.0	1.7
92.	9.8	8.6	7.5	6.5	5.5	4.7	3.9	3.2	2.7	2.2	1.8
91.	10.7	9.4	8.2	7.0	6.0	5.0	4.2	3.5	2.9	2.4	1.9
90.	11.6	10.2	8.9	7.6	6.5	5.4	4.5	3.8	3.1	2.5	2.1
89.	12.7	11.1	9.6	8.3	7.0	5.9	4.9	4.1	3.3	2.7	2.2
88.	13.8	12.1	10.5	8.9	7.6	6.3	5.3	4.4	3.6	2.9	2.4
87.	15.0	13.1	11.3	9.7	8.2	6.9	5.7	4.7	3.9	3.2	2.6
86.	16.4	14.3	12.3	10.5	8.8	7.4	6.1	5.1	4.2	3.4	2.8
85.	17.8	15.5	13.3	11.4	9.6	8.0	6.6	5.5	4.5	3.6	3.0
84.	19.4	16.9	14.5	12.3	10.3	8.6	7.1	5.9	4.8	3.9	3.2
83.	21.2	18.3	15.7	13.3	11.2	9.3	7.7	6.3	5.2	4.2	3.4
82.	23.0	19.9	17.0	14.4	12.1	10.0	8.3	6.8	5.6	4.5	3.7
81.	25.1	21.6	18.4	15.6	13.0	10.8	8.9	7.3	6.0	4.9	3.9
80.	27.3	23.5	20.0	16.9	14.1	11.7	9.6	7.9	6.4	5.2	4.2
79.	29.7	25.5	21.7	18.2	15.2	12.6	10.4	8.5	6.9	5.6	4.6
78.	32.3	27.7	23.5	19.7	16.4	13.6	11.2	9.2	7.5	6.1	4.9
77.	35.1	30.1	25.4	21.3	17.8	14.7	12.1	9.9	8.0	6.5	5.3
76.	38.2	32.6	27.6	23.1	19.2	15.8	13.0	10.6	8.6	7.0	5.6
75.	41.6	35.4	29.9	25.0	20.7	17.1	14.0	11.4	9.3	7.5	6.1
74.	45.2	38.4	32.3	27.0	22.4	18.4	15.1	12.3	10.0	8.1	6.5
73.	49.2	41.7	35.0	29.2	24.1	19.8	16.2	13.2	10.7	8.7	7.0
72.	53.5	45.2	37.9	31.5	26.0	21.4	17.5	14.2	11.5	9.3	7.5
71.	58.1	49.1	41.0	34.1	28.1	23.0	18.8	15.3	12.4	10.0	8.1
70.	63.2	53.2	44.4	36.8	30.3	24.8	20.3	16.4	13.3	10.8	8.7
69.	68.7	57.7	48.1	39.8	32.7	26.8	21.8	17.7	14.3	11.6	9.3
68.	74.6	62.6	52.0	43.0	35.3	28.8	23.5	19.0	15.4	12.4	10.0
67.	81.1	67.8	56.3	46.4	38.1	31.1	25.3	20.5	16.5	13.3	10.7
66.	88.1	73.5	60.9	50.2	41.1	33.5	27.2	22.0	17.8	14.3	11.5
65.	95.7	79.7	65.9	54.2	44.3	36.1	29.3	23.7	19.1	15.4	12.4
64.	103.9	86.4	71.3	58.5	47.8	38.8	31.5	25.5	20.5	16.5	13.3
63.	112.9	93.6	77.1	63.2	51.5	41.8	33.9	27.4	22.1	17.7	14.3
62.	122.6	101.4	83.4	68.2	55.5	45.1	36.5	29.4	23.7	19.1	15.3
61.	133.1	109.8	90.1	73.6	59.9	48.5	39.2	31.6	25.5	20.5	16.4
60.	144.4	119.0	97.4	79.4	64.5	52.2	42.2	34.0	27.4	22.0	17.6
59.	156.8	128.8	105.3	85.7	69.5	56.2	45.4	36.6	29.4	23.6	18.9
58.	170.1	139.5	113.8	92.5	74.9	60.6	48.8	39.3	31.6	25.3	20.3
57.	184.6	151.0	123.0	99.8	80.8	65.2	52.5	42.2	33.9	27.2	21.8
56.	200.3	163.5	132.9	107.7	87.0	70.2	56.5	45.4	36.4	29.2	23.4
55.	217.3	176.9	143.6	116.2	93.8	75.5	60.7	48.8	39.1	31.4	25.1
54.	235.7	191.5	155.1	125.3	101.0	81.3	65.3	52.4	42.0	33.7	27.0
53.	255.6	207.2	167.5	135.1	108.8	87.5	70.2	56.3	45.1	36.2	28.9
52.	277.2	224.2	180.9	145.7	117.2	94.1	75.5	60.5	48.5	38.8	31.1
51.	300.5	242.5	195.4	157.1	126.2	101.3	81.2	65.0	52.1	41.7	33.3
50.	325.7	262.3	210.9	169.4	135.9	108.9	87.3	69.9	55.9	44.7	35.8
49.	353.1	283.7	227.7	182.6	146.3	117.2	93.8	75.1	60.0	48.0	38.4
48.	382.6	306.8	245.8	196.9	157.6	126.1	100.8	80.6	64.5	51.5	41.2
47.	414.6	331.7	265.3	212.2	169.6	135.6	108.4	86.6	69.2	55.3	44.2
46.	449.2	358.6	286.4	228.7	182.6	145.9	116.5	93.1	74.3	59.4	47.4
45.	486.7	387.7	309.0	246.4	196.6	156.9	125.2	100.0	79.8	63.7	50.9
44.	527.1	419.0	333.4	265.5	211.6	168.7	134.6	107.4	85.7	68.4	54.6
43.	570.9	452.8	359.7	286.1	227.7	181.4	144.6	115.3	92.0	73.4	58.6
42.	618.2	489.3	388.0	308.2	245.1	195.1	155.4	123.8	98.8	78.8	62.9
41.	669.4	528.7	418.5	332.0	263.7	209.7	167.0	133.0	106.0	84.5	67.4
40.	724.6	571.1	451.4	357.6	283.8	225.5	179.4	142.8	113.8	90.7	72.4
39.	784.4	616.9	486.8	385.1	305.3	242.4	192.7	153.4	122.2	97.4	77.6
38.	848.9	666.3	524.9	414.7	328.5	260.6	207.1	164.7	131.1	104.5	83.3
37.	918.6	719.5	565.9	446.6	353.4	280.1	222.4	176.9	140.7	112.1	89.3
36.	994.0	776.9	610.1	480.9	380.1	301.1	239.0	189.9	151.1	120.3	95.8
35.	1075.3	838.7	657.6	517.7	408.8	323.6	256.7	203.9	162.1	129.1	102.8
34.	1163.2	905.4	708.8	557.3	439.7	347.8	275.7	218.9	174.0	138.5	110.3
33.	1258.1	977.3	763.9	599.9	472.9	373.8	296.1	235.0	186.8	148.6	118.3
32.	1360.6	1054.8	823.2	645.8	508.5	401.7	318.1	252.3	200.4	159.4	126.9
31.	1471.2	1138.3	887.0	695.0	546.8	431.7	341.6	270.9	215.1	171.0	136.1
30.	1590.6	1228.2	955.7	748.0	588.0	463.8	366.8	290.8	230.8	183.5	146.0
29.	1719.4	1325.1	1029.6	804.9	632.2	498.3	394.0	312.1	247.7	196.9	156.6
28.	1858.5	1429.5	1109.1	866.1	679.7	535.4	423.0	335.0	265.8	211.2	168.0
27.	2008.6	1542.0	1194.6	931.8	730.7	575.2	454.3	359.6	285.2	226.6	180.2
26.	2170.5	1663.1	1286.6	1002.5	785.4	618.0	487.8	386.0	306.1	243.0	193.3
25.	2345.2	1793.5	1385.5	1078.5	844.3	663.8	523.7	414.3	328.4	260.7	207.3

CORE TEMPERATURE (°C)	% RATED VOLTAGE 50.	55.	60.	65.	70.	75.	80.	85.	90.	95.	100.
105.	2.6	2.4	2.3	2.1	1.9	1.8	1.6	1.4	1.3	1.1	1.0
104.	2.8	2.7	2.5	2.3	2.1	1.9	1.7	1.6	1.4	1.2	1.1
103.	3.1	2.9	2.7	2.5	2.3	2.1	1.9	1.7	1.5	1.3	1.2
102.	3.4	3.2	3.0	2.7	2.5	2.3	2.1	1.8	1.6	1.4	1.3
101.	3.7	3.5	3.2	3.0	2.7	2.5	2.2	2.0	1.8	1.6	1.4
100.	4.1	3.8	3.5	3.2	3.0	2.7	2.4	2.2	1.9	1.7	1.5
99.	4.4	4.2	3.8	3.5	3.2	2.9	2.6	2.3	2.1	1.8	1.6
98.	4.9	4.5	4.2	3.9	3.5	3.2	2.8	2.5	2.2	2.0	1.7
97.	5.3	5.0	4.6	4.2	3.8	3.4	3.1	2.7	2.4	2.1	1.9
96.	5.8	5.4	5.0	4.6	4.2	3.7	3.4	3.0	2.6	2.3	2.0
95.	6.4	5.9	5.5	5.0	4.5	4.1	3.6	3.2	2.8	2.5	2.2
94.	7.0	6.5	6.0	5.4	4.9	4.4	3.9	3.5	3.1	2.7	2.3
93.	7.6	7.1	6.5	5.9	5.4	4.8	4.3	3.8	3.3	2.9	2.5
92.	8.4	7.7	7.1	6.4	5.8	5.2	4.6	4.1	3.6	3.1	2.7
91.	9.1	8.4	7.7	7.0	6.3	5.7	5.0	4.4	3.9	3.4	2.9
90.	10.0	9.2	8.4	7.6	6.9	6.1	5.4	4.8	4.2	3.7	3.2
89.	10.9	10.1	9.2	8.3	7.5	6.7	5.9	5.2	4.5	3.9	3.4
88.	12.0	11.0	10.0	9.1	8.1	7.2	6.4	5.6	4.9	4.3	3.7
87.	13.1	12.0	10.9	9.9	8.8	7.9	6.9	6.1	5.3	4.6	4.0
86.	14.3	13.1	11.9	10.7	9.6	8.5	7.5	6.6	5.7	5.0	4.3
85.	15.7	14.3	13.0	11.7	10.4	9.2	8.1	7.1	6.2	5.4	4.6
84.	17.2	15.6	14.2	12.7	11.3	10.0	8.8	7.7	6.7	5.8	5.0
83.	18.8	17.1	15.4	13.8	12.3	10.9	9.5	8.3	7.2	6.2	5.4
82.	20.5	18.6	16.8	15.0	13.4	11.8	10.3	9.0	7.8	6.7	5.8
81.	22.4	20.4	18.3	16.4	14.5	12.8	11.2	9.7	8.4	7.3	6.2
80.	24.5	22.2	20.0	17.8	15.8	13.9	12.1	10.5	9.1	7.8	6.7
79.	26.8	24.2	21.7	19.3	17.1	15.0	13.1	11.4	9.8	8.4	7.2
78.	29.3	26.5	23.7	21.0	18.6	16.3	14.2	12.3	10.6	9.1	7.8
77.	32.1	28.9	25.8	22.9	20.1	17.6	15.3	13.3	11.4	9.8	8.4
76.	35.1	31.5	28.1	24.9	21.9	19.1	16.6	14.4	12.4	10.6	9.0
75.	38.3	34.4	30.6	27.0	23.7	20.7	18.0	15.5	13.3	11.4	9.7
74.	41.9	37.5	33.3	29.4	25.7	22.4	19.4	16.8	14.4	12.3	10.5
73.	45.8	40.9	36.2	31.9	27.9	24.3	21.0	18.1	15.5	13.3	11.3
72.	50.0	44.6	39.4	34.7	30.3	26.3	22.7	19.5	16.7	14.3	12.2
71.	54.7	48.6	42.9	37.6	32.8	28.5	24.6	21.1	18.1	15.4	13.1
70.	59.7	53.0	46.7	40.9	35.6	30.8	26.6	22.8	19.5	16.6	14.1
69.	65.3	57.8	50.8	44.4	38.6	33.4	28.7	24.6	21.0	17.9	15.2
68.	71.3	63.0	55.3	48.2	41.8	36.1	31.0	26.6	22.7	19.3	16.4
67.	77.9	68.6	60.1	52.3	45.3	39.1	33.5	28.7	24.4	20.8	17.6
66.	85.1	74.8	65.4	56.8	49.1	42.3	36.2	30.9	26.3	22.4	18.9
65.	92.9	81.5	71.1	61.7	53.2	45.7	39.1	33.4	28.4	24.1	20.4
64.	101.5	88.8	77.3	66.9	57.6	49.5	42.3	36.0	30.6	25.9	21.9
63.	110.8	96.8	84.0	72.6	62.4	53.5	45.7	38.9	33.0	27.9	23.6
62.	121.0	105.4	91.4	78.8	67.6	57.8	49.3	41.9	35.6	30.1	25.4
61.	132.2	114.8	99.3	85.4	73.2	62.6	53.3	45.2	38.3	32.4	27.3
60.	144.3	125.1	107.9	92.7	79.3	67.6	57.5	48.8	41.3	34.9	29.4
59.	157.5	136.2	117.2	100.5	85.9	73.1	62.1	52.6	44.5	37.5	31.6
58.	171.9	148.3	127.4	109.0	92.9	79.0	67.0	56.7	47.9	40.4	34.0
57.	187.7	161.5	138.4	118.2	100.6	85.4	72.3	61.2	51.6	43.5	36.6
56.	204.8	175.8	150.3	128.1	108.9	92.3	78.1	65.9	55.6	46.8	39.4
55.	223.5	191.3	163.2	138.8	117.8	99.7	84.2	71.1	59.9	50.4	42.3
54.	243.9	208.2	177.2	150.5	127.4	107.7	90.9	76.6	64.5	54.2	45.5
53.	266.1	226.6	192.4	163.0	137.8	116.4	98.1	82.6	69.4	58.3	49.0
52.	290.3	246.5	208.9	176.6	149.1	125.7	105.8	89.0	74.7	62.8	52.7
51.	316.7	268.2	226.7	191.3	161.2	135.7	114.1	95.9	80.5	67.5	56.6
50.	345.4	291.8	246.1	207.2	174.4	146.5	123.1	103.3	86.6	72.6	60.9
49.	376.7	317.3	267.0	224.4	188.5	158.2	132.7	111.3	93.3	78.1	65.4
48.	410.8	345.1	289.7	243.0	203.8	170.8	143.1	119.9	100.4	84.0	70.3
47.	447.9	375.2	314.3	263.1	220.3	184.4	154.3	129.1	108.0	90.4	75.6
46.	488.3	408.0	340.9	284.9	238.1	199.0	166.4	139.1	116.3	97.2	81.3
45.	532.3	443.5	369.7	308.3	257.3	214.8	179.3	149.8	125.1	104.5	87.3
44.	580.2	482.0	400.9	333.7	278.0	231.7	193.3	161.3	134.6	112.4	93.9
43.	632.4	523.9	434.7	361.1	300.3	250.0	208.3	173.6	144.8	120.9	100.9
42.	689.2	569.2	471.2	390.7	324.5	269.7	224.5	187.0	155.8	129.9	108.4
41.	751.0	618.5	510.7	422.7	350.4	291.0	241.9	201.3	167.6	139.7	116.5
40.	818.2	671.9	553.5	457.2	378.5	313.8	260.6	216.7	180.3	150.2	125.2
39.	891.3	729.8	599.8	494.5	408.7	338.5	280.8	233.2	193.9	161.4	134.5
38.	970.9	792.6	649.9	534.8	441.3	365.0	302.4	251.0	208.6	173.5	144.5
37.	1057.5	860.7	704.1	578.2	476.4	393.5	325.8	270.1	224.3	186.5	155.2
36.	1151.6	934.6	762.7	625.2	514.3	424.3	350.9	290.7	241.2	200.5	166.7
35.	1254.0	1014.6	826.0	675.8	555.1	457.4	377.9	312.8	259.4	215.4	179.1
34.	1365.3	1101.3	894.5	730.4	599.1	493.0	406.9	336.6	278.9	231.5	192.4
33.	1486.3	1195.3	968.5	789.4	646.4	531.4	438.1	362.1	299.9	248.8	206.7
32.	1617.8	1297.1	1048.5	853.0	697.5	572.7	471.7	389.6	322.4	267.3	222.0
31.	1760.7	1407.4	1135.0	921.7	752.5	617.1	507.8	419.1	346.6	287.2	238.4
30.	1916.12	1526.8	1228.4	995.7	811.8	664.9	546.6	450.8	372.6	308.6	256.0
29.	2084.9	1656.2	1329.4	1075.5	875.6	716.4	588.4	484.8	400.5	331.6	275.0
28.	2268.2	1796.3	1438.4	1161.6	944.3	771.7	633.3	521.4	430.5	356.2	295.3
27.	2467.4	1948.0	1556.2	1254.5	1018.4	831.3	681.5	560.8	462.7	382.7	317.1
26.	2683.8	2112.2	1683.4	1354.6	1098.1	895.3	733.4	603.0	497.3	411.1	340.5
25.	2918.7	2289.9	1820.8	1462.5	1183.9	964.3	789.2	648.4	534.4	441.5	365.6

CORE TEMPERATURE (°C)	% RATED VOLTAGE										
	50.	55.	60.	65.	70.	75.	80.	85.	90.	95.	100.
115.	1.4	1.4	1.3	1.3	1.3	1.2	1.2	1.1	1.1	1.0	1.0
114.	1.5	1.5	1.5	1.4	1.4	1.3	1.3	1.2	1.2	1.1	1.1
113.	1.7	1.6	1.6	1.6	1.5	1.5	1.4	1.4	1.3	1.2	1.2
112.	1.8	1.8	1.8	1.7	1.7	1.6	1.5	1.5	1.4	1.4	1.3
111.	2.0	2.0	1.9	1.9	1.8	1.8	1.7	1.6	1.6	1.5	1.4
110.	2.2	2.2	2.1	2.0	2.0	1.9	1.9	1.8	1.7	1.6	1.5
109.	2.4	2.4	2.3	2.2	2.2	2.1	2.0	1.9	1.9	1.8	1.7
108.	2.7	2.6	2.5	2.5	2.4	2.3	2.2	2.1	2.0	1.9	1.8
107.	2.9	2.9	2.8	2.7	2.6	2.5	2.4	2.3	2.2	2.1	2.0
106.	3.2	3.1	3.1	3.0	2.9	2.8	2.6	2.5	2.4	2.3	2.2
105.	3.5	3.5	3.4	3.2	3.1	3.0	2.9	2.8	2.6	2.5	2.4
104.	3.9	3.8	3.7	3.6	3.4	3.3	3.2	3.0	2.9	2.7	2.6
103.	4.3	4.2	4.0	3.9	3.8	3.6	3.5	3.3	3.1	3.0	2.8
102.	4.7	4.6	4.4	4.3	4.1	4.0	3.8	3.6	3.4	3.3	3.1
101.	5.2	5.0	4.9	4.7	4.5	4.3	4.1	4.0	3.8	3.5	3.3
100.	5.7	5.5	5.3	5.1	4.9	4.7	4.5	4.3	4.1	3.9	3.6
99.	6.2	6.1	5.9	5.6	5.4	5.2	5.0	4.7	4.5	4.2	4.0
98.	6.9	6.6	6.4	6.2	5.9	5.7	5.4	5.1	4.9	4.6	4.3
97.	7.5	7.3	7.0	6.8	6.5	6.2	5.9	5.6	5.3	5.0	4.7
96.	8.3	8.0	7.7	7.4	7.1	6.8	6.5	6.1	5.8	5.5	5.1
95.	9.1	8.8	8.5	8.1	7.8	7.4	7.1	6.7	6.3	6.0	5.6
94.	10.0	9.7	9.3	8.9	8.5	8.1	7.7	7.3	6.9	6.5	6.1
93.	11.0	10.6	10.2	9.8	9.4	8.9	8.5	8.0	7.5	7.1	6.6
92.	12.1	11.7	11.2	10.7	10.2	9.7	9.2	8.7	8.2	7.7	7.2
91.	13.3	12.8	12.3	11.8	11.2	10.7	10.1	9.5	9.0	8.4	7.8
90.	14.6	14.1	13.5	12.9	12.3	11.7	11.0	10.4	9.8	9.1	8.5
89.	16.1	15.4	14.8	14.1	13.5	12.8	12.1	11.3	10.6	9.9	9.3
88.	17.7	17.0	16.2	15.5	14.7	14.0	13.2	12.4	11.6	10.8	10.1
87.	19.4	18.6	17.8	17.0	16.1	15.3	14.4	13.5	12.6	11.8	10.9
86.	21.4	20.5	19.6	18.6	17.7	16.7	15.7	14.7	13.8	12.8	11.9
85.	23.5	22.5	21.5	20.4	19.3	18.2	17.2	16.1	15.0	14.0	12.9
84.	25.8	24.7	23.6	22.4	21.2	20.0	18.7	17.5	16.4	15.2	14.1
83.	28.4	27.1	25.8	24.5	23.2	21.8	20.5	19.1	17.8	16.5	15.3
82.	31.3	29.8	28.4	26.9	25.4	23.8	22.3	20.9	19.4	18.0	16.6
81.	34.4	32.8	31.1	29.4	27.8	26.1	24.4	22.7	21.1	19.6	18.1
80.	37.8	36.0	34.1	32.3	30.4	28.5	26.6	24.8	23.0	21.3	19.6
79.	41.6	39.6	37.5	35.4	33.2	31.1	29.1	27.0	25.1	23.1	21.3
78.	45.8	43.5	41.1	38.8	36.4	34.0	31.7	29.5	27.3	25.2	23.1
77.	50.4	47.8	45.1	42.5	39.8	37.2	34.6	32.1	29.7	27.4	25.1
76.	55.4	52.5	49.5	46.5	43.6	40.7	37.8	35.0	32.3	29.7	27.3
75.	61.0	57.7	54.3	51.0	47.7	44.4	41.2	38.2	35.2	32.3	29.6
74.	67.1	63.4	59.6	55.9	52.2	48.5	45.0	41.6	38.3	35.1	32.2
73.	73.9	69.7	65.4	61.2	57.1	53.0	49.1	45.3	41.7	38.2	34.9
72.	81.4	76.6	71.8	67.1	62.4	57.9	53.5	49.3	45.3	41.5	37.9
71.	89.6	84.2	78.8	73.5	68.3	63.3	58.4	53.7	49.3	45.1	41.1
70.	98.6	92.5	86.5	80.5	74.7	69.1	63.7	58.5	53.6	49.0	44.6
69.	108.6	101.7	94.9	88.2	81.7	75.5	69.5	63.7	58.3	53.2	48.4
68.	119.6	111.8	104.1	96.7	89.4	82.4	75.8	69.4	63.4	57.8	52.5
67.	131.7	122.9	114.3	105.9	97.8	90.0	82.6	75.6	68.9	62.7	56.9
66.	145.0	135.1	125.4	116.0	106.9	98.3	90.0	82.3	74.9	68.1	61.7
65.	159.8	148.5	137.6	127.1	116.9	107.3	98.1	89.5	81.5	73.9	66.9
64.	176.0	163.3	151.0	139.2	127.9	117.1	107.0	97.4	88.5	80.2	72.6
63.	193.9	179.6	165.7	152.5	139.8	127.8	116.6	106.0	96.2	87.1	78.7
62.	213.7	197.5	181.9	167.0	152.9	139.5	127.0	115.4	104.5	94.5	85.2
61.	235.5	217.1	199.6	182.9	167.1	152.3	138.4	125.5	113.5	102.5	92.4
60.	259.5	238.8	219.0	200.3	182.7	166.2	150.8	136.5	123.3	111.2	100.1
59.	286.1	262.6	240.4	219.4	199.7	181.3	164.3	148.5	134.0	120.6	108.4
58.	315.4	288.9	263.8	240.3	218.3	198.7	178.9	161.5	145.5	130.8	117.5
57.	347.8	317.7	289.5	263.1	238.6	215.8	194.8	175.6	157.9	141.9	127.2
56.	383.5	349.5	317.7	288.1	260.7	235.4	212.2	190.9	171.5	153.8	137.8
55.	423.0	384.5	348.7	315.5	284.9	256.8	231.0	207.5	186.1	166.7	149.2
54.	466.6	423.0	382.7	345.5	311.3	280.0	251.5	225.5	202.0	180.7	161.5
53.	514.8	465.5	420.0	378.3	340.1	305.4	273.8	245.1	219.2	195.9	174.8
52.	568.1	512.2	461.0	414.2	371.6	333.0	298.0	266.4	237.9	212.2	189.2
51.	627.0	563.6	505.9	453.5	406.0	363.0	324.3	289.4	258.1	229.9	204.8
50.	692.1	620.3	555.2	496.5	443.4	395.7	352.8	314.4	279.9	249.1	221.6
49.	764.1	682.7	609.4	543.5	484.3	431.3	383.9	341.5	303.6	269.8	239.7
48.	843.8	751.5	668.8	594.9	529.0	470.1	417.6	370.9	329.3	292.2	259.3
47.	931.9	827.2	734.1	651.3	577.7	512.3	454.3	402.7	357.0	316.4	280.5
46.	1029.5	910.7	805.7	712.8	630.8	558.2	494.0	437.3	387.1	342.6	303.3
45.	1137.6	1002.7	884.3	780.2	688.7	608.2	537.2	474.7	419.6	370.9	328.0
44.	1257.2	1104.1	970.6	853.9	751.9	662.6	584.1	515.3	454.8	401.5	354.6
43.	1389.8	1215.8	1065.3	934.6	820.8	721.7	635.1	559.3	492.8	434.6	383.4
42.	1536.7	1339.1	1169.2	1022.8	896.0	786.0	690.3	606.9	534.0	470.3	414.4
41.	1699.6	1474.9	1283.3	1119.2	978.0	856.0	750.3	658.5	578.6	508.8	447.9
40.	1880.2	1624.7	1408.6	1224.7	1067.3	932.0	815.4	714.4	626.7	550.5	484.0
39.	2080.5	1790.0	1546.1	1340.0	1164.7	1014.8	886.0	774.9	678.8	595.5	522.9
38.	2302.9	1972.2	1697.1	1466.1	1270.8	1104.7	962.6	840.5	735.2	644.1	565.0
37.	2549.7	2173.3	1862.8	1604.0	1386.5	1202.4	1045.6	911.4	796.1	696.5	610.3
36.	2823.9	2395.2	2044.7	1754.8	1512.6	1308.6	1135.7	988.2	861.9	753.1	659.2
35.	3128.5	2640.1	2244.4	1919.6	1649.9	1424.1	1233.4	1071.4	933.0	814.3	712.0
34.	3467.2	2910.4	2463.7	2099.8	1799.6	1549.5	1339.3	1161.4	1009.9	880.3	768.8
33.	3843.8	3208.8	2704.4	2296.7	1962.6	1685.7	1454.1	1258.7	1093.0	951.5	830.1
32.	4262.8	3538.3	2968.6	2511.9	2140.2	1833.7	1578.5	1364.1	1182.7	1028.3	896.2
31.	4729.3	3902.3	3258.7	2747.2	2333.6	1994.5	1713.3	1478.1	1279.6	1111.2	967.4
30.	5248.9	4304.2	3577.3	3004.3	2544.1	2169.0	1859.4	1601.3	1384.3	1200.6	1044.2
29.	5828.0	4748.4	3926.9	3285.2	2773.4	2358.5	2017.7	1734.6	1497.4	1297.1	1126.9
28.	6473.9	5239.2	4310.9	3592.1	3023.0	2564.3	2189.1	1878.8	1619.5	1401.2	1216.1
27.	7194.5	5781.7	4732.4	3927.5	3294.7	2787.6	2374.8	2034.6	1751.3	1513.4	1312.1
26.	7999.2	6381.5	5195.2	4293.9	3590.5	3030.0	2575.9	2203.1	1893.7	1634.4	1415.6
25.	8898.3	7044.8	5703.4	4694.1	3912.3	3293.0	2793.7	2385.2	2047.3	1764.9	1527.1

% RATED VOLTAGE

CORE TEMPERATURE (°C)	50.	55.	60.	65.	70.	75.	80.	85.	90.	95.	100.
110.	1.4	1.4	1.4	1.3	1.3	1.2	1.2	1.2	1.1	1.1	1.0
109.	1.6	1.5	1.5	1.5	1.4	1.4	1.3	1.3	1.2	1.1	1.1
108.	1.7	1.7	1.6	1.6	1.5	1.5	1.4	1.4	1.3	1.3	1.2
107.	1.9	1.9	1.8	1.8	1.7	1.6	1.6	1.5	1.4	1.4	1.3
106.	2.1	2.0	2.0	1.9	1.9	1.8	1.7	1.6	1.6	1.5	1.4
105.	2.3	2.2	2.2	2.1	2.0	2.0	1.9	1.8	1.7	1.6	1.5
104.	2.5	2.5	2.4	2.3	2.2	2.1	2.1	2.0	1.9	1.8	1.7
103.	2.8	2.7	2.6	2.5	2.4	2.3	2.2	2.1	2.0	1.9	1.8
102.	3.0	3.0	2.9	2.8	2.7	2.6	2.5	2.3	2.2	2.1	2.0
101.	3.3	3.3	3.1	3.0	2.9	2.8	2.7	2.6	2.4	2.3	2.2
100.	3.7	3.6	3.5	3.3	3.2	3.1	2.9	2.8	2.7	2.5	2.4
99.	4.0	3.9	3.8	3.7	3.5	3.4	3.2	3.1	2.9	2.7	2.6
98.	4.4	4.3	4.2	4.0	3.8	3.7	3.5	3.3	3.2	3.0	2.8
97.	4.9	4.7	4.6	4.4	4.2	4.0	3.8	3.6	3.4	3.2	3.1
96.	5.4	5.2	5.0	4.8	4.6	4.4	4.2	4.0	3.8	3.5	3.3
95.	5.9	5.7	5.5	5.3	5.1	4.8	4.6	4.3	4.1	3.9	3.6
94.	6.5	6.3	6.0	5.8	5.5	5.3	5.0	4.7	4.5	4.2	3.9
93.	7.1	6.9	6.6	6.3	6.1	5.8	5.5	5.2	4.9	4.6	4.3
92.	7.8	7.6	7.3	7.0	6.6	6.3	6.0	5.7	5.3	5.0	4.7
91.	8.6	8.3	8.0	7.6	7.3	6.9	6.5	6.2	5.8	5.4	5.1
90.	9.5	9.1	8.7	8.4	8.0	7.6	7.1	6.7	6.3	5.9	5.5
89.	10.4	10.0	9.6	9.2	8.7	8.3	7.8	7.4	6.9	6.4	6.0
88.	11.4	11.0	10.5	10.0	9.5	9.0	8.5	8.0	7.5	7.0	6.5
87.	12.6	12.1	11.6	11.0	10.5	9.9	9.3	8.8	8.2	7.6	7.1
86.	13.8	13.3	12.7	12.1	11.4	10.8	10.2	9.5	8.9	8.3	7.7
85.	15.2	14.6	13.9	13.2	12.5	11.8	11.1	10.4	9.7	9.0	8.4
84.	16.7	16.0	15.3	14.5	13.7	12.9	12.1	11.4	10.6	9.8	9.1
83.	18.4	17.6	16.7	15.9	15.0	14.1	13.3	12.4	11.5	10.7	9.9
82.	20.3	19.3	18.4	17.4	16.4	15.5	14.5	13.5	12.6	11.7	10.8
81.	22.3	21.2	20.2	19.1	18.0	16.9	15.8	14.7	13.7	12.7	11.7
80.	24.5	23.3	22.1	20.9	19.7	18.5	17.3	16.1	14.9	13.8	12.7
79.	27.0	25.6	24.3	22.9	21.5	20.2	18.8	17.5	16.2	15.0	13.8
78.	29.7	28.2	26.6	25.1	23.6	22.1	20.6	19.1	17.7	16.3	15.0
77.	32.6	31.0	29.2	27.5	25.8	24.1	22.4	20.8	19.2	17.7	16.3
76.	35.9	34.0	32.1	30.2	28.2	26.3	24.5	22.7	20.9	19.3	17.7
75.	39.5	37.4	35.2	33.0	30.9	28.8	26.7	24.7	22.8	21.0	19.2
74.	43.5	41.1	38.6	36.2	33.8	31.5	29.2	26.9	24.8	22.8	20.8
73.	47.9	45.1	42.4	39.7	37.0	34.4	31.8	29.4	27.0	24.7	22.6
72.	52.7	49.6	46.5	43.5	40.5	37.5	34.7	32.0	29.4	26.9	24.6
71.	58.0	54.5	51.1	47.6	44.3	41.0	37.9	34.8	31.9	29.2	26.6
70.	63.9	59.9	56.0	52.2	48.4	44.8	41.3	37.9	34.7	31.7	28.9
69.	70.4	65.9	61.5	57.2	53.0	48.9	45.0	41.3	37.8	34.5	31.4
68.	77.5	72.4	67.5	62.6	57.9	53.4	49.1	45.0	41.1	37.4	34.0
67.	85.3	79.6	74.0	68.6	63.4	58.3	53.5	49.0	44.7	40.6	36.9
66.	94.0	87.5	81.3	75.2	69.3	63.7	58.3	53.3	48.6	44.1	40.0
65.	103.5	96.2	89.2	82.3	75.8	69.5	63.6	58.0	52.8	47.9	43.4
64.	114.0	105.8	97.9	90.2	82.9	75.9	69.3	63.1	57.4	52.0	47.0
63.	125.7	116.4	107.4	98.8	90.6	82.8	75.5	68.7	62.3	56.4	51.0
62.	138.5	128.0	117.9	108.2	99.1	90.4	82.3	74.8	67.7	61.2	55.2
61.	152.6	140.7	129.3	118.5	108.3	98.7	89.7	81.3	73.6	66.4	59.9
60.	168.2	154.7	141.9	129.8	118.4	107.7	97.7	88.5	79.9	72.1	64.9
59.	185.4	170.2	155.8	142.2	129.4	117.5	106.4	96.2	86.8	78.2	70.3
58.	204.4	187.2	171.0	155.7	141.5	128.2	115.9	104.6	94.3	84.8	76.1
57.	225.4	205.9	187.6	170.5	154.6	139.9	126.3	113.8	102.3	91.9	82.4
56.	248.5	226.5	205.9	186.7	169.0	152.6	137.5	123.7	111.1	99.7	89.3
55.	274.1	249.2	226.0	204.5	184.6	166.4	149.7	134.5	120.6	108.0	96.7
54.	302.4	274.1	248.0	223.9	201.7	181.5	163.0	146.2	130.9	117.1	104.6
53.	333.6	301.6	272.2	245.1	220.4	197.9	177.4	158.8	142.1	126.9	113.3
52.	368.1	331.9	298.7	268.4	240.8	215.8	193.1	172.6	154.1	137.5	122.6
51.	406.3	365.2	327.8	293.9	263.1	235.2	210.1	187.5	167.2	149.0	132.7
50.	448.5	402.0	359.8	321.7	287.4	256.4	228.7	203.7	181.4	161.4	143.6
49.	495.2	442.4	394.9	352.2	313.9	279.5	248.8	221.3	196.7	174.9	155.3
48.	546.8	487.0	433.4	385.5	342.8	304.6	270.6	240.3	213.4	189.4	168.0
47.	603.9	536.0	475.7	422.0	374.3	332.0	294.4	261.0	231.4	205.1	181.7
46.	667.2	590.1	522.1	461.9	408.8	361.7	320.2	283.4	250.8	222.0	196.6
45.	737.2	649.8	573.0	505.6	446.3	394.1	348.1	307.6	271.9	240.4	212.5
44.	814.7	715.5	628.9	553.4	487.3	429.3	378.5	333.9	294.7	260.2	229.8
43.	900.6	787.9	690.3	605.6	531.9	467.7	411.5	362.4	319.4	281.6	248.4
42.	995.8	867.7	757.7	662.8	580.6	509.4	447.3	393.3	346.1	304.7	268.5
41.	1101.4	955.8	831.6	725.3	633.7	554.7	486.2	426.7	374.9	329.7	290.2
40.	1218.4	1052.9	912.8	793.6	691.6	604.0	528.4	462.9	406.1	356.7	313.6
39.	1348.2	1159.9	1001.9	868.3	754.7	657.6	574.1	502.2	439.9	385.9	338.9
38.	1492.3	1278.0	1099.7	950.1	823.5	715.8	623.8	544.6	476.4	417.4	366.1
37.	1652.3	1408.4	1207.1	1039.4	898.5	779.2	677.6	590.6	515.9	451.4	395.5
36.	1829.9	1552.1	1325.0	1137.1	980.2	848.0	736.0	640.4	558.5	488.0	427.2
35.	2027.3	1710.8	1454.4	1243.9	1069.2	922.8	799.3	694.3	604.6	527.7	461.4
34.	2246.8	1886.0	1596.5	1360.7	1166.2	1004.1	867.9	752.6	654.4	570.4	498.2
33.	2490.8	2079.4	1752.5	1488.3	1271.8	1092.4	942.3	815.7	708.3	616.6	537.9
32.	2762.4	2292.9	1923.7	1627.8	1386.9	1188.3	1022.9	884.0	766.4	666.4	580.7
31.	3064.7	2528.7	2111.7	1780.2	1512.2	1292.4	1110.3	957.8	829.2	720.1	626.9
30.	3401.4	2789.2	2318.1	1946.8	1648.6	1405.5	1204.9	1037.7	897.1	778.0	676.6
29.	3776.7	3077.0	2544.7	2128.9	1797.2	1528.4	1307.5	1124.1	970.4	840.6	730.3
28.	4195.2	3395.1	2793.5	2327.8	1959.0	1661.7	1418.6	1217.5	1049.5	908.0	788.0
27.	4662.1	3746.7	3066.7	2545.1	2135.0	1806.4	1538.9	1318.5	1134.9	980.7	850.3
26.	5183.6	4135.3	3366.6	2782.5	2326.7	1963.5	1669.2	1427.7	1227.1	1059.1	917.3
25.	5766.2	4565.2	3695.9	3041.9	2535.2	2133.9	1810.3	1545.7	1326.7	1143.7	989.6

appendix H
A Critical Comparison of Ferrites with Other Magnetic Materials

Reprinted with permission of Magnetics, a Division of Spang Industries, Inc., Butler, Pa.

Basic Differences – Composition and Structure

The difference in properties and performance of ferrites as compared with other magnetic materials is due to the fact that the ferrites are oxide materials rather than metals. Ferromagnetism is derived from the unpaired electron spins in only a few metal atoms, these being Iron, Cobalt, Nickel, Manganese, and some rare earth elements. It is not surprising that the highest magnetic moments and the saturation magnetizations are to be found in the metals themselves or in alloys of these metals. The oxides, on the other hand, suffer from a dilution effect of the large oxygen ions in the crystal lattice. In addition, the net moment resulting from ferromagnetic alignment of the atomic spins is reduced because a different, less efficient type of exchange mechanism is operative. The oxygen ions do serve a useful purpose, however, since they insulate the metal ions and, therefore, greatly increase the resistivity. This property makes the ferrite especially useful at higher frequencies. The purpose of this paper is to list the various considerations which enter into the choice of a material for a specific application and to contrast pertinent ferrite properties with those of bulk metal or powdered metal materials.

Material Considerations Magnetic and Mechanical Properties

SATURATION MAGNETIZATION

As mentioned previously, the highest saturation values are found in the metals and alloys. Thus, if high flux densities are required in high power applications, the bulk metals, iron, silicon-iron and cobalt-iron are unexcelled. Since the flux in maxwells $\phi = BA$, where B = flux density in gauss and A = cross-sectional area in cm^2, obtaining high total flux in materials such as ferrites or permalloy powder cores can be accomplished only by increasing the cross-sectional area. Powdered iron has a fairly high saturation value, but exhibits low permeabilities.

CURIE TEMPERATURES

All magnetic materials lose their ferromagnetism at the Curie temperature. One overriding consideration for a magnetic material is that the Curie point of the material be well above the proposed operating temperatures. Table 1 lists the Curie Temperatures of the various materials. The Curie point depends only on composition and not on geometry. Even though some of the magnetic materials shown can be used at higher operating temperatures than others, very often the temperature limitations of the accessory items (wire insulation, potting or damping compound) can be more important; in this case,no practical advantage may be gained by the higher Curie point materials.

MAGNETIC LOSSES

The magnetic losses in an A.C. application can be represented by the familiar Legg equation:

$$R_m = \mu f L(ef + aB_m + c)$$

where:

- R_m = total core loss in ohms
- e = eddy current coefficient
- a = hysteresis coefficient
- c = residual loss coefficient
- μ = magnetic permeability
- f = frequency in hertz
- L = inductance in henries
- B_m = maximum flux density in gausses

Eddy current losses will increase quite rapidly with frequency. In bulk metals, these high frequency losses can be reduced by reducing the thickness of the material perpendicular to the flux flow. This is accomplished by using thin gage tapes or laminations or by powdering and insulating the particles. In ferrites, the same result is obtained by increasing the resistivity by many orders of magnitude. Thus, at the highest operating frequencies where further particle reduction is impractical, ferrites are the only available materials.

The hysteresis losses are proportional to the flux density and can be depicted as the area inside the hysteresis loop. High hysteresis losses are accompanied by the presence of unwanted harmonics. The nickel-iron (permalloy) alloys have low hysteresis losses and a great asset to the permalloy powder core is that these low losses are maintained with the accompanying reduction in eddy current losses.

The residual losses are not too well understood and perhaps represent an expression of our ignorance of the system. They apparently are tied in partially to absorption of energy from the system by gyromagnetic resonance.

A listing of the various losses in the materials under consideration is given in Table 1.

PERMEABILITY

Permeability is a function of composition and processing. The highest initial permeabilities (those measured at very low flux levels) are found in the nickel-iron alloys, particularly in supermalloy where the value is about 100,000. Powdered iron cores have low permeabilities (10-100) while permalloy powder cores are somewhat higher (15-550). Ferrites can be made over a wide range of permeabilities. The linear filter type permeabilities vary from 100-2000, while those used in power applications range from 3000-15,000. As the operating frequency increases, ferrites with lower permeabilities are used because these have distinctly lower losses in these regions. The permeabilities for a variety of materials are listed in Table 1.

FIGURE OF MERIT

A useful figure of merit for linear core materials is the μQ product. Values of this factor are tabulated in Table 1. At frequencies of 100 KHz and above, the value for ferrites is considerably above all other materials.

SQUARENESS

The squareness ratio is defined as the ratio of B_r to B_m and is especially important in memory work and switching core applications. Magnesium-manganese ferrites can be produced with extremely high squareness ratios. While some metal tape and bobbin cores possess similar high ratios, their higher cost and difficulty in miniaturization make the ferrites the material of choice in large scale memory applications. Thin film memories may be considered bulk metals and may become increasingly important but at present, literally billions of small ferrite memory cores are being produced. The importance of this phase is emphasized by the fact that the market value for computer magnetics is now equal in dollars to that of the power materials market.

BRITTLENESS

One drawback to the ferrite core as compared with metal cores is its brittleness. Being ceramic in nature, care must be exercised in the handling of these cores. Powder cores are also somewhat brittle and similar precautions are required. Although metal tape cores are not brittle, they nevertheless are sensitive to strain and mechanical shock, especially in the high permeability materials. Consequently, tape wound cores are often embedded in a damping compound which prevents the transfer of strain or shock to the cores.

HARDNESS

Ferrites are very hard materials as compared with the other materials under consideration. This property is especially useful in applications in which wear is a factor. Consequently, ferrite material is being used extensively in magnetic recorder head applications.

Geometry Considerations

FORMABILITY

The three types of materials - bulk metal,powdered metal and ferrite - are produced by widely varying techniques and consequently the available geometries also vary.

Bulk metals - These are produced mostly by standard metallurgical processes involving melting followed by hot and cold rolling. The sheet material produced is either slit and wound into tape or bobbin cores or punched into laminations. Photoetching, a new method of forming small complex parts, avoids costly tooling, and produces stress-free parts.

Powdered Iron and permalloy - These materials are always die-pressed into toroids or slugs, molypermalloy usually in toroids and powdered iron into slugs.

Ferrites - Because ferrites are produced by a ceramic technique, they can be made in a large number of shapes. Unlike the bulk metals, they can be molded directly, and unlike the powdered permalloy, they can be machined and ground to close tolerances after firing. Various forming processes for ferrites include die pressing,extrusion, hydrostatic pressing and hot pressing. The available shapes include toroids, E-I cores, U-I cores, pot cores, rods, tubes, beads and blocks.

TUNABILITY

An exact inductance is required in certain L-C circuits. If the shape of the inductor is toroidal, the inductance can be trimmed only by the addition or removal of turns, a time consuming and costly procedure. If a ferrite pot core is used, the tuning can be accomplished by means of a screw-type trimmer core which changes the effective air gap of the core. Threaded rods of powdered iron or ferrite materials are used extensively as tuning elements in slug tuned inductors.

WINDING CONSIDERATIONS

Winding turns on a toroid involves specialized equipment and the process involves winding each core separately in a relatively time consuming operation. The bobbins used in ferrite pot cores can be wound many at a time on a rather simple machine. This ease of winding constitutes an important advantage for ferrite pot cores.

MAGNETIC SHIELDING

If magnetic components are relatively close in a circuit, the fields produced by one component may effect the performance of other cores. One solution is to increase the space between components. This increases the overall size of the system. Another is to use a magnetic shield which increases weight and size. A ferrite pot core is inherently self-shielding by nature of the enclosed magnetic circuit.

Inductance Stability Considerations

TEMPERATURE STABILITY

In telecommunications circuitry (tuned L-C filters), the maintenance of a near-constant inductance as a function of temperature and time is most critical. One method of achieving this stability is by the insertion of an air gap. The gap may be distributed as in powder cores or localized as in gapped ferrite pot cores. Gapping also results in a reduction in the effective permeability but often this is not a serious limitation. In gapped ferrite cores, the temperature coefficient (T.C.) can be linear to match a capacitor with an equal but opposite T.C. (polystyrene) or relatively flat if a flat T.C. capacitor (silver-mica) is used.

$$\text{T.C.} = \frac{\Delta L}{L \Delta T}$$

where ΔL and ΔT are corresponding changes in inductance and temperature and L is inductance at a standard temperature.

Figure 1 illustrates the temperature characteristics of several ferrite materials.

Figure 1

FERRITE TOROIDS

INITIAL PERMEABILITY (μ_o) VS. TEMPERATURE

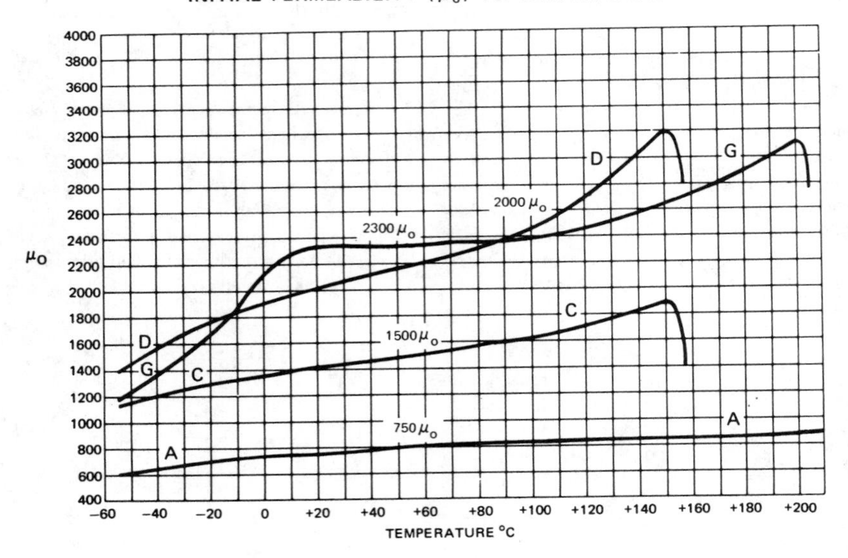

As pointed out, the use of an air gap greatly increases the temperature stability. The powder core toroid and ferrite pot core are thus used to good advantage. In the powder core, the T.C. is built into the toroid whereas in the pot core, the T.C. can be varied by changing the gap. However, in the latter, the effective permeability and therefore the inductance of the core is changed. By choice of the proper size core with the proper gap, the optimum inductance and T. C. can be obtained.

PERMEABILITY VS. A. C. FLUX DENSITY

It is often desirable to have a minimum change in permeability with A. C. excitation. Here again, the air gap in either permalloy powder cores or ferrites can be used to advantage. Figures 2, 3 and 4 show the relative change in inductance for a ferrite toroid, ferrite pot core, and permalloy powder toroid.

RELATIVE CHANGE OF INDUCTANCE WITH A. C. FLUX LEVEL FOR TOROIDS AND POT CORES

Figure 2

FERRITE TOROID

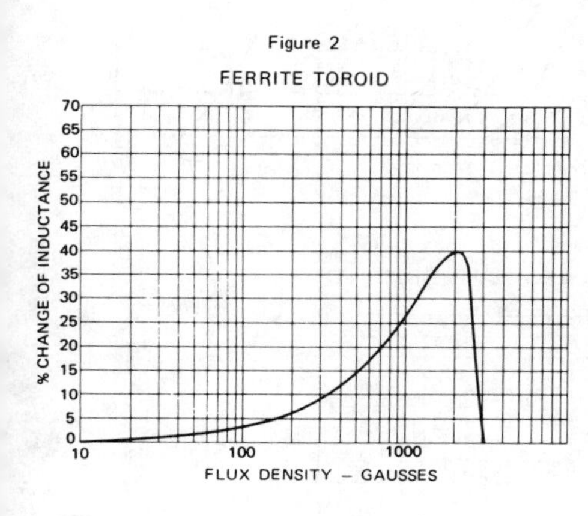

Figure 3

FERRITE POT CORE

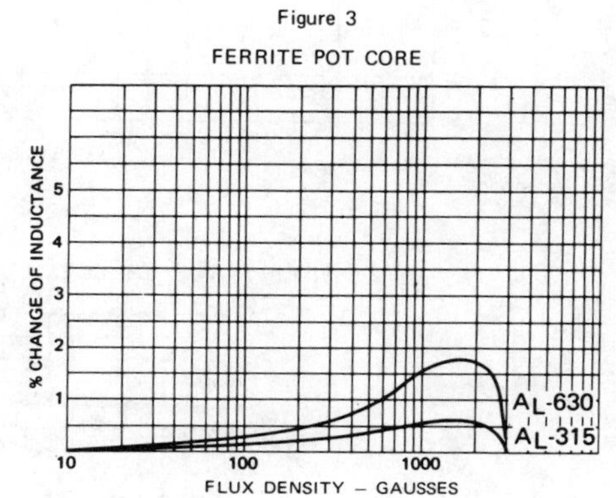

Figure 4

PERMALLOY POWDER TOROID

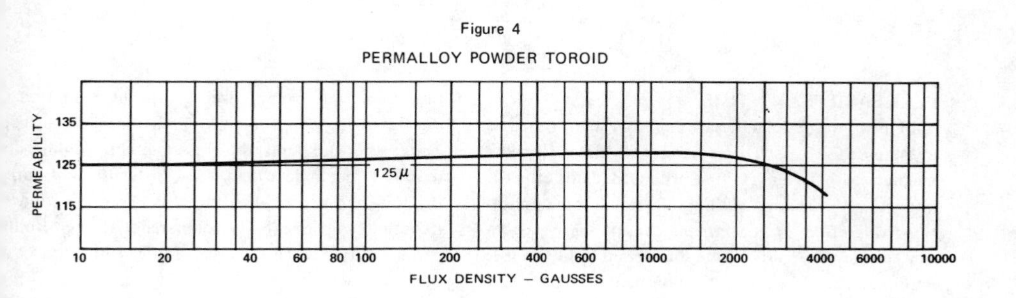

PERMEABILITY VS. D. C. BIAS

Often an A. C. circuit has a superimposed D. C. bias condition. Minimum variation of permeability with D. C. is desirable. Powder cores are especially resistant to these changes. Figure 5 shows typical variations of μ with D. C. bias for permalloy powder toroids. Gapped ferrite pot cores show a similar effect, shown in Figure 6.

Figure 5

INDUCTANCE VS. D. C. BIAS FOR PERMALLOY POWDER CORES

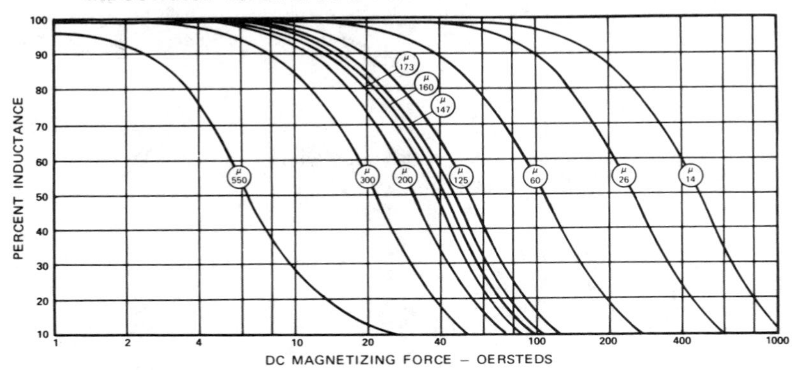

Figure 6

INDUCTANCE VS. D. C. BIAS FOR A FERRITE POT CORE

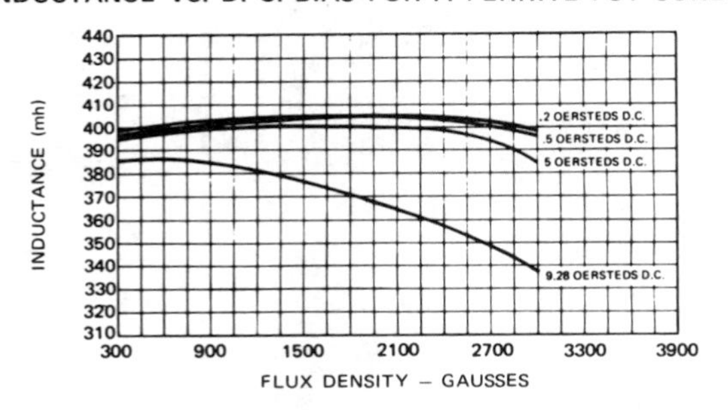

PERMEABILITY VS. TIME

In most magnetic materials there is a slight decrease in permeability with time after the material is demagnetized or after it is first produced. This effect is known as disaccommodation. In non-linear applications this effect is not too important. However, in low flux level circuits where a constant inductance is required, the effect must be considered. The effect is pronounced in low permeability materials and is negligible for high permeability materials. However, the effect can be minimized greatly by reduction o the effective permeability by insertion of an air gap Thus in powder cores, the change of permeability du to this effect is less than .1%. In ferrite pot cores the localized gap reduces the effect in proportion t the effective permeability compared with the toroid permeability. Since the effect is logarithmic, most o the decrease occurs in the first few days after firing If some aging of the cores occurs before usage, th change of inductance due to time will be negligibl

Application Considerations
Ferrite Advantages and Disadvantages

APPLICATION	ADVANTAGES	DISADVANTAGES
Low Frequency (< 1 KHz) **High Flux Applications** Generators Motors Power Transformers	• Ease of forming shapes allows possible use in inexpensive, high loss applications such as relays, small motors .	• Flux density low • Relative cost high • Limited size of parts
Medium Frequency (1-100 KHz) **Non-Linear High Flux Applications** Flyback Transformers Deflection Yokes Inverters Wide Band Transformers Recording Heads Pulse Transformers Memory Cores	• Cost much lower than Nickel-Iron alloys, especially thin tapes • Moderately high permeabilities available • Low losses, especially in upper half of this range • Inherent shielding in pot cores • Good wear resistance • Easily adapted to mass production	• Flux density lower than Nickel-Iron alloys • Permeabilities lower than Nickel-Iron alloys • Curie Temperature fairly low • Good mating surface necessary for high inductance • Smaller flux change than bobbin cores
Medium Frequency (1-100 KHz) **Low Flux, Linear Applications** Loading Coils Filter Cores Tuned Inductors Wide Band Transformers Antenna Rods	• Permeabilities higher than powdered iron or Permalloy cores • Gapped pot cores provide: 1. Adjustability 2. Stability - temperature, time, A. C. flux density, D. C. bias 3. Self-shielding • μQ Products higher than other materials • Wide choice of Inductance and Temperature Coefficient	• Low Curie point • Need precision grinding of air gap • Brittleness • Mounting hardware needed
Higher Frequencies (>200 KHz) **Low Flux, Linear Applications** Filters Inductors Tuning Slugs	• Low losses (especially eddy current) • Only Ferrites and powdered iron can operate at higher frequencies • Medium frequency advantages apply	• Permeability decreases with frequency • Medium frequency disadvantages apply • Poor heat transfer
Microwave Frequencies (>500 MHz)	• Low dielectric losses • Good gyromagnetic properties • Only bulk material available	

TABLE H.1 Properties of Soft Magnetic Materials

MATERIAL	Initial Perm. μ_o	B_{max} Kilogausses	Loss Coefficients		
			$e \times 10^6$	$a \times 10^3$	$c \times 10^3$
Fe	250	22	-	-	-
Si-Fe (unoriented)	400	20	870	120	75
Si-Fe (oriented)	1500	20	-	-	-
50-50 Ni Fe (grain-oriented)	2000	16	-	-	-
79 Permalloy	12,000 to 100,000	8 to 11	173	-	-
Permalloy powder	14 to 550	3	.01 to .04	.002	.05 to .1
Iron powder	5 to 80	10	.002 to .04	.002 to .4	.2 to 1.4
Ferrite-MnZn	750 to 15,000	3 to 5	.001	.002	.01
Ferrite-NiZn	10 to 1500	3 to 5	-	-	-
Co-Fe 50%	800	24	-	-	-

TABLE H.1 (Continued)

MATERIAL	Curie Temperature °C	Resistivity (ohm-cm)	$\mu_o Q$ at 100 KHz	Operating Frequencies
Fe	770	10×10^{-6}	-	60 - 1000 Hz
Si-Fe (unoriented)	740	50×10^{-6}	-	60 - 1000 Hz
Si-Fe (oriented)	740	50×10^{-6}	-	60 - 1000 Hz
50-50 Ni Fe (grain-oriented)	360	40×10^{-6}	-	60 - 1000 Hz
79 Permalloy	450	55×10^{-6}	8000 to 12,000	1 KHz - 75 KHz
Permalloy powder	450	1.	10,000	10 KHz - 200 KHz
Iron powder	770	10^4	2000 to 30,000	100 KHz - 100 MHz
Ferrite-MnZn	100 to 300	10 to 100	100,000 to 500,000	10 KHz - 2 MHz
Ferrite-NiZn	150 to 450	10^6	30,000	200 KHz - 100 MHz
Co-Fe 50%	980	70×10^{-6}		

List of Symbols

A	open-loop voltage gain
A_c	effective core area (cm^2)
A_p	area product (cm^4)
A_t	total surface area of magnetic component (cm^2)
A_w	wire area (cm^2)
$A_{w(B)}$	bare wire area (cm^2)
A_1, A_2	$N \times N$ square matrix
B	flux density (tesla or webers/m^2)
B	n-element column vector matrix
B_{ac}	alternating current flux density (tesla)
B_{dc}	direct current flux density (tesla)
B_1	$L \times M$ matrix
B_2	$L \times M$ matrix
C	output filter capacitor (farads)
C_1	$L \times N$ matrix
C_2	$L \times N$ matrix
D	steady-state ON time duty ratio = T_{ON}/T
D_H	maximum steady-state ON time duty ratio

D_L	minimum steady-state ON time duty ratio
D_m	optimum duty ratio for a given r_L to obtain maximum dc gain
D_o	steady-state "off" time duty ratio, continuous conduction mode = 1 - D
D_1	steady-state OFF time duty ratio for discontinuous conduction mode
D_2	steady-state dwell time duty ratio
d	dynamic ON time duty ratio = $D + \hat{d}$
$\hat{d}$	duty ratio perturbation
d_o	dynamic OFF time duty ratio = $D_o - \hat{d}$
esl	equivalent series inductance (henrys)
esr	equivalent series resistance (of output filter capacitor) (Ω)
f	frequency (Hz)
F	fringing flux factor
f, f_s	converter switching frequency (Hz)
G	coil length, window height (cm)
G_m	gain margin (dB)
G_{mH}	gain margin for high duty ratio, D_H (dB)
G_{mL}	gain margin for low duty ratio, D_L (dB)
H	magnetizing force (amp turns/cm)
$H_e(s)$	effective output filter transfer function
$H_i(s)$	input filter transfer function
I	unit matrix or identity matrix
I	current (A)
I_i	input current (A)
I_o	output current (A)
I_p	primary winding current (A)
I_s	secondary winding current (A)
J	current density (A/cm^2)

K	constant
K_e	electrical constant coefficient
K_g	core geometry coefficient
K_j	current density coefficient
K_u	window utilization factor
L	inductance (henry)
l_g	gap length (cm)
l_m	magnetic path length (cm)
M_c	control modulation function
M_i	pulse width modulator input port to output port transfer function
MLT	mean length turn (cm)
N	number, turns ratio, number of turns
n	number, turns ratio
nC	output filter damping capacitor $n \times C$ (farads)
N_p	number of primary turns
N_s	number of secondary turns
P_{cu}	copper loss (watts)
P_{fe}	core loss (watts)
P_i	input power (watts)
P_o	output power (watts)
P_t	apparent power (watts, volt-amp)
P_{Σ}	total loss (core and copper) (watts)
Q	quality factor of inductor or tuned circuit
R	load resistance (Ω)
R_p	primary winding resistance (Ω)
R_s	secondary winding resistance (Ω)
r	output filter damping resistor (Ω)
r_L	inductor loss equivalent resistance (Ω)

S_1 = (conductor area)/(wire area), where conductor area = copper area, wire area = copper area + insulation area

S_2 = (wound area)/(usable window area), where wound area = number of turns × wire area of one turn

S_3 = (usable window area)/(window area), where usable window area equals available window area minus unused area due to winding technique, and window area = available window area

S_4 = (usable window area)/(usable window area + insulation area)

s complex frequency variable $s = \sigma + j\omega$

T(s) loop gain (dB)

T, T_s period of converter switching frequency (sec) = $DT_s + D_oT_s$ for continuous conduction mode, = $DT_s + D_1T_s + D_2T_s$ for discontinuous conduction mode

VA volt-amperes

V_c steady-state control voltage (volts)

V_f diode forward voltage drop (volts)

V_i steady-state line input voltage (volts)

V_o steady-state output voltage (volts)

v_c dynamic control voltage = $V_c + \hat{v}_c$ (volts)

$\hat{v}_c$ control voltage perturbation (volts)

v_i dynamic line input voltage (volts)

$\hat{v}_i$ line input voltage perturbation (volts)

v_o dynamic output voltage = $V_o + \hat{v}_o$ (volts)

$\hat{v}_o$ output voltage perturbation (volts)

Vol volume (cm^3)

Vol_c core volume (cm^3)

Vol_g gap volume (cm^3)

W_a window area (cm^2)

$W_{a(eff)}$ effective window area (cm^2)

W_t	core weight (g)
X	steady-state value of x
X_c	reactance of capacitor C (Ω)
X_L	reactance of inductor L (Ω)
x	state vector, dimension N
$\dot{x}$	differential of state vector
$\hat{\dot{x}}$	perturbed differential of state vector x
y	output vector, dimension L
α	regulation constant
β	feedback network transfer coefficient
ζ	damping ratio, also used in a different context as resistance temperature correction factor
η	efficiency
μ_o	permeability of free space = $4\pi \times 10^{-7}$ henrys/m
μ_r	relative permeability of core
μ	product of μ_r and μ_o
μ_Δ	effective permeability
ρ	resistivity (Ω-cm)
σ	real part of complex frequency variable, s (nepers)
ϕ	total flux (weber)
φ_d	maximum phase lead (deg)
φ_g	maximum phase lag (deg)
φ_m	phase margin (deg)
φ_{mH}	phase margin at high duty ratio (deg)
φ_{mL}	phase margin at low duty ratio (deg)
ω	angular frequency, imaginary part of complex frequency variable, s (rad)
$a \cong b$	a is approximately equal to b
$a > b$	a is greater than b
$a \gg b$	a is very much greater than b

$a < b$	a is smaller than b
$a \ll b$	a is very much smaller than b
$a \equiv b$	a is identical to b
$a \neq b$	a is not equal to b
$r//R$	r in parallel with R

Bibliography

1. C. F. Wagner, Parallel Inverter with Resistance Load, *Electrical Engineering*, November 1935.
2. G. C. Uchrin and W. O. Taylor, A New Self-Excited Square-Wave Transistor Power Oscillator, *Proc. IRE 43*, January 1955. 1955.
3. G. H. Royer, A Switching Transistor DC to AC Converter Having an Output Frequency Proportional to the DC Input Voltage, *AIEE Trans. Communications and Electronics 74*, Pt. 1, July 1955.
4. J. L. Jensen, An Improved Square-Wave Oscillator Circuit, *IRE Trans. on Circuit Theory, CT-4*, September 1957.
5. A. J. Meyerhoff and R. M. Tillmaw, A High Speed Two-Winding Transistor-Magnetic-Core Oscillator, *IRE Trans. on Circuit Theory, CT-4*, 1957.
6. R. E. Morgan, A New Control Amplifier Using a Saturable Current Transformer and a Switching Transistor, Paper 58-858, AIEE Summer General Meeting and Air Transportation Conference, Buffalo, June 1958.
7. T. D. Towers, Practical Design Problems in Transistor DC/DC Converters and DC/AC Inverters, *Proc. IEE (London)*, *106B*, Suppl. 18, May 1959.
8. R. E. Morgan, A New Magnetic-Controlled Rectifier Power Amplifier with a Saturable Reactor Controlling On Time, Paper T-121, AIEE Special Technical Conference on Nonlinear Magnetics and Magnetic Amplifiers, Philadelphia, October 26-28, 1960.

9. G. A. Salters, A High Power DC-AC Inverter with Sinusoidal Output, *Electronic Engineering*, September 1961.
10. A. Kernick, J. L. Roof, and T. M. Heinrich, Static Inverter with Neutralization of Harmonics, *AIEE Trans. Appl. Ind.*, *81*, Pt. II, May 1962.
11. P. D. Corey, Methods for Optimizing the Waveform of Stepped-Wave Static Inverters, Paper 62-1147, *AIEE Summer General Meeting*, June 17, 1962.
12. R. W. Sterling et al., Multiple Cores Used to Simulate a Variable Volt-Second Saturable Transformer for Application in Self-Oscillating Inverters, IEEE International Conference on Nonlinear Magnetics, Washington, D.C., April 17-19, 1963.
13. T. Roddam, *Transistor Inverters and Converters*, Iliffe, London, 1963.
14. F. G. Turnbull, Selected Harmonic Reduction in Static DC-AC Inverters, Paper 63-1011, IEEE Summer General Meeting and Nuclear Radiation Effects Conferences, Toronto, June 16-21, 1963.
15. B. D. Bedford and R. G. Hoft, *Principles of Inverter Circuits*, Wiley, New York 1964.
16. R. E. Morgan, Bridge-Chopper Inverter for 400 CPS Sine Wave Power, *IEEE Trans. on Aerospace, 2*, April 1964.
17. R. E. Morgan, High-Frequency Time-Ratio Control with Insulated and Isolated Inputs, *IEEE Trans on Magnetics*, March 1965.
18. S. Lindena, The Current-Fed Inverter, A New Approach and a Comparison with the Voltage-Fed Inverter, *20th Power Sources Conference Proc.*, May 1966.
19. J. P. Vergez, Jr., and V. Glover, Low Power Solid-State Inverters for Space Applications, *WESCON 1966 Record.*
20. Y. Yu et al., Static DC to Sinusoidal AC Inverter Using Techniques of High Frequency Pulse-Width Modulation, *IEEE Trans. on Magnetics, MAG-3*, No. 3, September 1967.
21. R. J. Ravas et al., Staggered Phase Carrier Cancellation—A New Concept for Lightweight Static Inverters, *Suppl. to IEEE Trans. on Aerospace Elec. Syst.*, *AES-3*, No. 6, November 1967.
22. O. A. Kossov, Comparative Analysis of Chopper Voltage Regulators with LC Filters, *IEEE Trans. on Magnetics*, September 1968.
23. O. A. Kossov, Comparative Analysis of Chopper Voltage Regulators with LC Filters, *IEEE Trans. on Magnetics*, *MAG-4*, December 1968.

24. R. E. Morgan, Conversion and Control with High Voltage Transistors with Isolated Inputs, Paper 7.8, 1968 INTERMAG Conference, Washington, D.C., April 3-5, 1968.
25. J. R. Nowicki, *Power Supplies for Electronic Equipment*, Leonard Hill, London, 1971.
26. H. P. Hart and R. J. Kakalec, The Derivation and Application of Design Equations for Ferroresonant Voltage Regulators and Regulated Rectifiers, *IEEE Trans. on Magnetics*, March 1971.
27. T. G. Wilson et al., Regulated DC to DC Converter for Voltage Step-Up or Step-Down with Input-Output Isolation, *IEEE Fall Electronics Conference*, October 1971.
28. F. F. Judd and C. T. Chen, Analysis and Optimal Design of Self-Oscillating DC to DC Converters, *IEEE Trans. Circuit Theory, CT-18*, November 1971.
29. J. J. Pollack, Advanced Pulsewidth Modulated Inverter Techniques, *IEEE Trans. Ind. Appl., IA-8*, No. 2, March/April 1972.
30. F. C. Lee et al., Analysis of Limit Cycles in a Two-Transistor Saturable-Core Parallel Inverter, *IEEE Trans. on Aerospace and Elec. Syst., AES-9*, No. 4, July 1973.
31. D. H. Wolaver, Requirements on Switching Devices in DC to DC Converters, in *Power Electronics Specialists Conference Record*, 1973.
32. F. C. Lee and T. G. Wilson, Analysis of Starting Circuits for a Class of Hard Oscillators: Two-Transistor Saturable-Core Parallel Inverters, *IEEE Trans. on Aerospace and Elec. Syst., AES-10*, No. 1, January 1974.
33. F. E. Lukens, Linearization of the Pulse Width Modulated Converter, in *Power Electronics Specialists Conference Record*, June 1974.
34. F. C. Lee, Analysis of Transient Characteristics and Starting of a Family of Power Conditioning Circuits: Two-Transistor Saturable-Core Parallel Inverters, Ph.D. Dissertation, Duke University, Durham, N.C., 1974.
35. F. C. Lee and T. G. Wilson, Voltage-Spike Analysis for a Free-Running Parallel Inverter, in *1974 Digest of the INTERMAG Conference*, May 1974.
36. F. C. Lee and T. G. Wilson, Nonlinear Analysis of a Family of LC Tuned Inverters, *IEEE Trans. on Aerospace and Elec. Syst., AES-11*, No. 2, March 1975.
37. T. G. Wilson et al., Relationships Among Classes of Self-Oscillating Transistor Parallel Inverters, *IEEE Trans. on Aerospace and Elec. Syst., AES-11*, No. 2, March 1975.

38. D. E. Nelson and N. O. Sokal, Improving Load and Line Transient Response of Switching Regulators by Feed-Forward Techniques, in *POWERCON 2, the Second National Solid-State Power Conversion Conference Record*, 1975.
39. Y. Yu et al., Formulation of a Methodology for Power Circuit Design Optimization, in *Power Electronics Specialists Conference Record*, June 1976.
40. E. T. Calkin and B. H. Hamilton, A Conceptually New Approach for Regulated DC to DC Converters Employing Transistor Switches and Pulsewidth Control, *IEEE Trans. Ind. Appl.*, July/August 1976.
41. N. O. Sokal, Feed-Forward Control for Switching Mode Power Converters—A Design Example, in *POWERCON 3, the Third National Solid-State Power Conversion Conference Record*, 1976.
42. S. M. Ćuk, R. D. Middlebrook, A New Optimum Topology Switching DC to DC Converter, in *Power Electronics Specialists Conference Record*, 1977.
43. S. M. Ćuk and R. D. Middlebrook, Coupled Inductor and Other Extensions of a New Optimum Topology Switching DC to DC Converter, in *IEEE Ind. Appl. Society Annual Meeting, 1977 Record*, October 2-6, 1977.
44. S. M. Ćuk and R. Erickson, A Conceptually New High-Frequency Switched-Mode Amplifier Technique Eliminates Current Ripple, in *POWERCON 5, the Fifth National Solid State Power Conversion Conference Record*, May 1978.
45. S. M. Ćuk, Discontinuous Inductor Current Mode in the Optimum Topology Switching Converter, in *Power Electronics Specialists Conference Record*, June 1978.
46. S. M. Ćuk, Switching DC to DC Converter with Zero Input or Output Current Ripple, in *IEEE Ind. Appl. Society Annual Meeting, 1978 Record*, October 1-5, 1978.
47. Y. Yu et al., Development of a Standardized Control Module for DC-DC Converters, *NASA Contract Report NAS3-18918*, by TRW Defence and Space Systems, 1977.
48. A. I. Pressman, *Switching and Linear Power Supply, Power Converter Design*, Hayden, New York 1977.
49. R. D. Middlebrook and S. M. Ćuk, Isolation and Multiple Output Extensions of a New Optimum Topology Switching DC to DC Converter, in *Power Electronics Specialists Conference, 1978 Record*, June 1978.
50. F. Mahmoud and F. C. Lee, Analysis and Design of an Adaptive Multi-Loop Controlled Two-Winding Buck-Boost Regulator, Inter. Telecom. Energy Conference, 1979.

51. Y. Yu et al., Power Converter Design Optimization, *IEEE Trans. on Aerospace and Elec. Syst., AES-15* No. 3, May 1979.
52. R. Ruble and W. Treitel, A New Technique for Sine Synthesis Inverter Design, in *POWERCON 6, the Sixth National Solid-State Power Conversion Conference Record*, 1979.
53. L. Rensink et al., Design of a Kilowatt Off-Line Switcher Using a Ćuk Converter, in *POWERCON 6, the Sixth National Solid-State Power Conversion Conference Record*, May, 1979.
54. S. M. Ćuk, General Topological Properties of Switching Structures, in *Power Electronics Specialists Conference 1979 Record*.
55. S. P. Hsu, Problems in Analysis and Design of Switching Regulators, Ph.D. dissertation, California Institute of Technology, September 1979.
56. L. Rensink, Switching Regulator Configurations and Circuit Realizations, Ph.D. dissertation, California Institute of Technology, December 1979.
57. F. C. Lee and T. G. Wilson, State-Plane Analysis, Classification and Duality Relationships of Several Classes of Self-Oscillating Parallel Inverters, 1979 Int. Colloquium on Circuits and Systems, Taipei, Taiwan.
58. S. M. Ćuk and R. D. Middlebrook, Advances in Switched-Mode Power Conversion, Part 1, *Robotics Age, 1*, No. 2, Winter 1979.
59. S. M. Ćuk and R. D. Middlebrook, Advances in Switched-Mode Power Conversion, Part 2, *Robotics Age, 2*, No. 2, Summer 1980.
60. F. C. Lee et al., A Unified Design Procedure for a Standardized Control Module for DC-DC Switching Regulators, in *Power Electronics Specialists Conference, 1980 Record.*
61. C. J. Wu et al., Design Optimization for a Half-Bridge DC-DC Converter, in *Power Electronics Specialists Conference, 1980 Record.*
62. Y. Yu and F. C. Lee, Application Handbook for Standardized Control Module for DC-DC Converters, *NASA Contract Report NAS3-20102*, prepared jointly by TRW Defence and Space Systems and Virginia Polytechnic Institute and State University, 1980.
63. F. C. Lee and Y. Yu, An Adaptive Control Switching Buck Regulator Implementation, Analysis and Design, *IEEE Trans. on Aerospace and Elec. Syst.*, January 1980.
64. R. Erickson et al., Characterization and Implementation of Power MOSFETs in Switching Converters, *POWERCON 7*,

the Seventh National Solid-State Power Conversion Conference Record, March 24-27, 1980.
65. R. Redl and N. O. Sokal, Push-Pull Current-Fed Multiple Output DC-DC Power Converter with Only One Inductor and with 0 to 100% Switch Duty Ratio, in *Power Electronics Specialists Conference, 1980 Record.*
66. M. S. Makled and M. M. Fahmy, An Analytical Investigation of a Ferroresonant Circuit, *IEEE Trans. on Magnetics,* March 1980.
67. R. Redl and N. O. Sokal, Push-Pull Current-Fed Regulated Wide-Input-Range DC-DC Power Converter with Only One Inductor and with 0 to 100% Switch Duty Ratio: Operation at Duty Ratio Below 50%, in *Power Electronics Specialists Conference 1981 Record.*
68. V. J. Thottuvelil et al., Analysis and Design of a Push-Pull Current-Fed Converter, in *Power Electronics Specialists Conference 1981 Record.*
69. L. H. Dixon and C. J. Baranowski, Designing Optimal Multi-Output Converters with a Coupled-Inductor Current-Driven Topology, in *POWERCON 8, the Eighth National Solid-State Power Electronic Conference Record,* April 1981.
70. R. D. Middlebrook, Power Electronics: An Emerging Discipline, in *IEEE International Symposium on Circuits and Systems, 1981 Record.*
71. R. D. Middlebrook, Predicting Modulator Phase Lag in PWM Converter Feedback Loops, in *Powercon 8, the Eighth National Solid-State Power Electronics Conference Record,* April 1981.
72. J. N. Park and T. R. Zalom, A Dual Mode Forward/Flyback Converter, in *Power Electronics Specialists Conference, 1982 Record.*

MODELING

73. G. W. Wester and R. D. Middlebrook, Low-Frequency Characterization of Switched DC-to-DC Converters, in *Power Electronics Specialists Conference, 1972 Record.*
74. R. D. Middlebrook, Describing Function Properties of a Magnetic Pulse-Width Modulator, in *Power Electronics Specialists Conference, 1972 Record.*
75. T. G. Wilson and F. C. Lee, Analysis and Modeling of a Family of Two-Transistor Parallel Inverters, *IEEE Trans. on Magnetics, MAG-9,* No. 3, September, 1973.

76. R. D. Middlebrook, A Continuous Model for the Tapped Inductor Boost Converter, in *Power Electronics Specialists Conference, 1975 Record.*
77. D. J. Packard, Discrete Modeling and Analysis of Switching Regulators, Ph.D. dissertation, California Institute of Technology, May 1976.
78. R. D. Middlebrook and S. M. Ćuk, A General Unified Approach to Modeling Switching Converter Power Stages, in *Power Electronics Specialists Conference, 1976 Record.*
79. R. D. Middlebrook and S. M. Ćuk, Modeling and Analysis Methods for DC-to-DC Switching Converters, *IEEE International Semiconductor Power Converter Conference, 1977 Record.*
80. S. M. Ćuk and R. D. Middlebrook, A General Unified Approach to Modeling Switching DC-to-DC Converters in Discontinuous Conduction Mode, in *Power Electronics Specialists Conference, 1977 Record.*
81. Y. Yu et al., Modeling and Analysis of Power Processing Systems, in *NAECON 1977 Record.*
82. R. P. Iwen et al., Modeling and Analysis of DC-DC Converters with Continuous and Discontinuous Inductor Current, Second IFAC Symposium on Control in Power Electronics and Electrical Devices, Dusseldorf, October 1977.
83. Y. Yu et al., Modeling and Analysis of Power Processing Systems, *NASA Contract Report NAS3-19690*, by TRW Defence and Space Systems, November 1977.
84. R. P. Iwens et al., Generalized Discrete Time Domain Modeling and Analysis of DC-DC Converters, *IEEE Trans. Ind. Electron. Contr. Instrum., IECI-26*, No. 2, May 1979.
85. S. P. Hsu et al., Modeling and Analysis of Switching DC-to-DC Converters in Constant Frequency Current Programmed Mode, in *Power Electronics Specialists Conference, 1979 Record.*
86. Y. Yu, F. C. Lee, and J. Kolecki, Modeling and Analysis of Power Processing Systems (MAPPS), in *Power Electronics Specialists Conference, 1979 Record.*
87. F. C. Lee, Discrete Time Domain Modeling and Linearization of a Switching Buck Converter, International Symposium on Circuits and Systems, Tokyo, July 1979.
88. Y. Yu et al., Modeling of Switching Regulator Power Stages with and without Zero-Inductor-Current Dwell Time, *IEEE Trans. Ind. Electron. Contr. Instrum., IECI-26*, No. 3, August 1979.

89. R. D. Middlebrook, Modeling and Design of the Ćuk Converter, in *POWERCON 6, the Sixth National Solid-State Power Conversion Conference, 1979 Record.*
90. W. M. Polivka et al., State-Space Average Modeling of Converters with Parasitics and Storage Time Modulation, in *Power Electronics Specialists Conference 1980 Record.*
91. B. Erickson, S. Ćuk, and R. D. Middlebrook, Large Signal Modeling and Analysis of Switching Regulators, in *Power Electronics Specialists Conference, 1982 Record.*
92. V. J. Thottuvelil et al., Small Signal Modeling of a Push-Pull Current-Fed Converter, in *Power Electronics Specialists Conference, 1982 Record.*
93. D. J. Shortt and F. C. Lee, An Improved Switching Converter Using Discrete and Averaging Techniques, in *Power Electronics Specialists Conference, 1982 Record.*
94. R. D. Middlebrook, Power Electronics: Topologies, Modeling and Measurement, in *IEEE International Symposium on Circuits and Systems, 1981 Record.*

MAGNETIC COMPONENTS

95. C. R. Hanna, Design of Reactances and Transformers Which Carry Direct Current, *Trans. AIEE*, February 1927.
96. C. K. Hadlock and D. Lebell, Some Studies of Pulse Transformer Equivalent Circuits, *Proc. IRE*, January 1951.
97. R. Lee, *Electronic Transformers and Circuits*, 2nd ed., Wiley, New York, 1955.
98. I. Richardson, The Technique of Transformer Design, *Electro-Technology*, January 1961.
99. S. Lindena, Design of a Magnetic Voltage Stabilizer, *Electro-Technology*, May 1961.
100. N. R. Grossner, Pulse Transformer Circuits and Analysis 1 and 2, *Electro-Technology*, February/March 1962.
101. C. F. Wilds, Determination of Core Size in Pulse Transformer Design, *Electronic Engineering*, September 1961.
102. M. A. Cambre, A Generalized Approach to Transient Analysis of Wide Band Transformers, in *National Electronics Conference Record*, 1964.
103. H. B. Harms, Predicting Reliability of Electronic Transformers, in *National Electronics Conference Record*, 1964.
104. MIT Staff, *Magnetic Circuits and Transformers*, Massachusetts Institute of Technology, 1965.

105. S. Pro, Toroid Design Analysis, *Electro-Technology*, August 1966.
106. P. L. Dowell, Effects of Eddy Currents in Transformer Windings, *Proc. IEE, 113*, No. 8, August 1966.
107. E. C. Snelling, *Soft Ferrites—Properties and Applications*, Iliffe, London, 1969.
108. B. Castle, Optimum shapes for Inductors, *IEEE Trans. Parts, Materials and Packaging, PMP-5*, No. 1, March 1969.
109. F. C. Schwarz, An Unorthodox Transformer for Free-Running Inverters, *IEEE Trans. on Magnetics, MAG-5*, 1969.
110. S. Y. M. Feng and W. A. Sander, III, Optimum Toroidal Inductor Design Analysis, *20th Electronic Component Conference Proceedings*, 1970.
111. R. Lee and D. S. Stephens, Gap Loss in Current Limiting Transformers, *Electromechanical Design*, April, 1973.
112. R. Lee and D. S. Stephens, Influence of Core Gap in Design of Current Limiting Transformers, *IEEE Trans. on Magnetics*, September 1973.
113. J. R. Woodbury, Design of Imperfectly Coupled Power Transformers for DC-to-DC Conversion, *IEEE Trans. Ind. Elec. and Contr. Instrum., IECI-21*, No. 3, August 1974.
114. W. Dull, A. Kusko, and T. Knutrud, Designers' Guide to Current and Power Transformers, *EDN Magazine*, March 5, 1975.
115. V. B. Guizburg, The calculation of Magnetization Curves and Magnetic Hysteresis Loops for a Simplified Model of a Ferromagnetic Body, *IEEE Trans. on Magnetics*, March 1976.
116. A. K. Ohri et al., Design of Air-Gapped Magnetic Core Inductors for Superimposed Direct and Alternating Currents, *IEEE Trans. on Magnetics*, September 1976.
117. W. V. Manka, Design Power Inductors Step by Step, *Electronic Design*, December 20, 1977.
118. N. R. Grossner, The Geometry of Regulating Transformers, *IEEE Trans. on Magnetics*, March 1978.
119. W. A. Martin, Simplify Air Gap Calculating with a Hanna Curve, *Electronic Design*, April 12, 1978.
120. Col. W. T. McLyman, *Transformer and Inductor Design Handbook*, Dekker, New York, 1978.
121. Col. W. T. McLyman, *Cut Core Inductor Design Manual*, Arnold Catalog No. SC-142A, 1978.

122. Col. W. T. McLyman, *Magnetic Core Selection for Transformers and Inductors*, Dekker, New York, 1982.
123. T. Gross, Multistrand Litz Wire Adds "Skin" to Cut AC Losses in Switching Power Supplies, *Electronic Design*, February 1, 1979.
124. T. Konopinski and S. Szuba, Limit the Heat in Ferrite Pot Cores for Reliable Switching Power Supplies, *Electronic Design*, June 7, 1979.
125. T. Gross, A Little Understanding Improves Switching Inductor Designs, *EDN Magazine*, June 20, 1979.
126. A. J. Mas, Design and Performance of Power Transformers with Metallic Glass Cores, *Powerconversion International*, July/August 1980.
127. P. E. Thibodeau, The Switcher Transformer: Designing It in One Try for Switching Power Supplies, *Electronic Design*, September 1, 1980.
128. C. J. Wu et al., Minimum Weight EI Core and Pot Core Inductor and Transformer Designs, *IEEE Trans. on Magnetics*, September 1980.
129. S. A. Chin et al., Design Graphics for Optimizing the Energy Storage Inductor for DC-to-DC Power Converters, in *Power Electronics Specialists Conference, 1982 Record.*
130. J. R. Leehey et al., DC Current Transformer, in *Power Electronics Specialist Conference, 1982 Record.*
131. W. E. Rippel and Col. W. T. McLyman, Design Techniques for Minimizing the Parasitic Capacitance and Leakage Inductance of Switched Mode Power Transformers, in *POWERCON 9, the Nineth National Solid-State Power Electronic Conference Record*, 1982.

CONTROL THEORY

132. H. S. Black, Stabilized Feedback Amplifiers, *Bell System Technical Journal*, January 1934.
133. H. W. Bode, Relations Between Attenuation and Phase in Feedback Amplifier Design, *Bell System Technical Journal*, 1940.
134. H. W. Bode, *Network Analysis and Feedback Amplifier Design*, Van Nostrand Reinhold, New York, 1945.
135. B. C. Kuo, *Automatic Control Systems*, Prentice-Hall, Englewood Cliffs, N.J., 1962.
136. L. R. Poulo and S. Greenblatt, Research Investigations on Feedback Techniques and Methods for Automatic Control,

Contract ECOM-0520-F, Bose C2rp., Natick, Mass., April 1969.

137. C. Desoer and E. Kuh, *Basic Circuit Theory*, McGraw-Hill, New York, 1969.
138. D. E. Combs, Stability Analysis of a Pulse-Width Controlled DC to DC Regulated Converter Using Linear Feedback Control System Technique, *Nat. Elec. Conf. Record, 26*, 1970.
139. J. J. D'Azzo and C. H. Houpis, *Linear Control System Analysis and Design: Conventional and Modern*, McGraw-Hill, New York, 1975.
140. R. P. Iwens et al., Time Domain Modeling and Stability Analysis of an Integral Pulse Frequency Modulated DC to DC Power Converter, in *Power Electronics Specialists Conference, 1975 Record.*
141. C. Griffin, Optimizing the PWM Converter as a Closed Loop System, in *POWERCON 4, the Fourth National Solid-State Power Electronics Conference Record*, 1977.
142. H. D. Venable and S. R. Foster, Practical Techniques for Analyzing, Measuring and Stabilizing Feedback Control Loops in Switching Regulators and Converters, in *POWERCON 7, the Seventh National Solid-State Power Electronics Conference Record*, 1980.

INPUT FILTER

143. J. J. Beiss and Y. Yu, A Two-Stage Input Filter with Nondissipatively-Controlled Damping, in *INTERMAG Conference Record*, April 1971.
144. Y. Yu and J. J. Beiss, Some Design Aspects Concerning Input Filters for DC-DC Converters, in *Power Conditioning Specialists Conference Record*, 1971.
145. N. O. Sokal, System Oscillations from Negative Input Resistance at Power Input Port of Switching-Mode Regulator, Amplifier, DC/DC Converter or DC/AC Inverter, in *Power Electronics Specialists Conference, 1973 Record.*
146. R. D. Middlebrook, Input Filter Considerations in Design and Application of Switching Regulators, *IEEE Ind. Appl. Soc. Annual Meeting, 1976 Record.*
147. R. D. Middlebrook and S. M. Ćuk, Design Techniques for Preventing Input Filter Oscillations in Switched-Mode Regulators, in *POWERCON 5, the Fifth National Solid-State Power Electronics Conference, 1978 Record.*

148. F. C. Lee and Y. Yu, Input Filter Design for Switching Regulators, *IEEE Trans. on Aerospace and Elec. Syst., AES-15,* No. 5, September 1979.
149. T. K. Phelps and W. S. Tage, Optimizing Passive Input Filter Design, in *POWERCON 6, the Sixth National Solid-State Power Conversion Conference, 1979 Record.*
150. S. S. Kelkar and F. C. Lee, A Novel Input Filter Compensation Scheme for Switching Regulators, in *Power Electronics Specialists Conference, 1982 Record.*
151. S. S. Kelkar and F. C. Lee, Adaptive Feedforward Input Filter Compensation for Switching Regulators, in *POWERCON 9, the Ninth National Solid-State Power Conversion Conference, 1982 Record.*

CALCULATOR PROGRAMS

152. B. K. Murdock, *Handbook of Electronic Design and Analysis Procedures Using Programmable Calculators,* Van Nostrand Reinhold, New York, 1979.
153. C. McIntyre, SR-52 Solves Network Equations by Finding Complex Determinant, *Electronics,* May 12, 1977.
154. F. M. Lilienstein, Analyze Switcher Stability, Bandwidth and Gain with a Programmable Calculator, *Electronic Design,* June 7, 1979.
155. A. B. Przedpelski, Eliminate Bandwidth Calculation Drudgery with a Universal Calculator Program, *Electronic Design,* October 11, 1979.
156. W. A. Geckle, Compute S-Function/Time-Domain Response Quickly with a Programmable Calculator, *Electronic Design,* December 6, 1979.
157. F. W. Hauer, Speed Ferromagnetic Inductor Designs with a Programmable Calculator, *Electronic Design,* December 20, 1979.
158. C. Gyles, Analyze Complex Linear Networks with a Building-Block Calculator Program, *Electronic Design,* April 26, 1980.
159. C. J. McCluskey, TI-59 Calculator Analyzes Complex Ladder Networks, *Electronic Design,* May 10, 1980.
160. F. Cornelissen, TI-59 Solves Network Equations Using Complex Matrices, *Electronics,* July 31, 1980.
161. B. K. Erickson, Ladder Network Calculations, *IEEE Trans. Cons. Electronics, CE-26,* November 1980.
162. G. West, Use a Programmable Calculator to Ease Transformer Design, *EDN Magazine,* November 24, 1982.

MEASUREMENT

163. V. E. Legg, Magnetic Measurements at Low Flux Densities Using the Alternating Current Bridge, *Bell System Technical Journal*, January 1936.
164. R. A. Homan, DC Power System Dynamic Impedance Measurements, in *National Electronics Conference Record*, October 1964.
165. R. D. Middlebrook, Measurement of Loop Gain in Feedback Systems, *International Journal of Electronics, 38*, No. 4, 1975.
166. R. D. Middlebrook, Improved Accuracy Phase Angle Measurement, *International Journal of Electronics, 40*, No. 1, 1976.
167. P. C. Todd, Automating the Measurement of Converter Dynamic Properties, in *POWERCON 7, the Seventh National Solid-State Power Conversion Conf., 1980 Record.*
168. F. Barzegar et al., Using Small Computers to Model and Measure Magnitude and Phase of Regulator Transfer Functions and Loop Gain, in *POWERCON 8, the Eighth National Solid-State Power Conversion Conference, 1981 Record.*

COMPUTER-AIDED DESIGN ANALYSIS AND SIMULATION

169. B. A. Wells et al., Analog Computer Simulation of a DC-to-DC Flyback Converter, *Suppl. to IEEE Trans. on Aerospace and Elec. Syst., AES-3*, November 1967, pp. 399-409.
170. S. Y. M. Feng et al., A computer Aided Design Procedure for Flyback Step-Up DC-to-DC Converters, *IEEE Trans. on Magnetics, MAG-8*, No. 3, September 1972.
171. D. Y. Chen et al., Computer Aided Design and Graphics Applied to the Study of Inductor Energy Storage DC-to-DC Electronic Power Converters, *IEEE Trans. Aerospace and Elec. Syst., AES-9*, No. 4, July 1973.
172. W. A. Schnider, Verify Network Frequency Response with This Simple BASIC Program, *EDN Magazine*, October 5, 1977.
173. Y. Yu, Computer Aided Analysis and Simulation of Switched DC-DC Converters, in *1978 IEEE Southeaston Proceedings*, April 1978.
174. R. Keller, Closed-Loop Testing and Computer Analysis Aid Design of Control Systems, *Electronic Design*, November 22, 1978.

175. N. P. Episcopo and R. P. Massey, Computer Predicted Steady State Stability of Pulse-Width-Controlled DC/DC Converters, in *POWERCON 6, the Sixth National Solid-State Power Conversion Conference, 1979 Record.*
176. F. C. Lee and Y. Yu, Computer Aided Analysis and Simulation of Switched DC-DC Converters, *IEEE Trans. Ind. Appl., IA-15,* No. 5, September/October 1979.
177. V. G. Bello, Computer Modeling of Pulse-Width Modulators Simplifies Analysis of Switching Regulators, *Electronic Design,* January 18, 1980.
178. G. H. Warren, Computer Aided Design Program Supplies Low-Pass Filter Data, *EDN Magazine,* August 20, 1980.
179. J. E. Crowe, Mains Hold-up Performance in Switched Mode PSU's, *Electronic Engineering,* November 1980.
180. E. Niemeyer, Network Analysis Program Runs on Small Computer System, *EDN Magazine,* February 4, 1981.
181. V. G. Bello, Computer Program Adds SPICE to Switching Regulator Analysis, *Electronic Design,* March 5, 1981.
182. V. G. Bello, Using the SPICE 2 CAD Package for Easy Simulation of Switching Regulators in Both Continuous and Discontinuous Conduction Modes, in *POWERCON 8, the Eighth National Solid-State Power Conversion Conference, 1981 Record.*
183. H. T. Meyer, Matrix Statements Define Complex Variables, Perform Complex Math in BASIC, *Electronic Design,* July 23, 1981.
184. S. Hageman, Program Analyzes Six-Element Active RC Networks, *Electronic Design,* January 7, 1982.
185. W. N. Waggener, Analyze Complex Circuits with a Matrix Inversion Program, *EDN Magazine,* March 17, 1982.

PATENTS

186. H. W. Bode, Amplifier, U.S. Pat. No. 2,123,178, July 12, 1938.
187. F. C. Schwarz, Analog Signal to Discrete Time Interval Converter (ASDTIC), U.S. Pat. No. 3,659,184, April 25, 1972.
188. E. T. Calkin, B. H. Hamilton, and F. C. Laporta, Regulated DC-to-DC Converter with Regulated Current Source Driving a Nonregulated Inverter, U.S. Pat. No. 3,737,755, June 5, 1973.

189. H. D. Venable, Regulated DC-to-DC Converter, U.S. Pat. No. 3,925,715, December 9, 1975.
190. P. W. Clarke, Converter Regulation by Controlled Overlap, U.S. Pat. No. 3,938,024, February 10, 1976.
191. P. Kotlarewsky, Master-Slave Voltage Regulator Employing Pulse Width Modulation, U.S. Pat. No. 4,174,534, November 13, 1979.
192. S. M. Ćuk and R. D. Middlebrook, DC-to-DC Switching Converter, U.S. Pat. No. 4,184,197, January 15, 1980.
193. S. M. Ćuk, Push-Pull Switching Power Amplifier, U.S. Pat. No. 4,186,437, January 29, 1980.
194. S. M. Ćuk, DC-to-DC Switching Converter with Zero Input and Output Current Ripple and Integrated Magnetics Circuits, U.S. Pat. No. 4,257,087, March 17, 1981.
195. G. E. Bloom and A. Eris, DC-to-DC Converter, U.S. Pat. No. 4,262,328, April 14, 1981.
196. S. M. Ćuk and R. D. Middlebrook, DC-to-DC Converter Having Reduced Ripple Without Need for Adjustments, U.S. Pat. No. 4,274,133, June 16, 1981.

SNUBBER NETWORKS

197. E. T. Calkin and B. H. Hamilton, Circuit Techniques for Improving the Switching Loci of Transistor Switches in Switching Regulators, in *IEEE Ind. Appl. Society Conference Record 1972.*
198. F. C. Lee and T. G. Wilson, Voltage Spike Analysis for a Free-Running Parallel Inverter, *IEEE Trans. on Magnetics, MAG-10,* No. 3, September 1974.
199. J. M. Peter, ed., *The Power Transistor in Its Environment,* Thomson-CSF Semiconductor Division Publication, 1978.
200. F. C. Lee and T. G. Wilson, Nonlinear Analysis of Voltage Spike Supression Networks for a Free-Running Parallel Inverter, in *IEEE Ind. Appl. Annual Meeting, 1979 Record.*
201. W. McMurray, Selection of Snubbers and Clamps to Optimize the Design of Transistor Switching Converters, in *Power Electronics Specialists Conference, 1979 Record.*
202. W. J. Shaughnessy, LC Snubber Networks Cut Switcher Power Losses, *EDN Magazine,* November 20, 1980.

MATHEMATICS

203. H. F. Baker, On the Integration of Linear Differential Equations, *Proc. London Math. Soc., 34,* 1902, pp. 347-360; *35,* 1903, pp. 333-374; *second series 2,* 1904, pp. 293-296.
204. B. Van der Pol, Forced Oscillations in a Circuit with Nonlinear Resistance, *Phil. Mag., 7-3,* 1927, pp. 65-80.
205. W. J. Cunningham, *Nonlinear Analysis,* McGraw-Hill, New York, 1958.
206. S. Austen Stigant, *The Elements of Determinants, Matrices and Tensors for Engineers,* MacDonald and Co. (Publishers) Ltd., 1959.
207. R. Bellman, *Perturbation Techniques in Mathematics, Physics and Engineering,* Holt, Rinehart and Winston, New York.
208. A. V. Fiacco and G. P. McCormick, Computational Algorithm for the Sequential Unconstrained Minimization Technique for Nonlinear Programming, *Management Science, 10,* July 1964, pp. 601-617.
209. D. A. Pierre, *Optimization Theory with Applications,* Wiley, 1969, pp. 36-43.
210. J. Abadie and J. Carpenter, Generalization of the Walte Reduced Gradient Method to the Case of Nonlinear Constraints, in *Optimization,* R. Fletcher, ed., Academic Press, New York, 1969.
211. W. C. Mylander, R. L. Holmes, and G. P. McCormick, A Guide to SUMT-Version 4, *Paper RAC-P-63,* Research Analysis Corp., October 1971.
212. *Motorola Application Note AN460.*
213. S. Ćuk and R. D. Middlebrook, Modeling Analysis and Design of Switching Converters, *Report No. NASA CR-135174, Contract No. NAS3-19690 and NAS3-20102.*
214. Loop Gain Measurements with HP Wave Analyzers. *Hewlett-Packard Application Note 59.*
215. Low Frequency Gain Phase Measurements. *Hewlett-Packard Application Note 157.*

Index

DATE DUE

WITHDRAWN